Jason Hoppe

Adobe Illustrator CC

A Complete

Course

and

Compendium

of Features

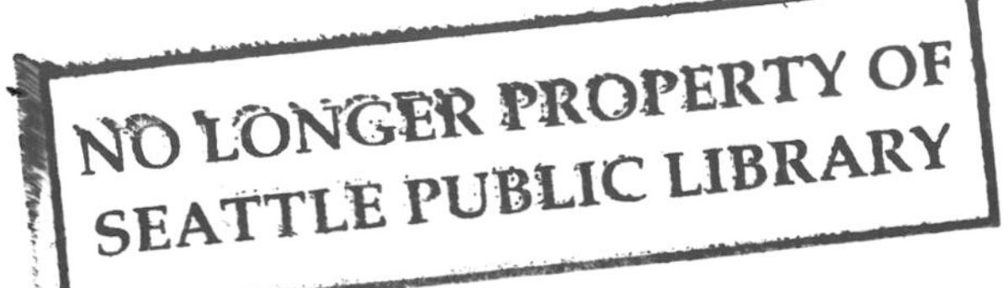

rockynook

Adobe Illustrator: A Complete Course and Compendium of Features
Jason Hoppe
www.jasonhoppe.com

Project editor: Maggie Yates
Project manager: Lisa Brazieal
Marketing manager: Mercedes Murray
Copyeditor: Maggie Yates
Interior design and layout: Jason Hoppe
Cover design: Steve Laskevitch
Indexer: James Minkin

ISBN: 978-1-68198-531-2
1st Edition (1st printing, April 2020)

Rocky Nook, Inc.
1010 B Street, Suite 350
San Rafael, CA 94901
USA

www.rockynook.com

Distributed in the UK and Europe by Publishers Group UK
Distributed in the U.S. and all other territories by Ingram Publisher Services

Library of Congress Control Number: 2018949104

Printed in China

About the Author

Jason Hoppe is an Adobe Certified Expert and Adobe Certified Instructor in Photoshop, InDesign, Illustrator, and Acrobat. His 20 years of teaching and working in the advertising world have led him to writing this book.

Born in Japan and raised in Rome, NY, Jason went to school at Mohawk Valley Community College to get a degree in Advertising, Design, and Production. He earned his BFA in (Swiss) Graphic Design at Fredonia State University where he fell in love with the Mac and the early versions of Photoshop, Illustrator, and Quark. While off for the summers, he did production work for the local newspaper, the *Rome Daily Sentinel.*

Having had enough of the snow, salted roads that ate up his cars, and hot, humid summers, he left New York for Seattle in the fall of 1994. He signed up with Mac Temps and landed his first job at a small advertising agency. His love for production brought him to several agencies in Seattle. He taught Quark and Photoshop classes at The Art Institute of Seattle for seven years. In 2000, he began teaching at the School of Visual Concepts. In 2010 he quit his agency life and teamed up with Craig Swanson of CreativeTechs and became the founding instructor of CreativeLive, having done more than 250 videos on all things creative, including pumpkin carving.

Gaining his Adobe Certifications in in Photoshop, InDesign, Illustrator, and Acrobat, he met Steve Laskevitch and Carla Fraga, owners of LuminousWorks, an Adobe authorized training center in Ballard, WA. Teaching software and design classes there for seven years, Steve accidentally suggested he write a book on Illustrator. Jason also teaches in the Design program and Visual Media program at Seattle Central College on all things creative and production.

On the personal side, Jason has always been a fixer. He started customizing lawn mowers at an early age while mowing lawns to save for college. When he turned 14 he bought his first car, a 1964 Corvair, and the addiction began. He collects, fixes, and restores cars as his passion and hobby, having collected over 200 cars from the 1940s to the 1990s.

Jason also loves to build, renovate, and remodel houses, especially ones with multi-car garages to store his cars. He lives in Seattle with his husband and two hairless cats.

This book is his first of many—he found that writing is not only fun, but also exciting.

Photo: Michelle Smith-Lewis

Acknowledgments

Thanks to Steve Laskevitch for accidentally proposing that I write this book on Illustrator. Seeing the InDesign and Photoshop books was writing prompted him to ask me if I wanted to write the Illustrator book in the series. And here it is!

My mom, Flo Hoppe, artist and world-renowned basket maker, immersed us in art from an early age. Mom made the best christmas ornaments and silkscreened our christmas cards. She brought us to the library every week to get books and records since we grew up with no television. She helped us make all our Halloween costumes and knitted all our hats and mittens for the frigid winters in upstate New York. Mom makes the best cards for every occasion and I have saved every one she has ever sent me. No matter what, she always supported and encouraged me in all that I did. She was the person who made me fall in love with pop-up books and paper folding. Thanks mom!

Virginia Jorgensen, amazing illustrator and absolute hoot of a human being, encouraged me to go into graphic design in 1988. She and her husband Ed were illustrators with an amazing flair for expressing the character of everything they illustrated. They are both greatly missed.

Bob Clarke, teacher at Mohawk Valley Community College, who taught me and encouraged me in the beginning of my education. His infinite patience and encouragement helped me so much on my journey.

Vera Beggs, my 3rd grade teacher at Turin Road School in Rome, NY. She showed me what a great teacher is and how much that shapes a student at an early age. And she and her husband drove Pontiacs which was another big plus in my book.

All the teachers out there who make teaching and learning their life's work. Thank you for helping shape, encourage, and support all the students out there.

My husband, Greg, who puts up with all my car exploits and adventures and never-ending stream of boxes, packages and parcels filled with car parts that arrive weekly.

Jason Hoppe
Seattle, March 2020

Contents

About the Author v
Acknowledgments vii

Introduction 1

The Course

1 Build with a Solid Base 5

Before You Begin 6
- General 6
- Selection & Anchor Display 7
- Type 8
- Units 8
- User Interface 9
- GPU Performance 10

Configuring the Workspace 11
- Choose an Initial Workspace 11
- Create a New Workspace 12

Projects: An Introduction to Illustrator 13

Lesson A: Create a New Artboard 14
Lesson B: Hand Tool and Zoom Tool 15
Lesson C: Create an Apple 16
Lesson D: Create a Spray Can 19

Project: Colors and Gradients 24

Lesson A: Selecting and Applying Color . . . 25

Lesson B: Gradients . . . 28

Project: Building Weather Icons 30

Lesson A: Water Droplet . . . 31

Lesson B: Moon and Stars . . . 33

Lesson C: Cloud . . . 35

Lesson D: Rain . . . 37

Lesson E: Sun . . . 39

Lesson F: Snowflake . . . 41

Lesson G: Wind . . . 44

Lesson H: Adding Texture . . . 47

Project: Building Kitchen Icons 50

Lesson A: Measuring Cup . . . 51

Lesson B: Coffee Pot . . . 53

Lesson C: Spatula and Spoon . . . 54

Lesson D: Boiling Water . . . 57

Lesson E: Toaster with Toast . . . 59

Lesson F: Egg and Avocado . . . 61

Project: Creating Plaid Fabric 62

Lesson A: Create Plaid Fabric . . . 63

Project: Building a Sewn Patch 68

Lesson A: Create a Sewn Patch . . . 69

Project: Big Build—Camping Gear in an Outdoor Scene 77

Lesson A: Create a Hot Air Balloon . . . 78
Lesson B: Create a Tent . . . 83
Lesson C: Create a Campfire . . . 86
Lesson D: Create Pine Trees . . . 89
Lesson E: Create a Backpack . . . 91
Lesson F: Assemble the Scene . . . 94

Project: Output 96

Lesson A: Asset Export . . . 96
Lesson B: Saving and Printing Files . . . 97
Lesson C: Printing . . . 98

The Compendium

1 Preferences and Workspaces 101

Preferences . . . 102
Document-Specific and Global . . . 102
Selection & Anchor Display . . . 105
Type . . . 107
Units . . . 109
Guides & Grid . . . 110
Smart Guides . . . 111
Slices . . . 113
Hyphenation . . . 113
User Interface . . . 115
Performance . . . 117
File Handling & Clipboard . . . 118
Appearance of Black . . . 120

Workspaces . . . 121
Panel Locations . . . 121
Choose A More Useful Initial Workspace . . . 121
Creating a New Column of Panels . . . 121

Customizing Menus and Keyboard Shortcuts 122
What's on the Menu? 122
Keys to Success 123
Creating a New Document 124
Presets 124
Artboards 126
Artboard Panel 126
Artboard Tool 127
Artboard Options 127
Layout Artboards 128
Convert to Artboards 128
Duplicate Artboards 128
Printing Artboards 128
Layers Panel 129
About Layers 130
Layer Panel Overview 130
Layer Management Best Practices 131
Naming 132
For Segregating Content 133
For Protecting Content 133
Creating New Layers 133
Layer Options 134
Panel Options 134
Layer Order 135
Moving Objects to Layers 135
Reordering Layers 135
Selecting Objects 136
Locating Objects 136
Editing Groups 136
Editing Clipping Masks 137
Consolidate Layers and Groups 137
Release Items to Separate Layers 138

2 Shape Creation 139

Drawing Vector Shapes 140
- Rectangle 140
- Ellipse 140
- Polygon 141
- Star 141
- Editing Corners/Corner Widgets 142
- Drawing Lines 143
- Line Segment 143
- Arc Segment 144
- Spiral 144
- Rectangular Grid 145
- Polar Grid 146

Editing Shapes and Lines 148
- Rotate Shapes 148
- Resize Shapes 148
- Duplicate Shapes 148
- Flip Shapes 148
- Properties Panel 149
- Transform a Rectangle 150
- Transform an Ellipse/Circle 151
- Creating a Simple Pie Chart 151
- Transform a Polygon 151
- Transform a Star 152
- Transform a Line 152

Precision Editing 152

Grouping Items 154
- Selecting Objects 154
- Isolation Mode 155
- Locking/Unlocking 155
- Symbols Versus Groups 156

3 Advanced Construction 157

Pathfinder Panel 158
- Shape Modes Section 158
- Pathfinders Section 160
- Expand Compound Shapes 162
- Pathfinders Options 164

Shape Builder Tool . . . 164
Delete Shapes . . . 164
Add Shapes . . . 164
Delete Lines . . . 165

Appearance Panel . . . 166
Stroke Panel . . . 167
Opacity . . . 170
Blend Modes . . . 171
Fill and Stroke Color . . . 171

Live Paint . . . 172
Paint with Live Paint . . . 173
Live Paint Objects with Open Edges . . . 174
Expand a Live Shape . . . 174

Image Trace . . . 175
Black and White Mode . . . 176
Applying Color to a Black and White Vector . . . 177
Grayscale Mode . . . 178
Color Mode . . . 178

4 Editing and Transformation 179

Editing and Transformation . . . 180

Corner Widgets . . . 181
Editing Corners . . . 182

Live Shapes . . . 185

Shape Editing . . . 187
Direct Selection Tool . . . 187
Outline Mode . . . 191
Joining Lines . . . 191
Curvature Tool . . . 192
Pen Tool . . . 194
Pencil Tool . . . 199

Transform Shapes and Lines 201
- Movement 201
- Rotation 201
- Scaling 202
- Reflection/Flip 203
- Shearing 204
- Direct Selection Tool 204
- Group Selection Tool 205
- Free Transform Tool 205
- Scaling Corners 205
- Scaling Strokes & Effects 207

Paths 208
- Open Paths Versus Closed Paths 208
- Outline Stroke/Expand 208
- Offset Path 210
- Join Paths 210
- Divide Paths 210
- Compound Paths 211

Blend 215
- Blend Options 215
- Editing the Blend Spine 216

Symbols 221
- Symbol Tools 222

Patterns 224
- Creating Patterns 224
- Editing and Scaling Patterns 226

Alignment and Distribution 227
- Guides and Grids 227
- Guides 228
- Smart Guides 229
- Grids 230

Align Panel 231
- Distribute 232
- Spacing 233
- Precise Positioning and Sizing 233
- Make It Pixel Perfect 235

5 Effects and Graphic Styles 237

Opacity and Blending Modes . . . 228
- Show Transparency . . . 228
- Knockout Group . . . 239
- Opacity Masks . . . 240
- Blending Modes . . . 241
- Opacity . . . 242

Effects Menu . . . 243
- Raster and Vector Effects . . . 243

Expand/Expand Appearance . . . 246
- Outline Path . . . 247

Graphic Styles . . . 248
- Create a Graphic Style . . . 248

Effects Expanding . . . 250

Clipping Masks / Draw Inside . . . 252
- Create a Clipping Mask . . . 252

6 Type and Text 255

- Working with Type . . . 256
- Typefaces, Categories, and Styles: . . . 256

Positioning, Spacing, and Measuring . . . 258
- Measuring . . . 258
- Baseline . . . 258
- Kerning . . . 258
- Tracking . . . 259
- Leading/Line Spacing . . . 259

The Anatomy of a Letter . . . 260
- X-Height . . . 260
- Descender . . . 260
- Ascender . . . 260

Text Creation . . . 261
- Point Type/Paragraph Type . . . 261

Start Typing . . . 262
- Creating a Text Box . . . 262
- Text in a Container . . . 262
- Type on a Path . . . 262

Setting Character Attributes . . . 263
- Fonts . . . 263
- Size . . . 263
- Leading . . . 264
- Kerning . . . 264
- Tracking . . . 265
- Scaling . . . 265
- Baseline Shift . . . 265
- Character Formatting Styles . . . 265
- Language . . . 266
- Display . . . 266
- Setting Paragraph Attributes . . . 267

Import Text . . . 268

Manage Text . . . 269
- Resize a Text Area . . . 269
- Selecting Text . . . 270

Text Options . . . 271
- Area Type Options . . . 272
- Type on a Path Options . . . 273
- Thread Text/Linking Text . . . 274
- Remove or Break Threads . . . 274
- Wrap Text Around an Object . . . 275

Touch Type Tool . . . 276

Outline Fonts . . . 277

Exporting Text . . . 277

Adobe Fonts . . . 278
- Activating Fonts . . . 279
- Find/Replace Missing Fonts . . . 279
- Packaging Fonts . . . 279

Paragraph and Character Styles . . . 281
- Paragraph Style Attributes . . . 281
- Character Style Attributes . . . 281
- Create Paragraph Styles . . . 281
- Applying a Paragraph Style . . . 282
- Editing a Paragraph Style . . . 283
- Style Overrides . . . 283
- Redefine a Style . . . 283
- Delete a Style . . . 283
- Load Styles . . . 283

Character Styles . . . 284
Creating, Applying, and Editing . . . 285
Graphic Styles . . . 286
Create a Graphic Style . . . 286
Save a Graphic Style . . . 286
Merge Two or More Existing Graphic Styles . . . 287
Styles Applied to a Group . . . 287
Graphic Styles on Active Type . . . 288
Graphic Style Libraries . . . 289
Save a Graphics Style Library . . . 289
Rename a Graphic Style . . . 293
Delete a Graphic Style . . . 290
Break the Link to a Graphic Style . . . 290
Replace Graphic Style Attributes . . . 290
7 Working with Color 291
The Basics . . . 292
RGB . . . 292
CMYK . . . 292
Process Colors Versus Spot Colors . . . 292
Color Myths, Theory, and Management . . . 294
Ask an Expert About Color . . . 294
Grasping at Light . . . 294
Devices and Their Disappointing Limitations . . . 295
So What Should I Do? . . . 296
Profiles . . . 297
The Flexibility of RGB . . . 298
The Useful Rigidity of CMYK . . . 299
Final Advice . . . 300
Swatches Panel . . . 301
Swatches Panel Options . . . 305
Creating New Swatches . . . 302
Global and Non-Global Colors . . . 305
Color Modes . . . 305
Spot Colors . . . 306

Editing Colors . . . 307
- Swatch Panel . . . 307
- Color Panel . . . 307
- Color Picker . . . 307
- HEX Colors . . . 307
- Tints . . . 308
- Opacity . . . 308

Sampling Color . . . 309
- Eyedropper . . . 309
- Adding Sampled Colors . . . 311

Color Theme Panel . . . 312
- Create . . . 312
- Explore . . . 313
- My Themes . . . 314

Color Guide Panel . . . 316
- Color Harmonies . . . 316
- Color Guide Options . . . 317

Gradient Panel . . . 319
- Apply a Predefined Gradient . . . 319
- Editing and Creating Gradients . . . 319
- Edit Colors . . . 320
- Gradient Types . . . 321
- Gradient Tool . . . 322
- Apply a Gradient to a Stroke . . . 324
- Freeform Gradients . . . 326

Color Libraries in AI . . . 329
- Spot Colors . . . 329
- Saving Swatches . . . 329
- AI or ASE files . . . 329
- Importing Swatches . . . 330

Color in CC Libraries . . . 331
- Creative Cloud Color Storage and Usage . . . 332

Recolor Artwork . . . 334
- Recolor Artwork Panel . . . 334
- Assign Colors . . . 335
- Editing Existing Colors . . . 336

8 Output 337

PDF . . . 338
Presets . . . 338
General Options . . . 339
Compression . . . 340
Marks and Bleeds . . . 341
Output . . . 342
Advanced . . . 343
Security . . . 343
Summary . . . 343

Asset Export . . . 344
Adding Assets . . . 344
Export Settings . . . 344
File Formats . . . 344
Export Assets . . . 347
Updating Assets . . . 347

Print . . . 348

Package . . . 349
A Copy of Everything . . . 349

Appendix 351

Illustrator Keyboard Shortcuts . . . 352
Tools . . . 352
Menu Commands . . . 354
Miscellaneous Shortcuts . . . 359

Index 361

Introduction

Welcome to Illustrator!

In this book, you will be working your way through a full course curriculum that will expose you to all of the essential features and functions of Adobe Illustrator. Along the way, you'll learn the concepts and vocabulary of graphic design and page layout.

Between several larger projects are chapters of lessons. In those lessons, each action that I'd like you to try is marked with an arrow icon:

➡ This is what an action looks like.

The surrounding paragraphs explain some of the why and how. For greater depth, the second section of this book is a Compendium of those features and functions, providing the "deep dive" needed for true mastery of this powerful application. Throughout the Course section, I will suggest readings in the Compendium section. Although you will be able to complete the entire course without them, I think if you do those readings you'll find yourself regularly nodding and muttering, "oh, that's why it works that way."

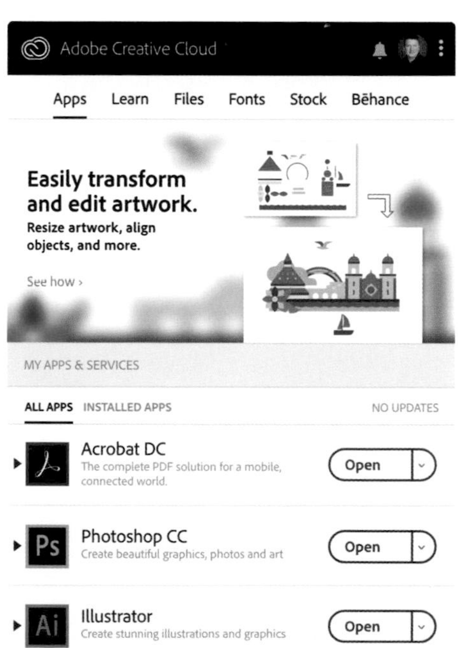

To follow along with the projects and lessons in this book, you'll need the files. Launch your favorite web browser and go to rockynook.com/illustratorCandC, answer a simple question, and download the files. Put that files somewhere convenient (and memorable).

Have you installed Illustrator yet? If you work for a company with an enterprise license, it's likely your IT people have installed it for you. We will be using the Creative Cloud app as our hub for launching Adobe applications and accessing the services that come with a Creative Cloud (CC) license. This app also checks to make sure your software license is up-to-date, so it should remain running whenever you use your creative applications. I use the CC app's Preferences to have it launch on startup and auto-update software so I don't have to worry about it.

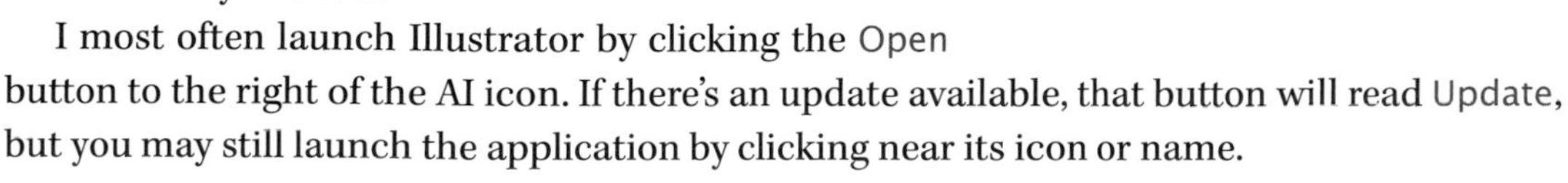

I most often launch Illustrator by clicking the Open button to the right of the AI icon. If there's an update available, that button will read Update, but you may still launch the application by clicking near its icon or name.

THE COURSE

1 Build with a Solid Base

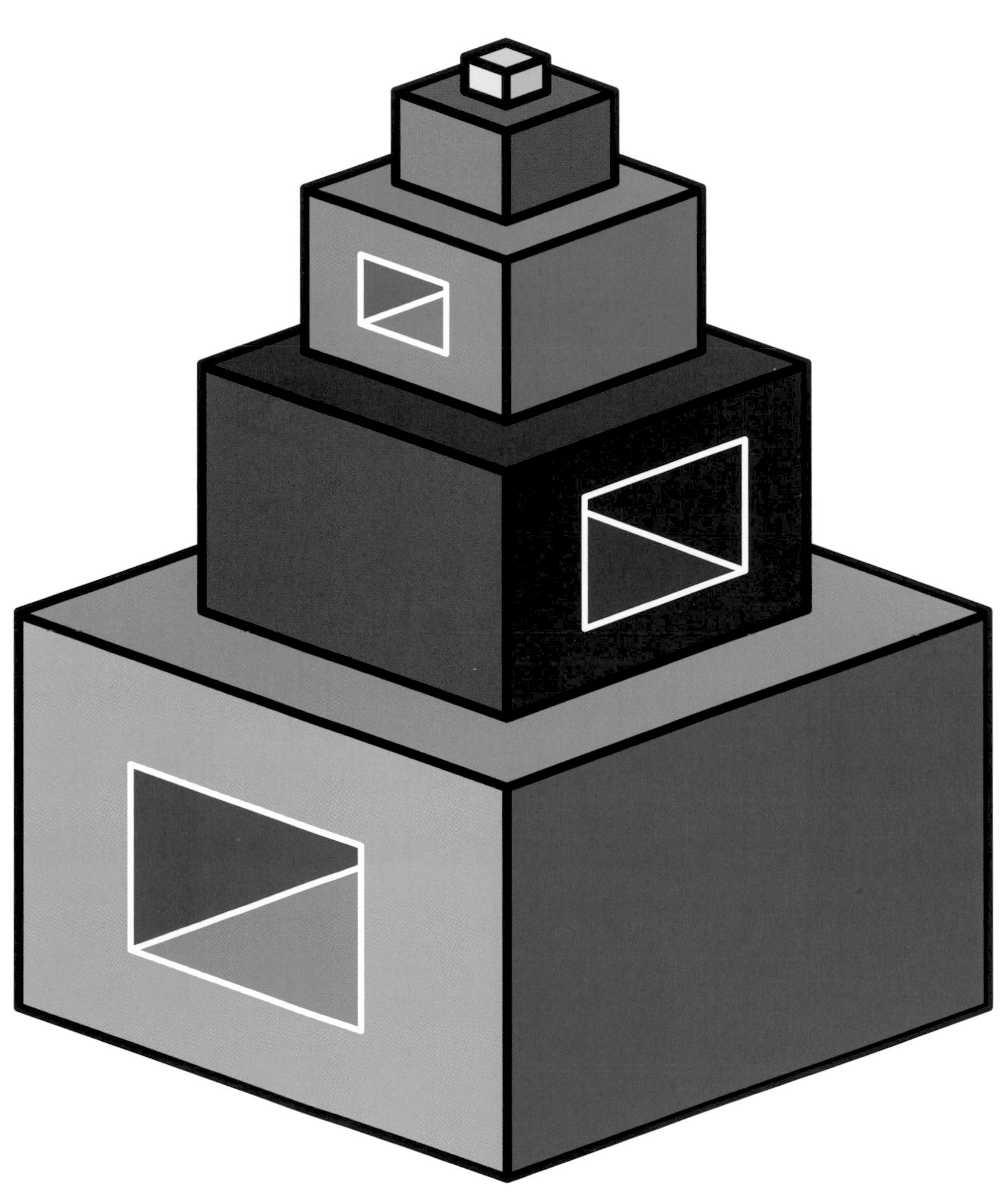

Before You Begin

Before you start creating in Illustrator, there are some preference you may want to set to ensure success with all you create. Setting your preferences before you open any documents will ensure that all the documents you create will have the same preferences. Many of the preferences you'll set are specific to each document, but there's no way to distinguish them from those that are globally applied to the application. The full discussion about customizing Illustrator can be found in the first chapter of the compendium, Preferences and Workspaces, but this first section will walk you through the basic customizations that you will likely want to set.

General

Let's open the Preferences menu. On a Mac, use the Illustrator CC menu; on a PC go to Edit > Preferences. The Preferences shortcut is ⌘-K/Ctrl-K.

Set the Keyboard Incerement value to a low increment. I would use 1 point or 1 mm. Currently the document is in points as a unit of measure so each time you nudge an object with the keyboard arrows, it will advance in this increment. You can change the units later if points is not what you prefer.

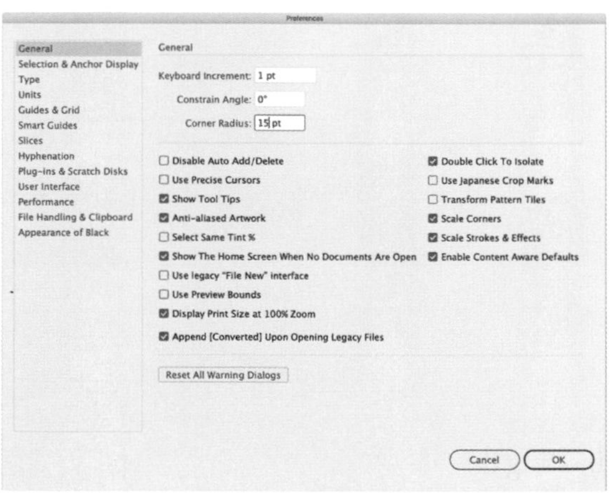

Show The Home Screen When No Documents Are Open

The Home screen appears when this setting is checked, and displays your recent documents and new document presets. This screen is not necessary since you get another screen that looks nearly the same once you create a new document.

Selection & Anchor Display

A few setting here will make your creations easier to manage, select, move, and edit. Many of the boxes are checked by default, which is good as they are helpful.

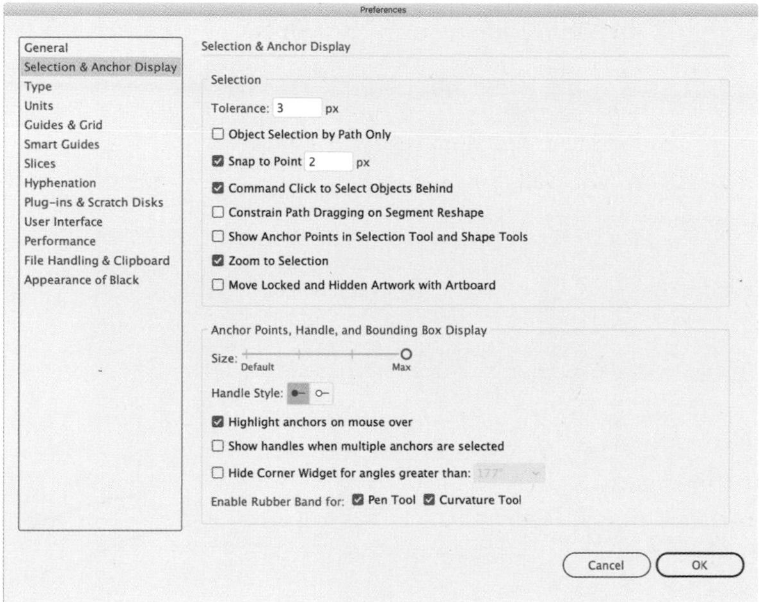

Anchor Points, Handle, and Bounding Box Display

Adjusting this value makes the anchor points, handles, and points larger based on the sliding scale. Slide the slider to make the points larger and find a size that works for you.

Show Handles When Multiple Anchors Are Selected

If you are using curves, then you will be working with pull handles. By default, the handles only show on the point you have selected. Check this box and all the points on your selected object will show handles.

Type

Size/Leading

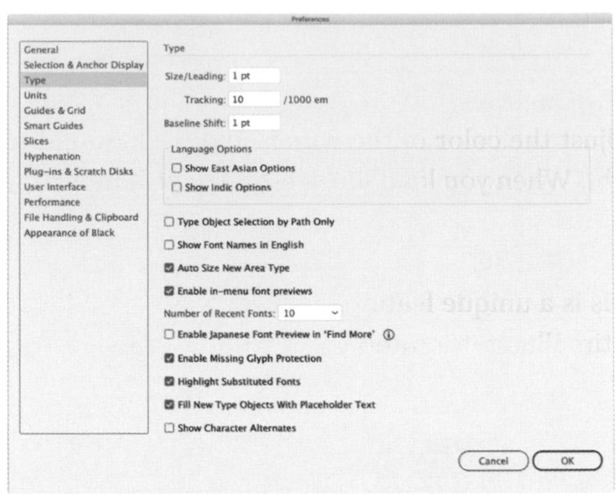

Since I am a huge fan of shortcuts, I change these three settings from their defaults. I set the Size/Leading keyboard shortcuts to adjust in 1 pt increments when using the Shift+⌘+. (Mac) / Shift+Ctrl+. (PC) to increase the type size and Shift+⌘+,/ Shift+Ctrl+, to decrease type size.

Tracking

Here I set the Tracking defaults to adjust in 10/1000 em, using ⌥+→ /**Alt**+→ and ⌥+← /**Alt**+← to increase and decrease the kerning and tracking.

Baseline Shift

I change the defaults of 2 pt to 1 pt increments when using the ⌥+↑ /**Alt**+↑ and ⌥+↓ /**Alt**+↓ to increase and decrease the baseline shift.

Units

General

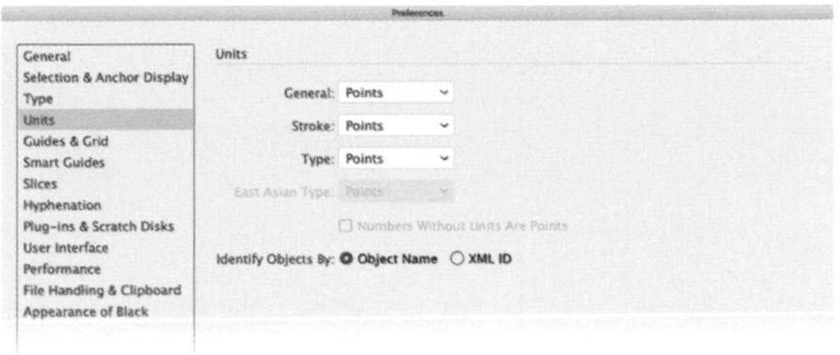

This is used for measuring the size and position of objects, the size of the artboard, and the size of the rulers. Choose whichever unit of measurement you're most comfortable using.

Stroke, Type

Points is the standard unit of measure for specifying the stroke weight and type size.

User Interface

Brightness

Adjust the color of the panels and background to be dark, medium dark, medium light, or light. When you load Illustrator, the default is a dark background.

UI Scaling

This is a unique feature that I have not seen before and I like it. This allows you to scale the entire Illustrator interface, including tools, cursors, and menu fonts from small to large.

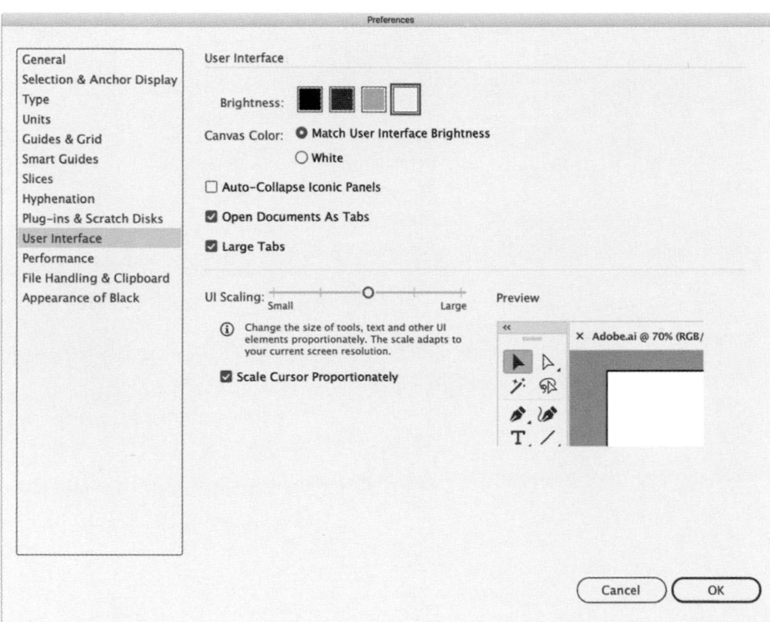

Preferences & Workspaces
Shape Creation
Advanced Construction
Editing & Transformation
Effects & Graphic Styles
Type & Text
Working with Color
Output

GPU Performance

To better control zooming to specific objects, I recommend disabling Animated Zoom. I prefer to use the Zoom tool in a precise way by clicking and dragging over the object I want to zoom in on rather than sliding the Zoom tool over the object to control the zoom size.

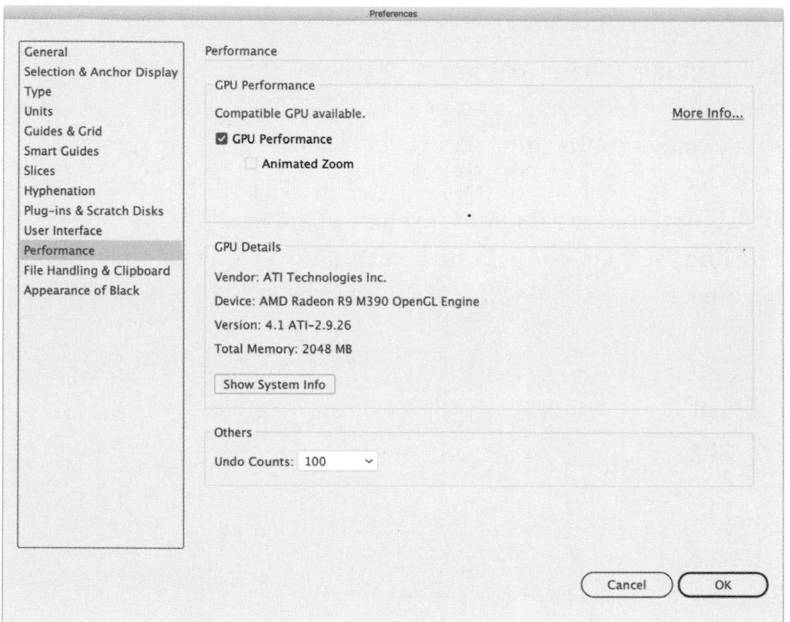

Configuring the Workspace

Choose an Initial Workspace

I start off with Essentials Classic Workspace because it offers all the basic panels and tools. I add panels as needed when I am working. In the upper-right corner of the application, you'll see the Workspace menu. Choose the workspace that best fits your creative goal, and the panels that Illustrator thinks are the most useful will appear.

One feature I find very annoying is that in some workspaces, the toolbar will become streamlined and many of the tools will not be visible. How frustrating that can be! To make things look "cleaner" the tools are moved to the Edit section at the bottom of the toolbar. Look for the three dots at the bottom of the toolbar; that is where all the other tools are hiding.

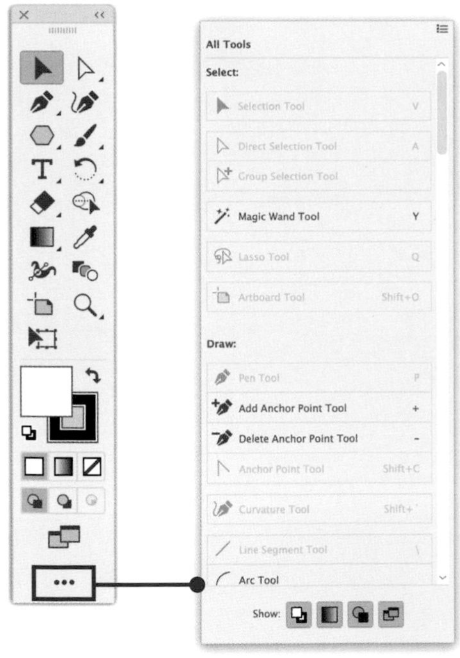

The "streamlined" toolbar has fewer tools. Click on the three dots to display the list of hidden tools.

Click on the tool to add it back to the toolbar.

When you choose a workspace, all the panels relevant to that workspace will be visble. At the top of the stack of panels is a small button with << in it. When clicked, it expands the panels so you can see their contents. Clicking it again collapses the panels to icons.

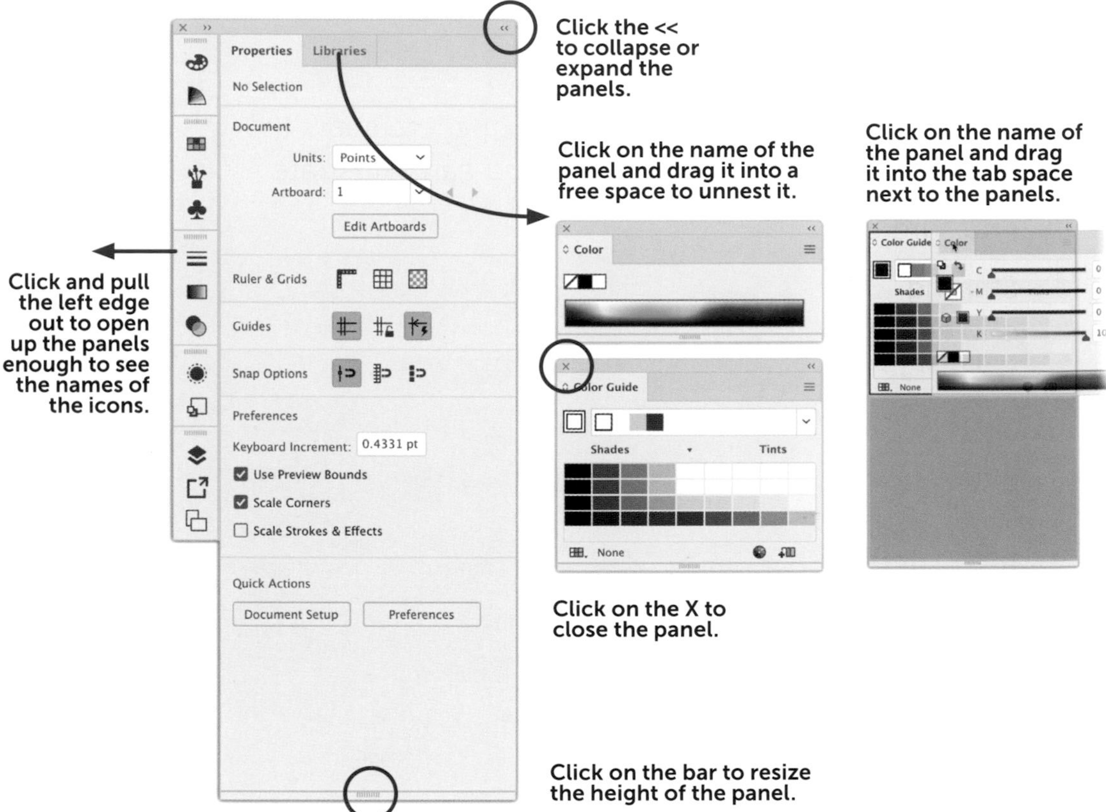
Click the << to collapse or expand the panels.
Click on the name of the panel and drag it into a free space to unnest it.
Click on the name of the panel and drag it into the tab space next to the panels.
Click and pull the left edge out to open up the panels enough to see the names of the icons.
Click on the X to close the panel.
Click on the bar to resize the height of the panel.
Properties
Libraries
No Selection
Document
Units:
Points
Artboard:
1
Edit Artboards
Ruler & Grids
Guides
Snap Options
Preferences
Keyboard Increment:
0.4331 pt
Use Preview Bounds
Scale Corners
Scale Strokes & Effects
Quick Actions
Document Setup
Preferences
Color
Color Guide
Shades
Tints
None

You can adjust each panel's height by grabbing the bar along the bottom edge—watch for the two-headed arrow, then drag the bar to resize the panels.

To unnest panels, click on the panel's tab and drag the panel to a free space. Nest panels together by clicking on the panel tab and dragging it into another tab. The panel will become transparent and a blue outline will appear when you are in the right nesting zone.

Create a New Workspace

When the panels appear to be just as you'd like them (for now), capture that arrangement by returning to the Workspace menu and choosing New Workspace.... Name it — I'm going with "Real Essentials." If any of those panels go missing, or if there is a mess of panels in the way, you can choose Reset Real Essentials from that menu to recall your preferred setup.

Projects: An Introduction to Illustrator

I have heard from so many people over the years that they are not creative; they can't draw and can't illustrate.

You sure can!

Starting with the basics of shape creation tools and color choices, this book will get you up to speed with the tools that will push your creativity to its potential.

These lessons walk you through the basics of creating a new document, working with basic shape building tools, adding some color, and learning new creation techniques. If you want more, several of these lessons will show you just how much Illustrator has to offer.

Get ready to create and wow yourself!

Lesson A: Create a New Artboard

Lets start with a new document and get used to using artboards. Don't call them pages! You're creating in Illustrator, so artboards they are.

- First, launch Illustrator. Use the Creative Cloud app—that's your hub for all Adobe apps. Once the program is running, you can create a new document either by going to File > New > Document... or by clicking on the Create New... button on the welcome screen. If you use the menu method, you'll notice a keyboard shortcut that does the job, too: on a Mac, it's ⌘-N (hold down the Command key and type "n"), and on Windows, it's Ctrl-N (hold down the Ctrl key and type "n"). Hereafter, I'll indicate shortcuts in that order for Mac and Windows, respectively, like this: ⌘-N/Ctrl-N.
- In the New Document window, you can set your preferred unit of measurements, as well as the number of artboards and their size and orientation. If you don't know how many or what size artboards you need, you can change the number and size of your artboards at any time in the creation process. Click on the Artboard tool in the toolbar. The shortcut is ⇧+O.
- Click on the artboard with the Artboard tool to make pull handles appear at the corners and midpoints. Pull a handle to resize the artboard. To exit artboard editing, click on any other tool in the toolbar.
- Double-click on the Artboard tool to call up the options menu. Here you can name the artboard and edits its size and orientation. Check out the helpful hint at the bottom of the panel showing a quick way to duplicate an artboard using the Artboard tool. You can also use the Artboard tool to draw more artboards in your document.

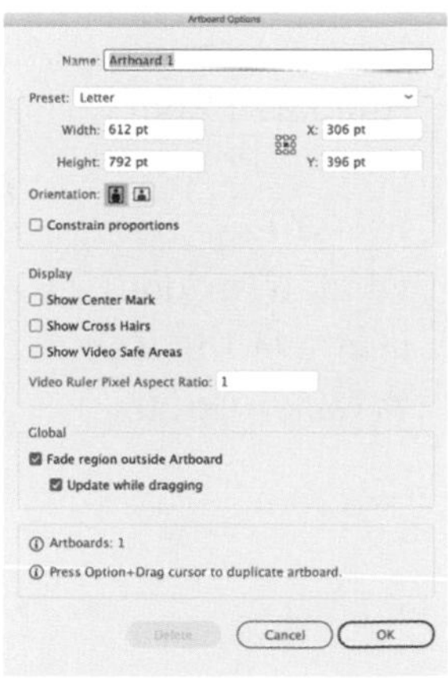

- Open the Artboard panel by choosing Window > Artboards. In this panel you can see the artboards listed. You can click on the dropdown menu and choose options such as New Artboard, Duplicate Artboard, or Artboard options. The Artboard panel shows all the active artboards. Double-click on the artboard's number to locate it. Double-click on the name to rename the artboard. Double-click on the artboard icon to open the artboard options. Click on the New Artboard icon at the bottom of the panel to create a new artboard. Click on the name of the artboard then click on the Trash Can icon at the bottom of the panel to delete it.

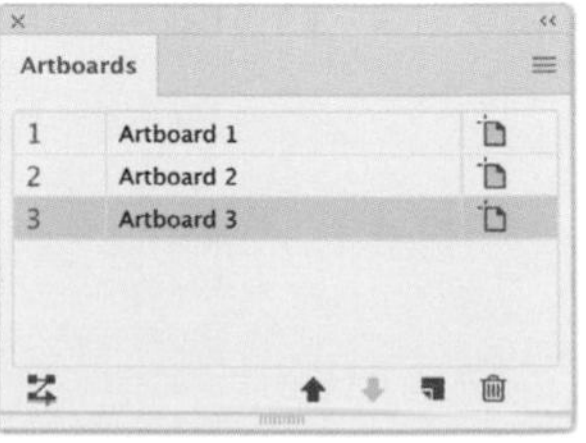

Lesson B: Hand Tool and Zoom Tool

Quick navigation is essential to getting proficient in Illustrator. The Hand tool and Zoom tool are keys to that navigation, and there are a few trick to use them efficiently. Scroll bars take too long to navigate and I never use them. I like these shortcuts instead.

Hand Tool

- Find the Hand tool in the toolbar or use H as a shortcut.
- Click anywhere on the document to grab and move the window around.
- Hold the Space Bar. This will activate the Hand tool. Let go and you are back to the last tool you were using. Note: If you try that when you are typing, you'll get spaces! In this case there is a shortcut to the shortcut: ⌥/Alt will show the Hand tool when you are in type editing.

Zoom Tool

- Find the Zoom tool in the toolbar or use Z as a shortcut.
- Click anywhere on the document to zoom in. No mystery there.
- With Animated Zoom turned off in Preferences (Preferences > Performance > Uncheck Animated Zoom) you can click and drag the Zoom tool over any object and the area you define will zoom full screen and be centered in the window.
- Everything is big. Now what? Double-click on the Hand tool to fit the artboard to the window, or double-click on the Zoom tool to show the artboard as actual size. You can also use the Zoom In and Out commands under the View menu: ⌥ (Mac)/Alt (PC), ⌘ (Mac)/Ctrl (PC).

Zoom In	⌘+
Zoom Out	⌘–
Fit Artboard in Window	⌘0
Fit All in Window	⌥⌘0
Actual Size	⌘1

Lesson C: Create an Apple

One things that I stress in any Illustrator class I teach is you do not need to know how to draw. We create with basic shapes. You will see the magic happen once we start creating. Start a New document and lets begin.

Drawing an Oval

- **1** Draw an oval using the Ellipse tool (L). Select it with the Selection tool (V).
- **2** Duplicate the shape using Option + click (Mac)/ Alt + click (PC) and drag it.
- **3** Select the first oval and open the Properties panel. Rotate the shape 5 degrees. Select the other shape and rotate it -5 degrees.

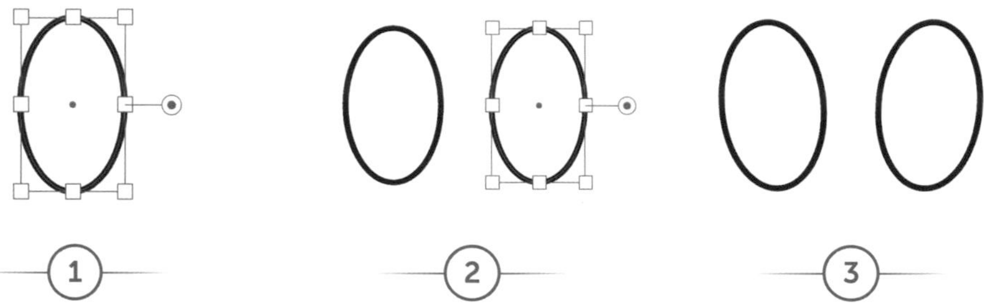

Pathfinder Tool

- **4** Select both ovals with the Selection tool (V).
- **5** Open the Pathfinder panel and choose the Unite option. This makes the two shapes into one shape.
- **6** Select the apple and adjust the height by using the pull handles.

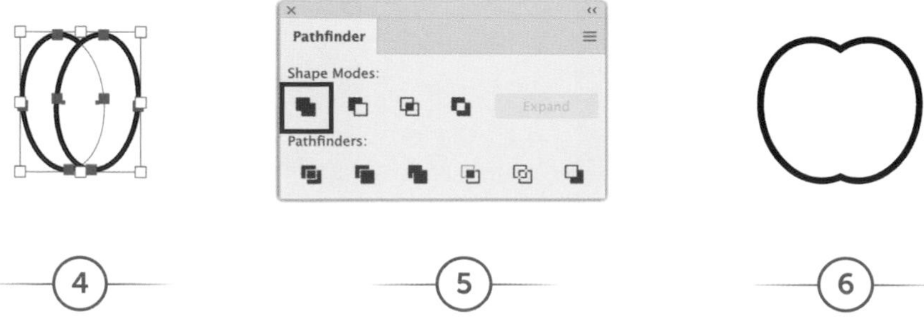

7 Create a circle with the Ellipse tool, and hold Shift while drawing to make it a perfect circle. Duplicate the circle, and place one halfway on top of the other.

8 Choose the Intersect option in the Pathfinder panel. This makes the two shapes into one shape by keeping only the piece of the shape made by the intersection of the two circles.

9 You just made a leaf without drawing it.

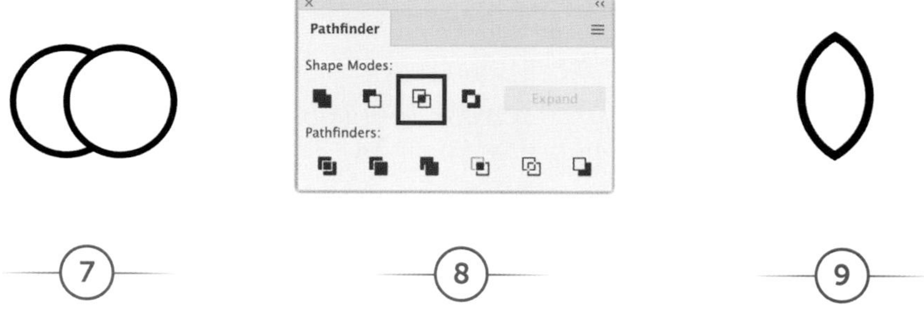

Selecting Color

10 Open the Swatches panel, Window > Swatches. Select the apple. Click on the fill icon and choose red. Click on the stroke icon and select a darker red.

11 Select the leaf and fill it with a green. Add a darker green stroke.

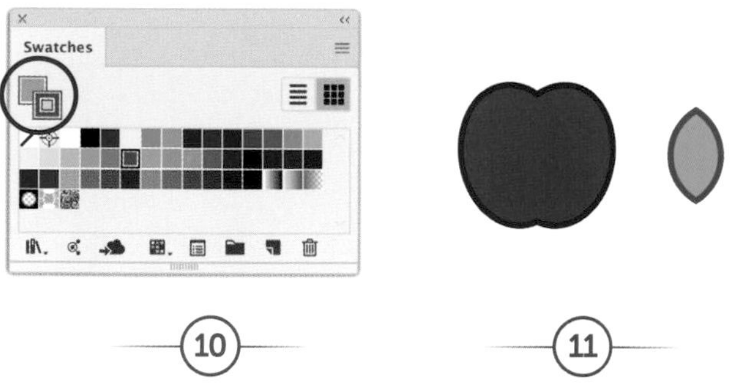

Arc Tool: Stem and Highlight

- **12** Select the Arc tool. Hold Shift and draw an Arc.
- **13** Open the Stroke panel. Window > Stroke. If the Stroke panel is small, click on the panel dropdown menu and choose Show Options. Select the arc and cap the end. (I call it hotdogging the ends.)
- **14** Choose a brown stroke for the arc to make the stem.

- **15** Assemble the apple, leaf, and stem. Rotate the leaf by selecting it and hovering over (but not touching) the outside pull corner, and hold Shift when you rotate the shape to constrain it to a 45-degree angle. You may need to scale the arc as well. Open the Transform panel and uncheck the Scale Stroke & Effects checkbox to keep the stroke from scaling when you scale the object.
- **16** Select the stem, then duplicate with Option + click (Mac) / Alt + click (PC) and dragging it. Change the color of the stroke to white and add it to the apple as a highlight "noodle."

Lesson D: Create a Spray Can

Let's build a spray can with a nice flow of spray coming out of it.

Drawing Basic Shapes and Lines

1 Draw a rectangle using the Rectangle tool (M). Apply an 8 pt stroke to it. Set the stroke weight in the Stroke panel or in the Properties panel. Both panels are under the Window menu. In the Stroke panel set the corners to be Round Join, which rounds the four corners of the stroke.

2 Draw a line using the Line Segment tool (\); hold Shift while drawing to keep the line constrained in one direction. In the Stroke panel set the End Caps to be rounded. Make the line the width of the rectangle and the same 8 pt stroke weight. Duplicate the line and move one to the top and one to the bottom of the rectangle.

3 Draw an oval using the Ellipse tool (L). Make it the same width as the rectangle and the same 8 pt stroke weight.

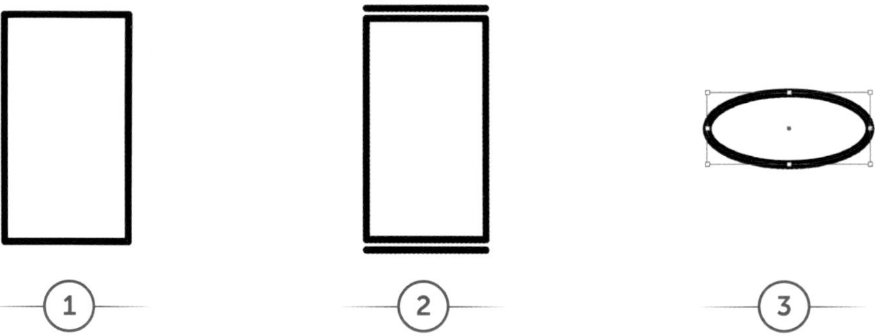

Pathfinder Tool

- 4 Draw a rectangle over the oval so the top of the rectangle is in the middle of the oval.
- 5 Select both shapes, then use the Pathfinder panel to select the Minus Front Shape mode to remove the lower half of the oval.
- 6 This becomes the cap of the spray can.

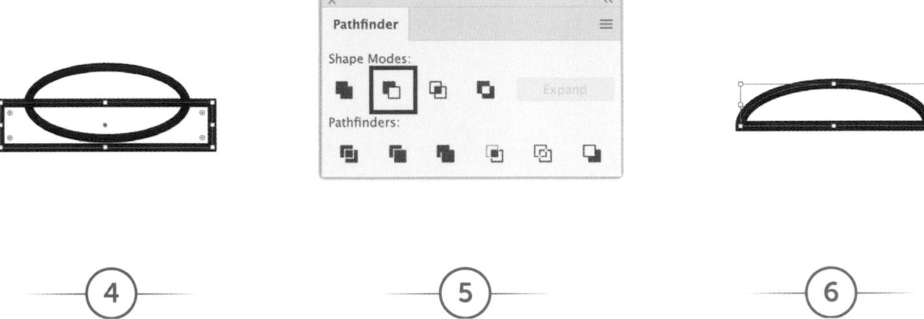

4 5 6

- 7 Draw a small rectangle on top of the oval for the spray tip.
- 8 Use the Direct Selection tool to select the two top corners, and pull the corner widget targets in toward the center slightly to round the corners.
- 9 Select the two lines, the oval, and the spray head, and change the stroke color to light gray. Choose a color for the spray can body and add a stroke and fill color (they should be the same color, in this case, to blend the lines of the rectangle into the shape).

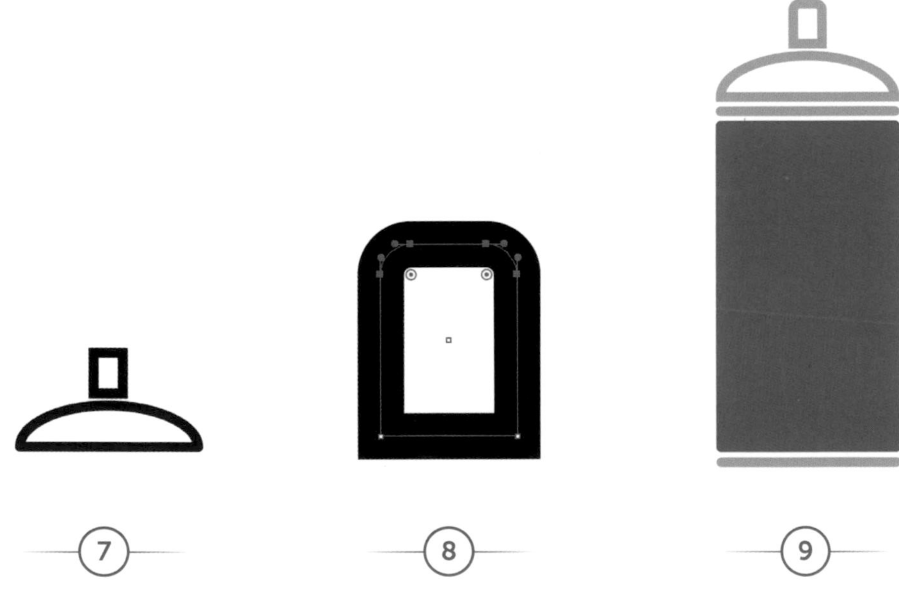

7 8 9

Add Color and Highlights

10 Move the shapes together so they touch. Change the spray head to a darker color to create contrast.

11 Add highlights to the can by drawing a line and applying white to the stroke. Use the Arc tool to draw an arc for the highlight noodle on the dome of the can. Round the ends of the lines and arc in the Stroke panel and round cap the ends.

12 Select the tip and choose Object > Path > Outline Path to convert the stroke to a shape. In the Pathfinder panel choose Unite to merge the shapes and the tip together.

Pathfinder

- **13** Draw a circle and move it to one side of the spray head.
- **14** Select both the spray head and the circle, then use the Pathfinder panel Minus Front mode to remove the dot and create a spray nozzle.
- **15** Draw a small circle to the right of the spray head, and fill it with the same color as the can.

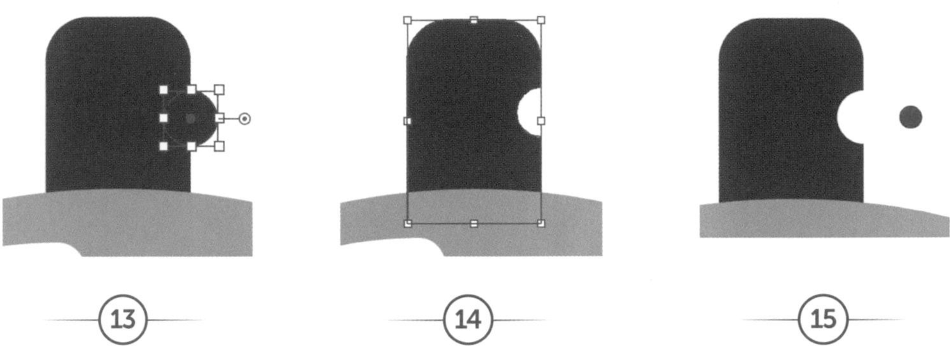

Duplicate and Repeat

- **16** Using the Selection tool, Option+click and drag (Mac) or Alt+click (PC) the circle to duplicate it.
- **17** Use Command+D (Mac) or Ctrl+D (PC) to duplicate it several times. Select all the dots and choose Object > Group.
- **18** Select the group, then select the Rotate tool (R) and double-click on it to open the Rotate dialog box. Set the angle to 10 degrees and click Copy (clicking OK will not copy it).

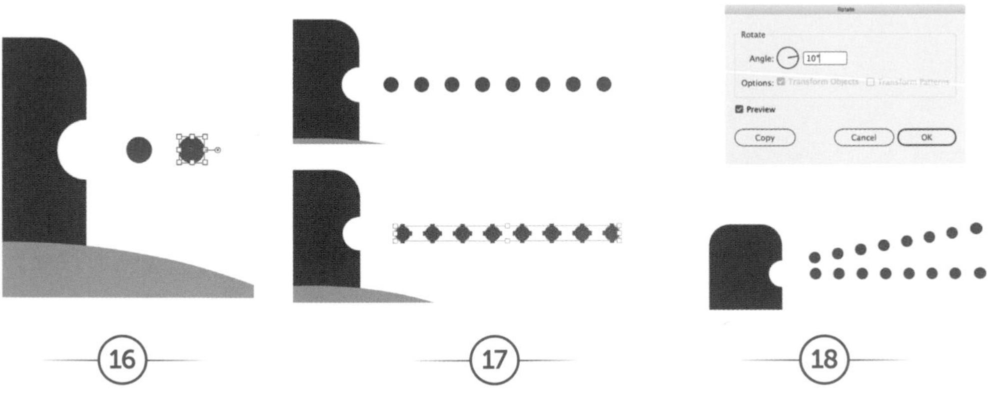

Final Assembly

- **19** Repeat the process of selecting the first grouped line. Open the Rotate dialog box and rotate the first set to -10 degrees and click Copy.
- **20** Select each of the three spray lines and space them equally.
- **21** Admire your finished spray can. Nice job!

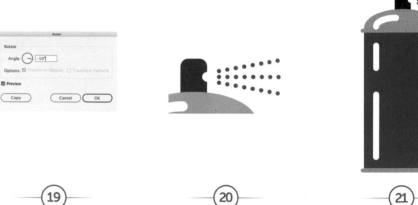

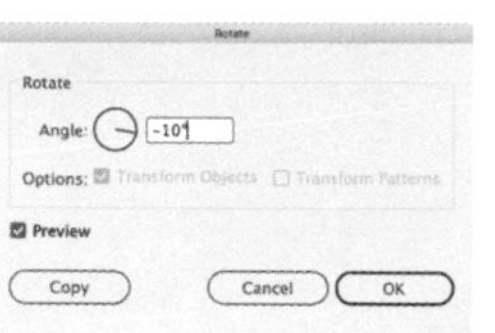

21

Project: Colors and Gradients

Illustrator has many options for altering the color and the shade of the color you're using. These projects will show you how to create and select colors and color schemes; and how to make those colors change on a gradient.

Lesson A: Selecting and Applying Color

Select Color

1 Draw a shape to apply a fill color and stroke color to it.

2 Open the Swatches panel from the Window menu. With the object selected, click on the stroke or fill icon in the upper left of the Swatches panel, then select a color.

3 Open the Properties panel from the Window menu. Here you can click on the fill or stroke color by clicking on the color thumbnail. You can also edit the stroke weight here (something you cannot do in the Swatches panel).

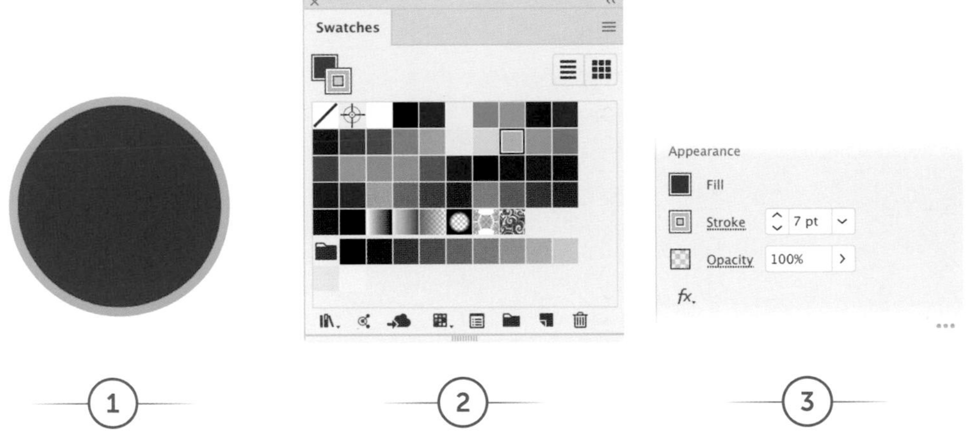

4 In the toolbar at the bottom are the stroke and fill icons that are larger and easier to click on. You can switch the fill and stroke colors by clicking on the double-ended arrow in the upper right of the fill/stroke icons (X). Click the box with the red slash under the icons to set the selected fill or stroke to none (/).

5 Double-click on either the fill or stroke icon to open the Color Picker to choose a color from the color spectrum or the different color models.

6 Open the Color panel from the Window menu to pick from the color spectrum or from the sliders.

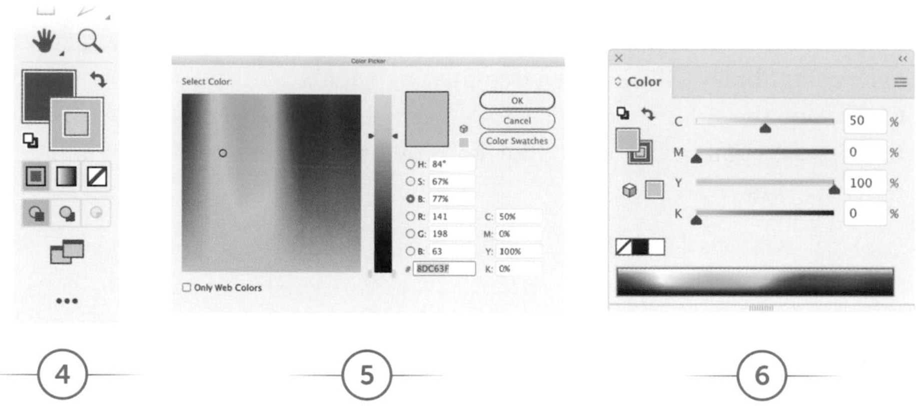

7 To turn a color into a global color that can be tinted, double-click on the color in the Swatches panel and check the Global checkbox.

8 Once a color is a global color, you can create a tint of it in the Color panel by sliding the tint slider or entering in a tint value.

9 Global colors appear with a white folded edge in the Swatches panel and offer advantages that non-global colors do not.

10 Explore more color options at the bottom left of the Swatches panel in the Color Libraries.

11 Color Libraries are built into Illustrator so you can choose from a large list of colors. Each color library will open in a free-floating panel.

12 Check out the Adobe Color Themes panel from the Window menu and click the Explore tab. You can search color harmonies based on your input value. This is a compact version of color.adobe.com.

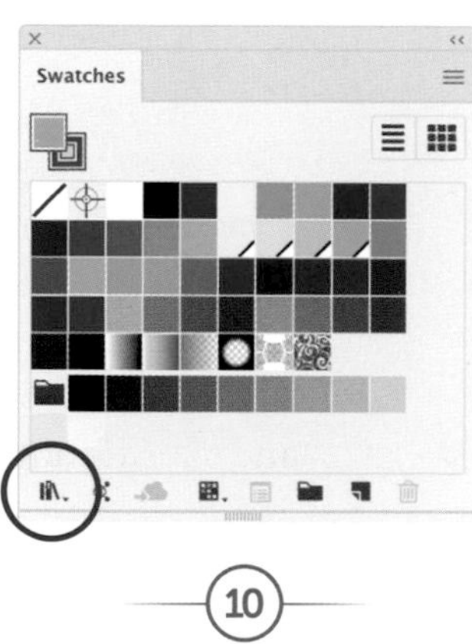

10

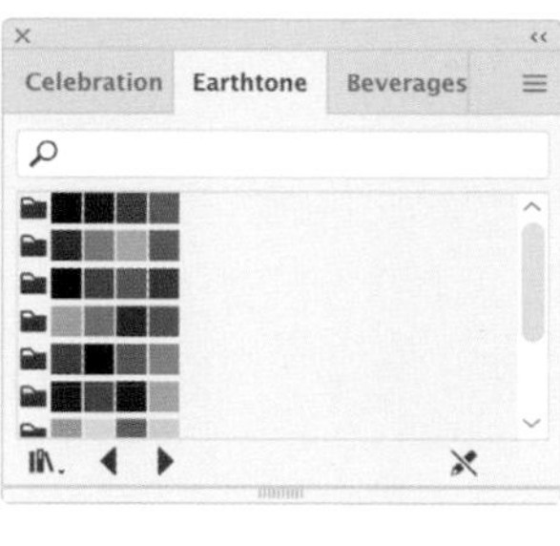

11

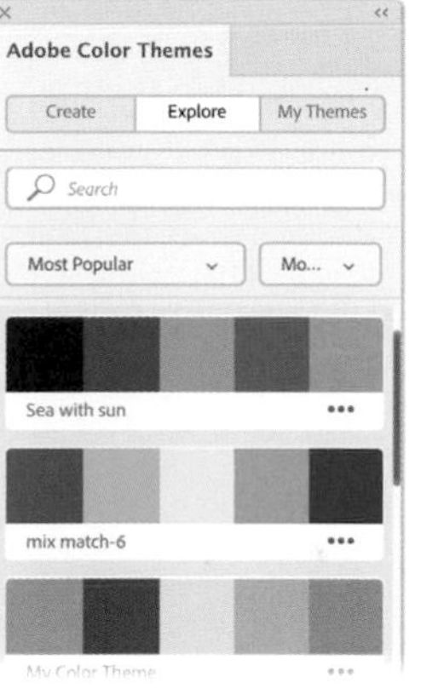

12

Lesson B: Gradients

Gradient Panel

- 1 Open the Gradient panel: Window > Gradient.
- 2 Draw a circle with the Ellipse tool (L). Hold Shift while drawing to constrain it to a perfect circle.
- 3 Select the circle and click on the fill icon in the Gradient panel. Click on the Gradient thumbnail to apply the black-and-white gradient to the fill of the circle.

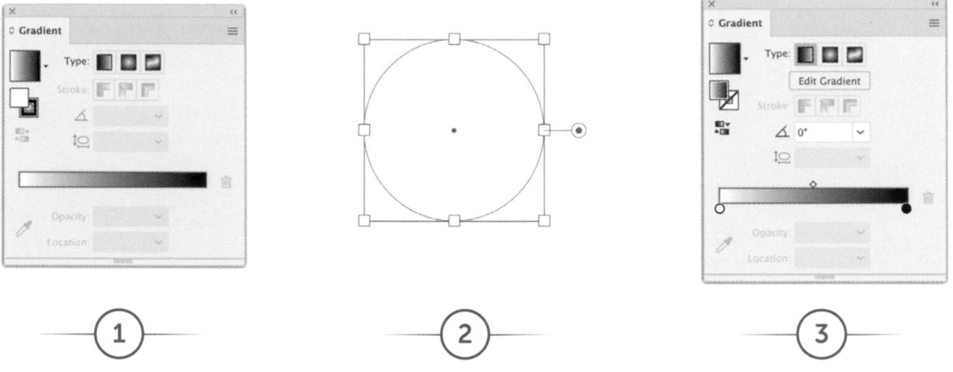

Editing Gradients

- 4 Double-click on the color stop below the gradient ramp. This will open the Swatches panel under the ramp.
- 5 Choose a color from the Swatch list.
- 6 Select the other color stop; double-click and choose white.

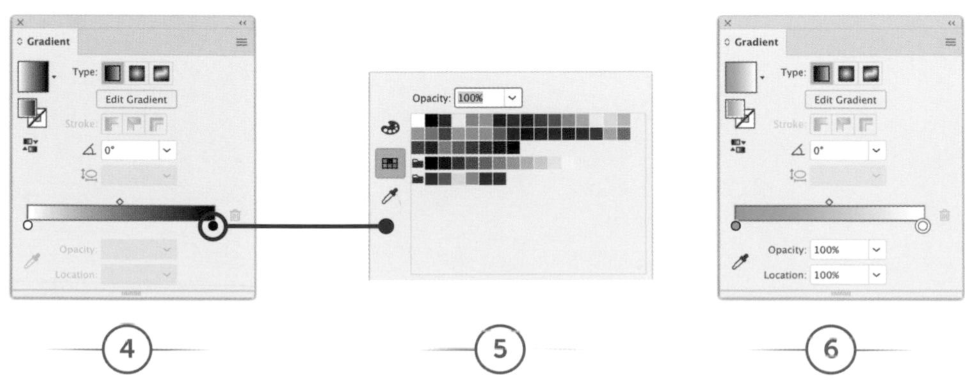

Offset Path

- **7** This is how the circle looks with the gradient applied to the fill.
- **8** Choose Object > Path > Offset Path. Set the Offset amount to -15 and click OK. This creates an exact duplicate of the shape inside (because of the negative number). A positive number will create a shape larger than the original.
- **9** Select the inside shape and click on the Reverse Gradient button in the Gradient panel to flip the gradient direction.

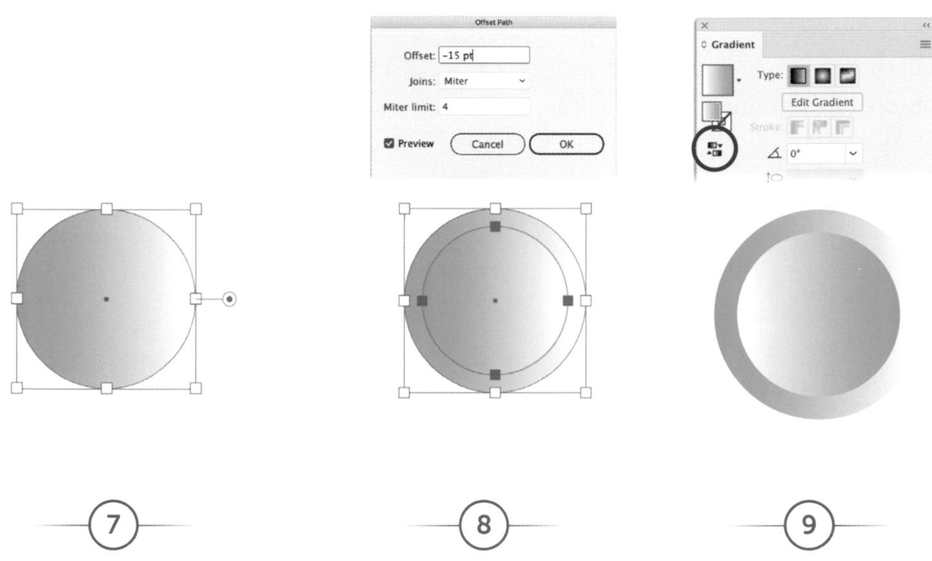

Project: Building Weather Icons

Lesson A: Water Droplet

Drawing a Circle

1 Draw a circle with the Ellipse tool (L). Hold Shift while drawing to constrain it to a perfect circle.

2 With the Direct Selection tool (A), click on the top of the circle to activate the top anchor.

3 Use the up arrow key to move the point up. Hold Shift+up arrow to move it ten times as fast. You've created and egg!

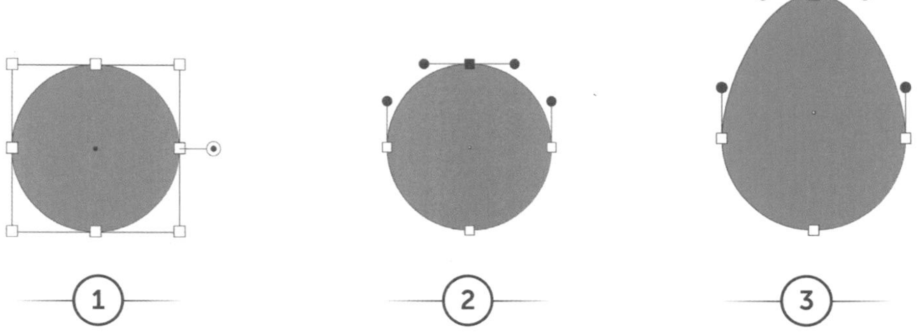

Convert Anchor Point

4 Open the Properties panel. Window > Properties. There are two type of points; Smooth and Corner. The egg shape is a rounded point, so we need to make this a corner point to make it into a raindrop.

5 Select the Corner point icon to convert the anchor point.

6 Well done—you just created a raindrop.

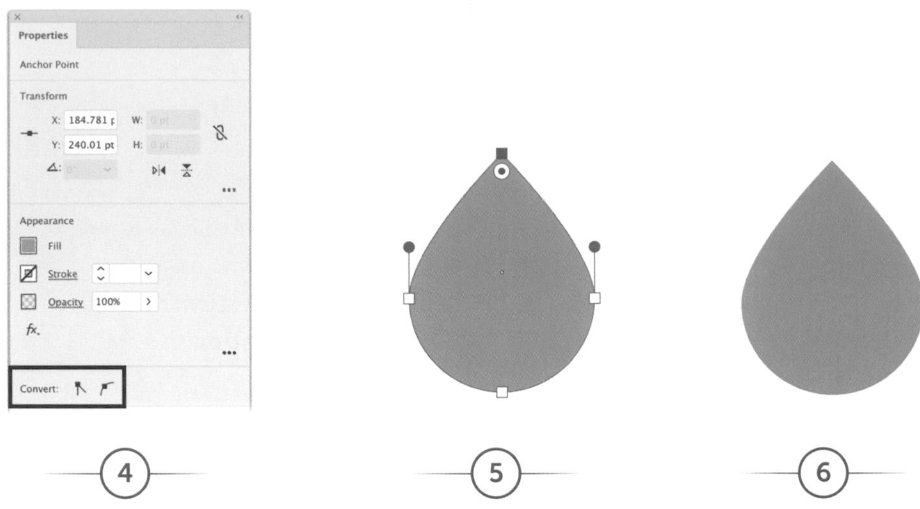

Arc Tool—Add a Highlight Noodle

- **7** Select the Arc tool. Hold Shift and draw an arc. Make the stroke white.
- **8** Open the Stroke panel: Window > Stroke. Select the arc and round cap the end.
- **9** Admire your handy work. A raindrop with a highlight noodle. The hotdogged ends make all the difference.

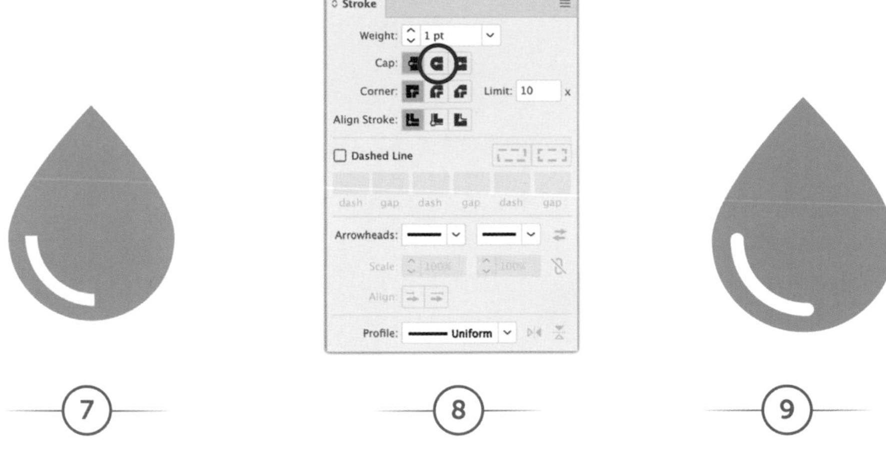

Lesson B: Moon and Stars

Pathfinder Tool/Shape Tool Alternative

1 Draw a circle with the Ellipse tool (L). Hold Shift while drawing to constrain it to a perfect circle.

2 Duplicate the circle by using the Selection tool (V). Option+click and drag (Mac) or Alt+click and drag (PC) the shape. Hold Shift while duplicating the shape to keep it directly in line with the original shape.

3 In the Pathfinder panel, choose minus front mode, which deletes the front-most object, leaving the moon shape.

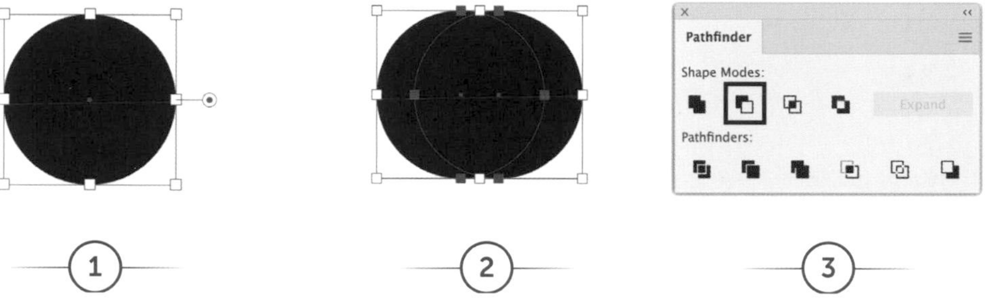

4 An alternate method to deleting the shapes to reveal the moon is using the Shape Builder tool (Shift + M). Select both shapes with the Selection tool, choose the Shape Builder tool, and hold Option (Mac) or Alt (PC). This converts the Shape Builder tool to subtract whatever part of the shape you drag over. Drag over the shapes you want to delete, and a mesh overlay appears showing what will be deleted.

5 The moon rises from these shapes.

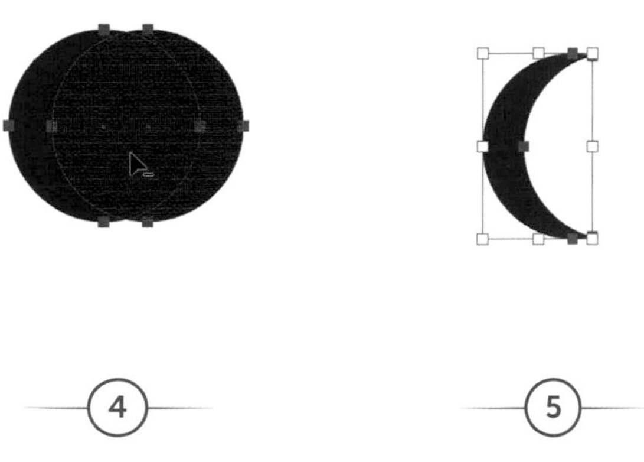

Star Tool Settings and Editing

1 Select the Star tool from the toolbar and click on your artboard to open the Star tool dialog box.

2 Radius 1 is measured from the center to the inside point. Radius 2 is measured from the center to the outside point

3 That is how to create a star using the Star dialog box.

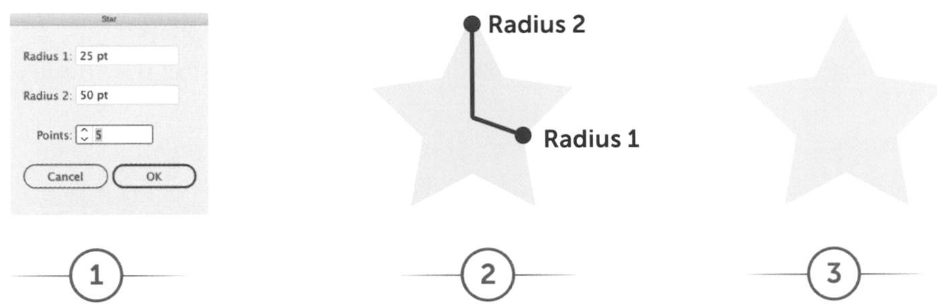

4 An alternate way to draw a star using the Star tool is to manually change the sides and point length while drawing. Use the up arrow to add points, and the down arrow to remove points. Don't let go of the mouse or you will be moving the shape, not editing it.

5 To make the points longer, hold Command (Mac) or Ctrl (PC) and pull out from the center. To make the points shorter, hold Command (Mac) or Ctrl (PC) and pull into the center.

6 The star you just created will be the new default star when you draw. Change the default by clicking on the artboard with the Star tool or manually create a new star with different attributes.

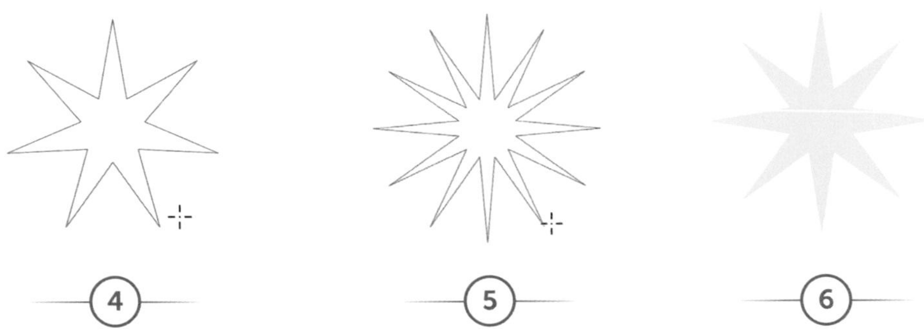

Lesson C: Cloud

Drawing Circles

1 Draw a circle with the Ellipse tool (L). Hold Shift while drawing to constrain it to a perfect circle.

2 Duplicate the circle by using the Selection tool (V) and Option+click and drag (Mac) or Alt+click and drag (PC) the shape. Hold Shift while duplicating the shape to keep it directly in line with the original shape.

3 Use Command+D (Mac) or Ctrl+D (PC) to duplicate the shape several times.

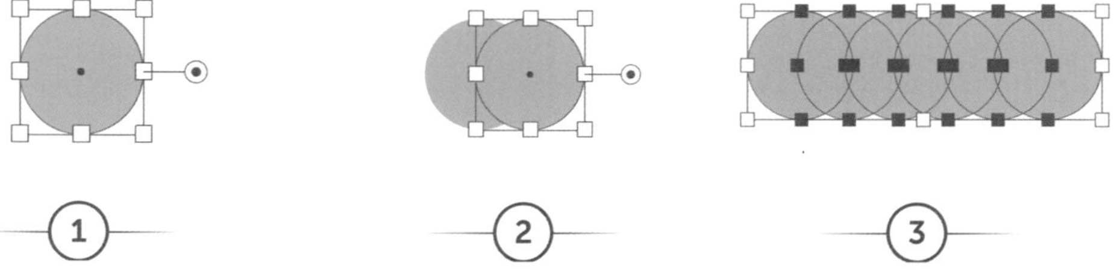

Duplication and Pathfinder

4 Duplicate the first row of circles, moving them halfway up on the first line. Repeat this for four lines. Delete a few circles from each line, ending up with one at the top. Select all the circles with the Selection tool.

5 Apply the Unite mode from the Pathfinder panel to create one shape.

6 You've creates one solid cloud from several circles.

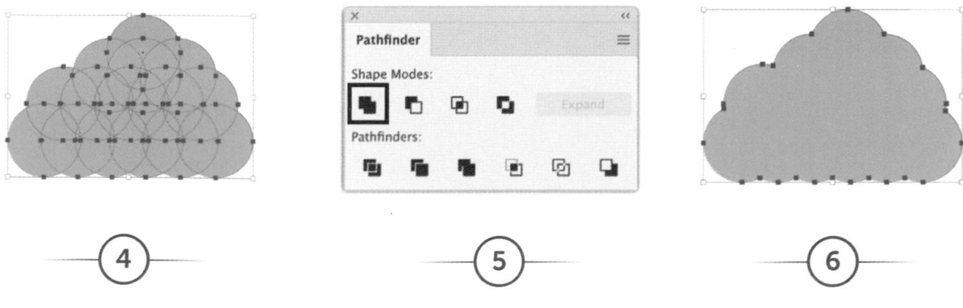

Pathfinder Tool/Corner Widgets

- 7 Draw a rectangle and move it over the base of the clouds. Select the clouds and the rectangle and choose the Minus Front mode from the Pathfinder panel.
- 8 Select the lower two corners of the rectangle with the Direct Selection tool: this will show the corner widget targets.
- 9 Pull the corner widget targets in toward the center of the cloud to round them.

7 8 9

Lesson D: Rain

Line Tool

1 Draw a vertical line with the Line tool (\). Hold Shift while drawing to constrain it to a straight line. Round cap the ends in the Stroke panel.

2 Duplicate the line by using the Selection tool (V) and Option+click and drag (Mac) or Alt+click and drag (PC) the shape. Hold Shift while duplicating the shape to keep it directly in line with the original shape.

3 Use Command+D (Mac) / Ctrl+D (PC) to duplicate it several times.

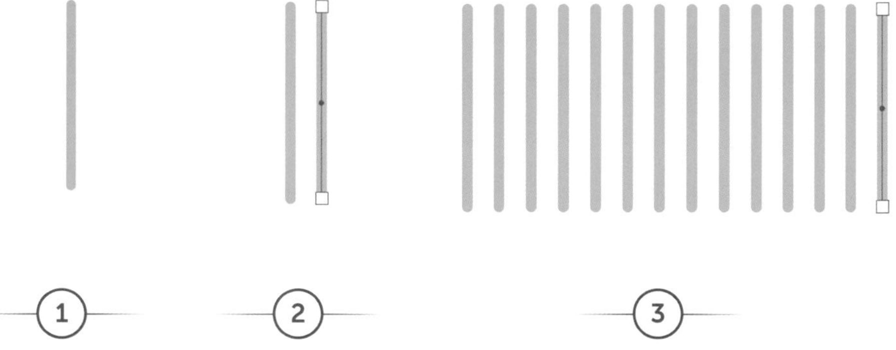

Shear Tool

4 Use the Direct Selection tool (A) to select the bottom point of the first line. Hold Shift+up arrow to move the end point up. Select the lower point of the next line and hold Shift+ down arrow (twice) to move the point down. Repeat this process, staggering the length of each line. Select all the lines and choose Object > Group.

5 Select the group of lines with the Selection tool. Double-click the Shear tool in the toolbar to open the Shear settings. Set the Shear Angle to be -10 degrees, Axis to Horizontal. Click OK.

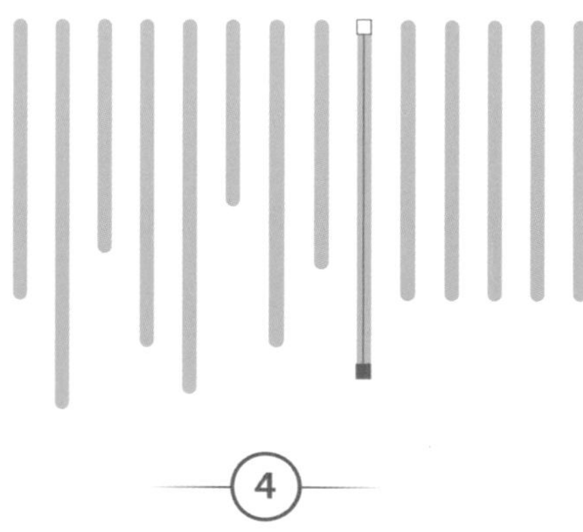

4

5

6 Apply a larger amount of shear to the group of lines to make it look like harder rain.

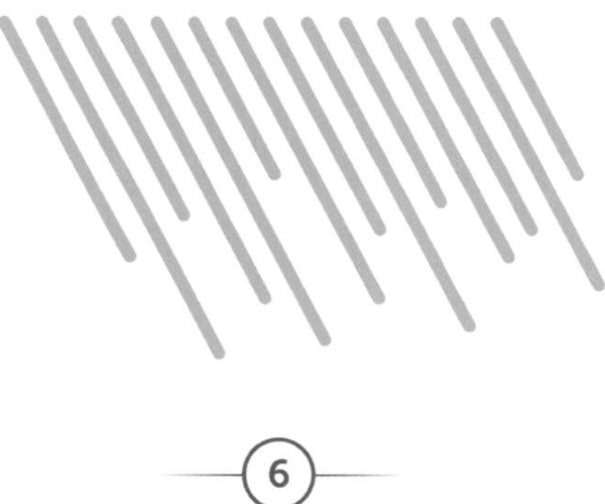

6

Lesson E: Sun

Lines and Circles

1 Draw a circle with the Ellipse tool (L). Hold Shift while drawing to constrain it to a perfect circle.

2 Draw a vertical line with the Line tool (\). Hold Shift while drawing to constrain it to a straight line. Round cap the ends in the Stroke panel.

3 Move the line so it is vertically centered on the circle. It does not need to touch the circle but it won't change the process if it does.

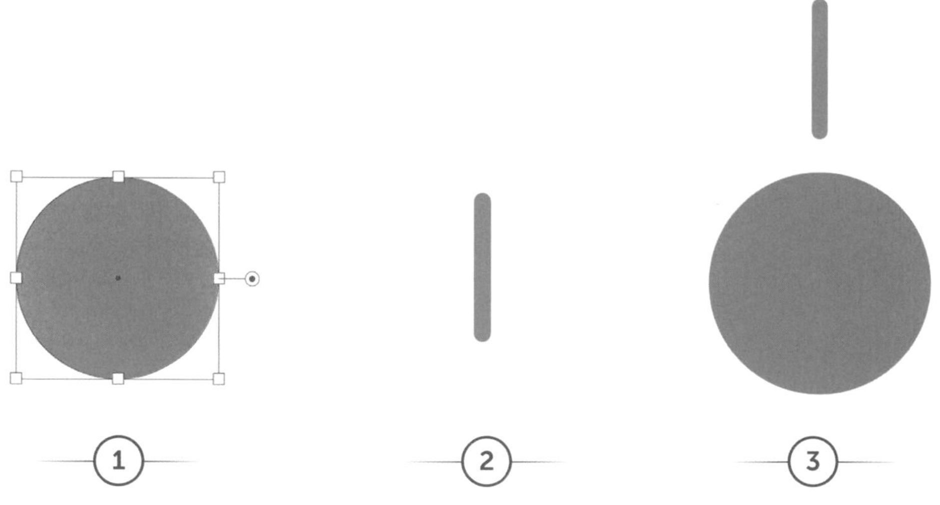

Rotate Tool

4 Select the vertical line, then select the Rotate tool (R).

5 The Rotate tool will show as crosshairs. Position the crosshairs in the center of the circle. The word center should show when you find it. If the center does not show, go to View > Smart Guides to turn Smart Guides on. Hold Option (Mac) / Alt (PC) (the crosshairs will show three dots to the right of it) and click on the center of the circle to open the Rotate tool dialog box.

6 In the Rotate tool dialog box, enter 360 (the number of degrees in a circle) divided by the number of rays you want for your sun. You could do this with a calculator and then enter that value but you can calculate directly in the entry field, with no need for a calculator! Click Copy; don't click OK.

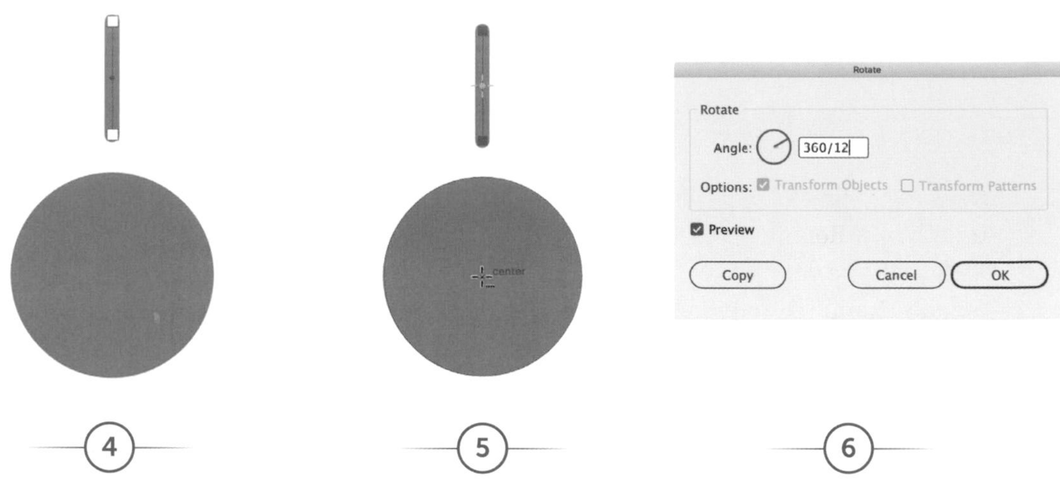

Duplicate

- 7 Selecting Copy in the Rotate tool dialog box will create a copy of the line in the amount that was calculated in the entry field: in this example, 360°/12 = 30°.
- 8 Use Command+D (Mac) / Ctrl+D (PC) to duplicate the Copy command and create multiple lines around the sun.

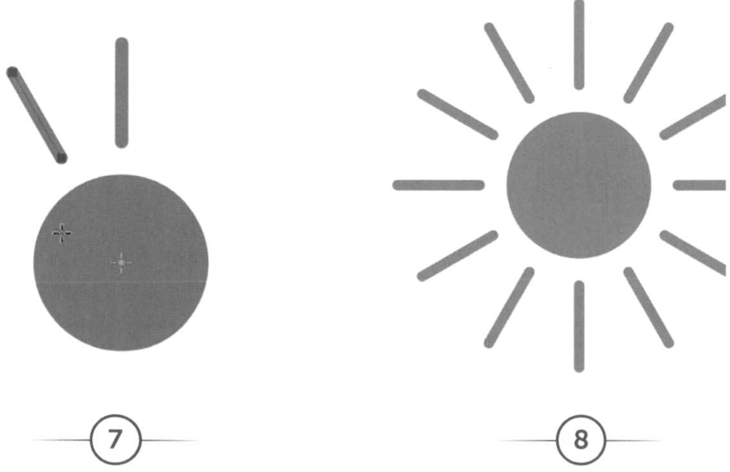

Lesson F: Snowflake

Lines, Shapes, and Endcaps

1 Draw a vertical line with the Line tool (\). Hold Shift while drawing to constrain it to a straight line. Round cap the ends in the Stroke panel.

2 Draw a square with the Rectangle tool (M). Hold Shift while drawing to constrain it to a square. Rotate the square 45 degree and position it on the line.

3 Draw a circle with the Ellipse tool (L). Hold Shift while drawing to constrain it to a perfect circle and position it at the top of the line.

4 Draw a square with the Rectangle tool (M). Hold Shift while drawing to constrain it to a square. Rotate the square 45 degree.

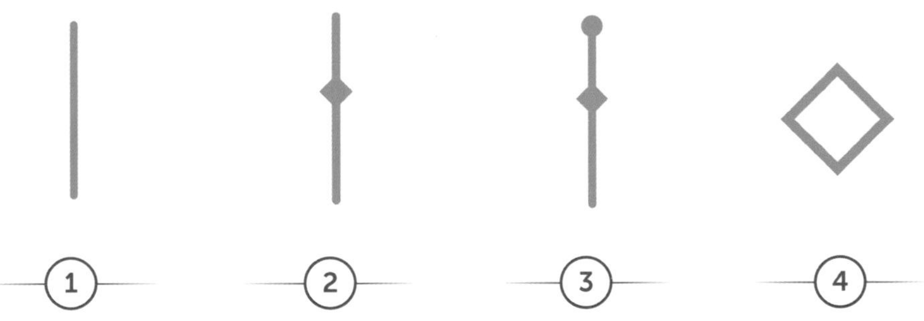

Lines and Circles

5 Select the top point of the rotated square with the Direct Selection tool (A) and delete the point. This will create a vee-shaped line.

6 Position the vee on the line between the square and the circle.

7 Duplicate the vee, scale it smaller and move it up the line.

8 Duplicate the vee twice and position them lower on the line, one to the left and one to the right to form "arms."

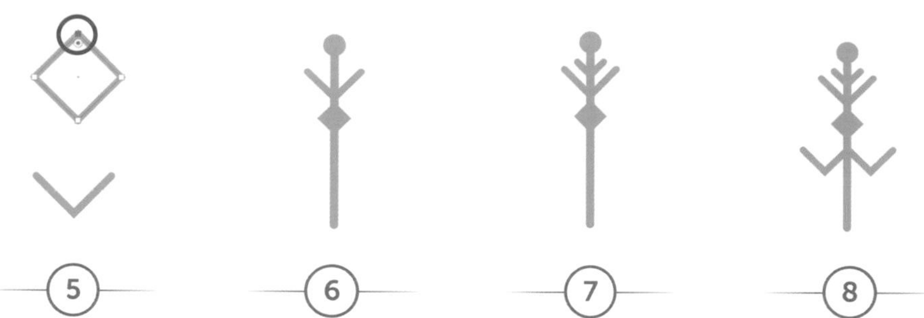

Rotate Tool

- **9** Select all the shapes and group them together with Object > Group. Select the Rotate tool (R) and Option + click (Mac) / Alt + click (PC) at the bottom point of the creation. The crosshair cursor will show four dots indicating a dialog box will open upon click.
- **10** With the Rotate dialog box open, enter in the number of duplicate shapes to create. Easiest way is to take a 360-degree circle and divide it into the number of duplicates. This field is your calculator: 360/6 will produce one object every 60 degrees.
- **11** Press the Copy button. The object will copy and rotate around the origin point at the specified 60 degrees.

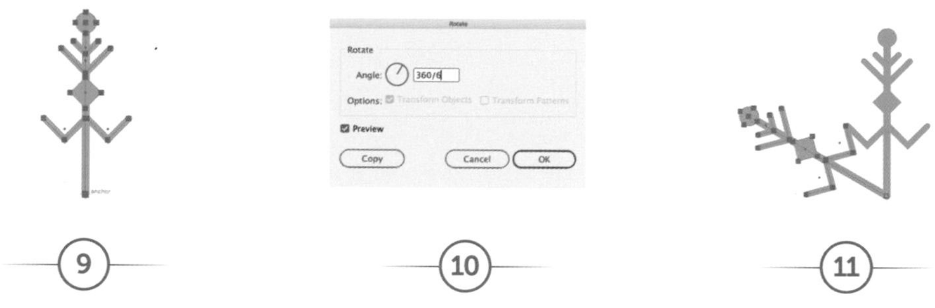

12 To repeat the Copy command and duplicate the arms around the origin point, use Command+D (Mac) / Ctrl+D (PC) to create multiple arms.

13 Draw a polygon for the center of the snowflake. Use the diamond on the upper-right side of the polygon to change the number of sides to six.

14 Move the polygon to the center of the snowflake.

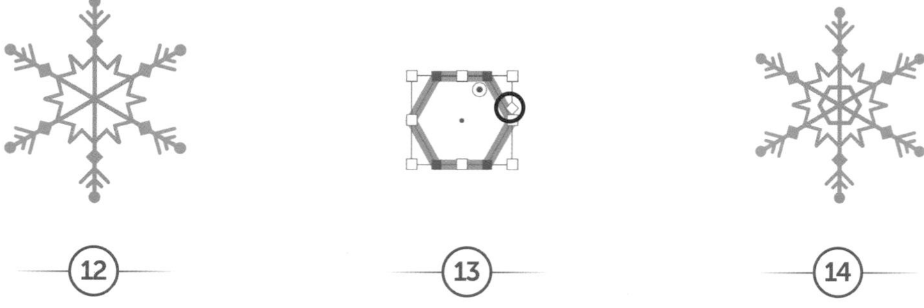

Lesson G: Wind

Offset Path

- 1 Draw a circle with the Ellipse tool (L). Hold Shift while drawing to constrain it to a perfect circle.
- 2 Select the circle and choose Object > Path > Offset path. Set the Offset to -20. Click OK.
- 3 This creates a smaller circle inside the original circle.

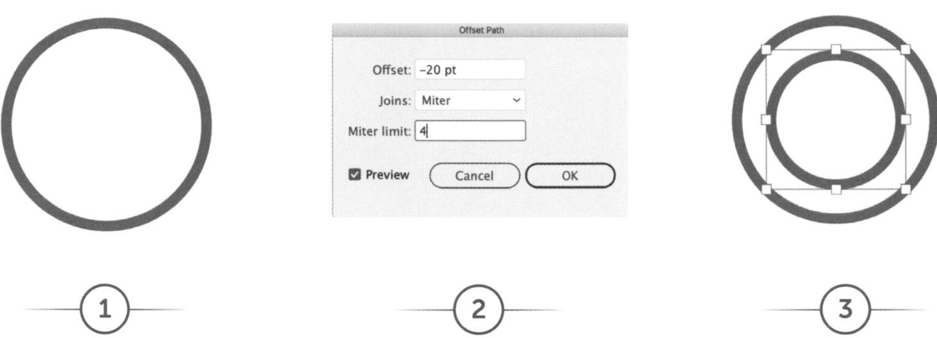

Direct Selection and Delete

- 4 With the Direct Selection tool (A), select the left anchor point on the outer circle and delete it.
- 5 This will delete both arcs that connect to the anchor point.
- 6 Direct Select the lower-left arc on the inner circle.

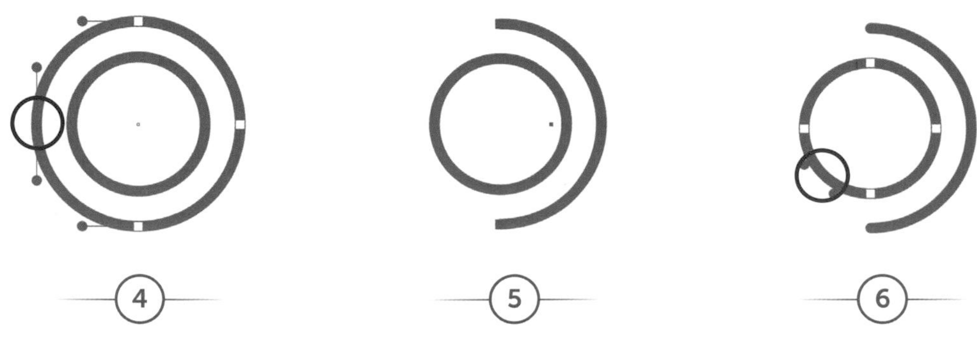

Pen Tool

- 7 Click the inner circle with the arc deleted.
- 8 Select the Pen tool (P). Click on the bottom anchor point of the outer circle, and an anchor point box will appear.
- 9 Hold Shift and click to the left, drawing a line that is connected to the circle. Holding Shift will draw a straight line.

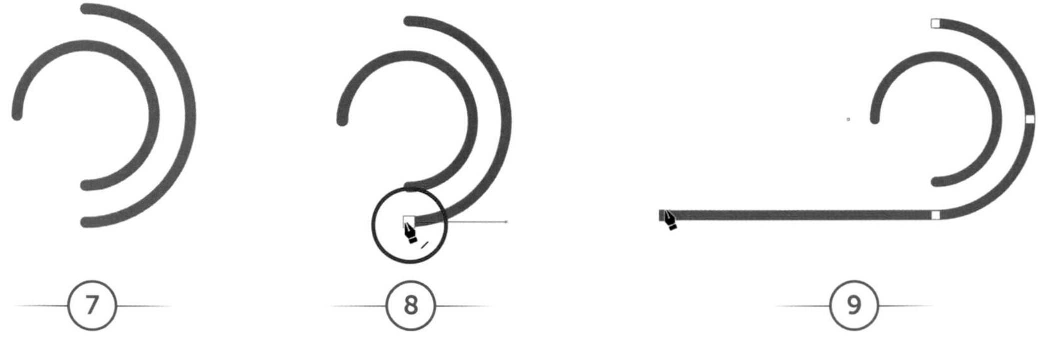

- 10 Click on the bottom anchor point of the inner circle with the Pen tool, and an anchor point box will appear.
- 11 Hold Shift and click to the left, drawing a line that is connected to the circle.

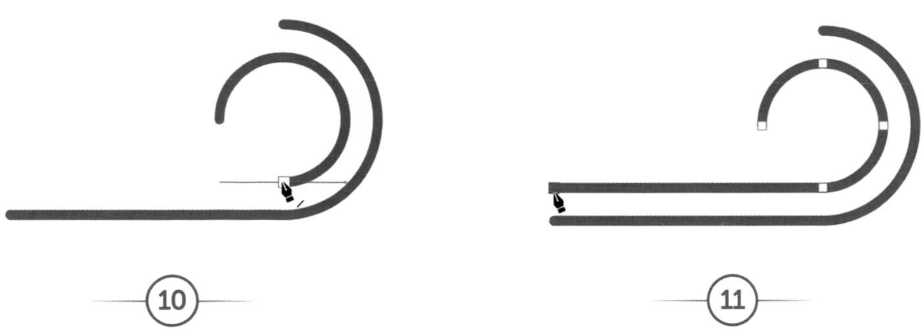

Duplicate, Flip, and Scale

- **12** Select the smaller inside arc. Duplicate the shape by using the Selection tool (V) and Option+click and drag (Mac) / Alt+click and drag (PC) the shape.
- **13** In the Properties panel, click on the Flip Vertical button to flip the shape.
- **14** Position it below the other two shapes.

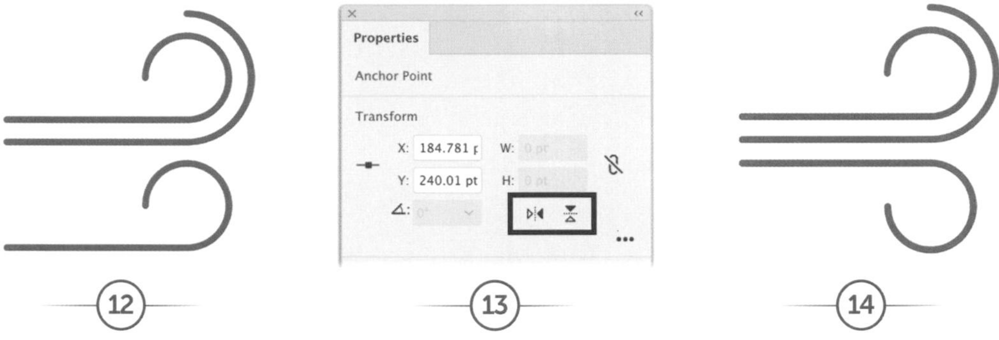

- **15** Open the Transform panel. Uncheck the Scale Stroke & Effects box to prevent the stroke from scaling down with the shape.
- **16** Select the shape, hold Shift, and pull a corner handle in toward the center to scale the shape smaller. Let the wind blow!

Lesson H: Adding Texture

By applying pictures of texture to our creations, we can give them dimension. The Image Trace function will trace to turn images into vector shapes so they can be applied to our creations. There are a few tricks to making texture work in all the shapes, so follow these steps carefully.

Image Trace

1 Create a new Document. Find an image of the texture you want to convert to a vector. Choose File > Place to put the image into the Illustrator document.

2 Open the Image Trace panel from the Window menu. Select the image and choose the Black and White preset. Click the Preview button to see the results.

3 Adjust the Threshold slider to achieve the desired results. If you see nothings, slide the slider toward More. If it is nearly solid black, move the slider toward Less.

1

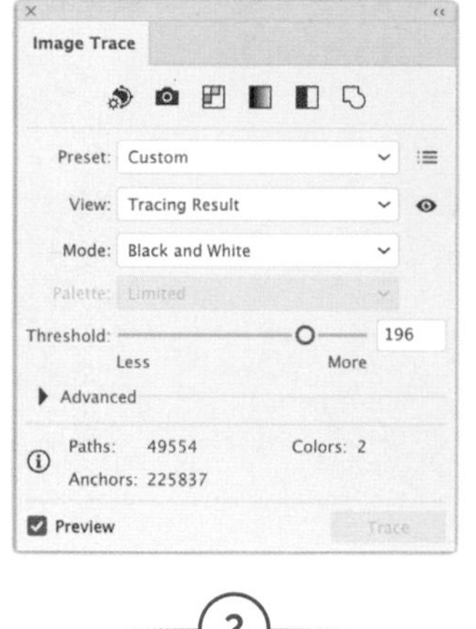

2

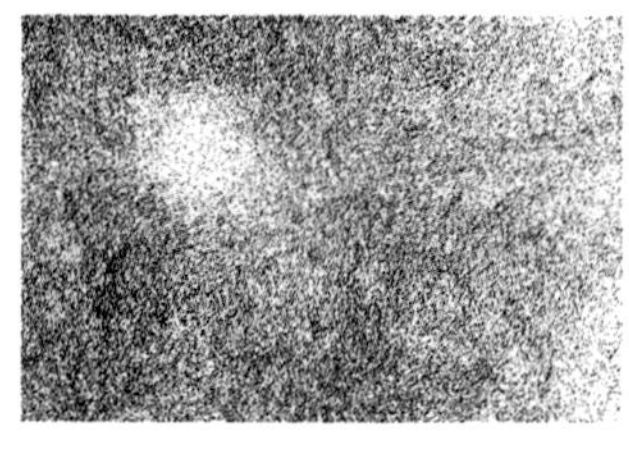

3

4 Click on the Advanced section to open up further options. Choose Ignore White so the white areas do not become shapes.

5 In the Properties panel, click Expand from the Quick Actions section to convert the image into vector. This is a close-up of the vector conversion.

6 When the vector shapes are selected, it can look like a sea of anchor points.

Draw Inside

- **7** Select the vector texture and fill it with a darker blue than was used for the clouds. Copy the texture.
- **8** Select the cloud and choose Draw Inside mode at the bottom of the toolbar.
- **9** Dashed lines will appear around the corners of the object (this is a much easier way to do a Clipping Mask).

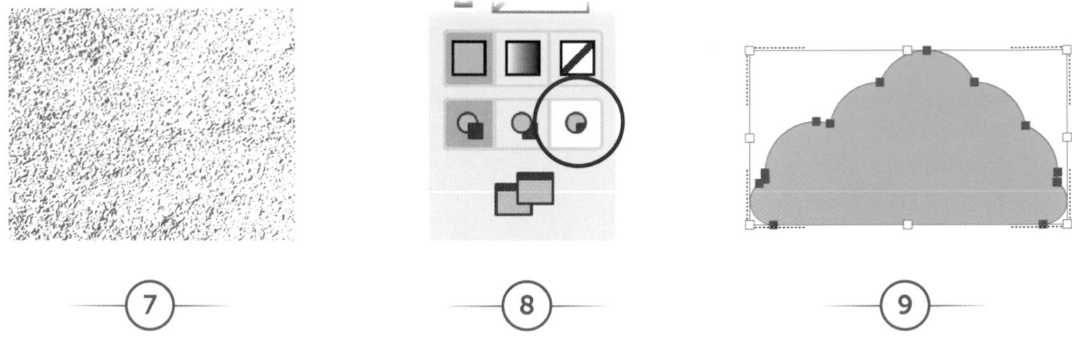

10 Paste the texture into the cloud shape.

11 Click the Draw Normal mode at the bottom of the toolbar to exit the Draw Inside mode.

12 The dashed lines will disappear and the shape will be back in Draw Normal mode.

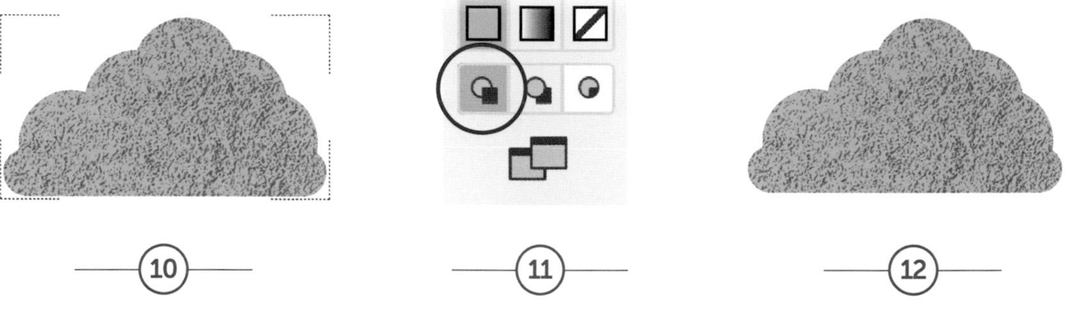

Outline Paths

13 Paths cannot have anything drawn inside them since they are not shapes. We need to convert any lines to shapes to make this work.

14 Select all the paths and choose Object > Path > Create Outlines. This will make the paths shapes. Before you can use the Draw Inside feature, we need to convert this to "one" shape. Illustrator calls this a compound path. Choose Object > Compound Path > Make.

15 Click the Draw Inside mode at the bottom of the toolbar and paste in the vector texture. Click the Draw Normal mode to exit the Draw Inside mode.

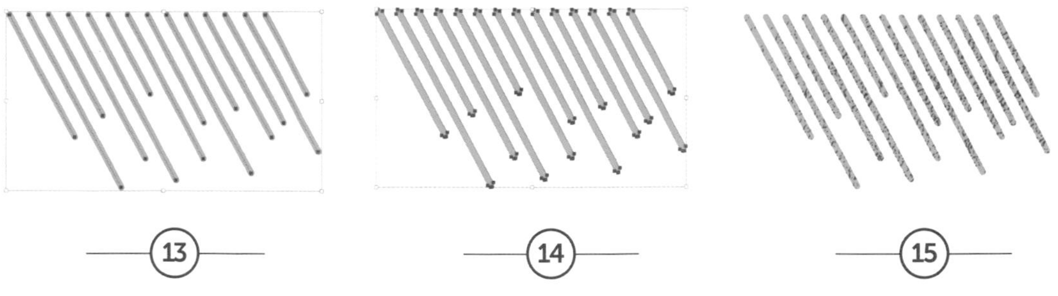

Project: Building Kitchen Icons

Using a grid structure to help build icons can be of great help. With the Snap to Grid feature activated, you can keep lines and points snapped to the grid for easy icon creation and consistency. Lets look at an example: I set this up in millimeters so everything is based on units of 10.

Lesson A: Measuring Cup

Set up a Grid

1 Set the general units in your preferences to be millimeters.

2 Set the grid layout in preferences: Gridline every: 100 mm; Subdivisions: 10. This creates heavier grid lines with lighter grid lines for the subdivisions.

3 Select Show Grid to show the grid and turn on Snap to Grid under the View menu.

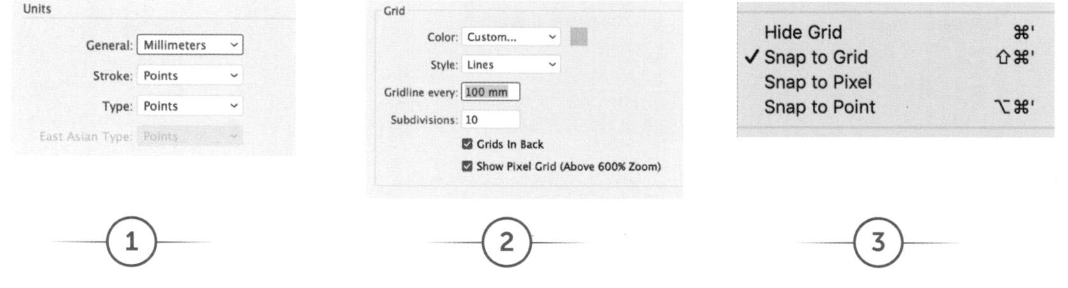

Snap to Grid

1 Draw a square, 10 units by 10 units, 20 pt stroke, no fill.

2 Select the lower two corners with the Direct Selection tool (A) and set the corner radius in the Properties panel to be 10 mm.

3 Create the handle with the Pen tool (P), click two units away from square, then four units up, two units to the left. To exit the Pen tool, click on the Selection tool (V).

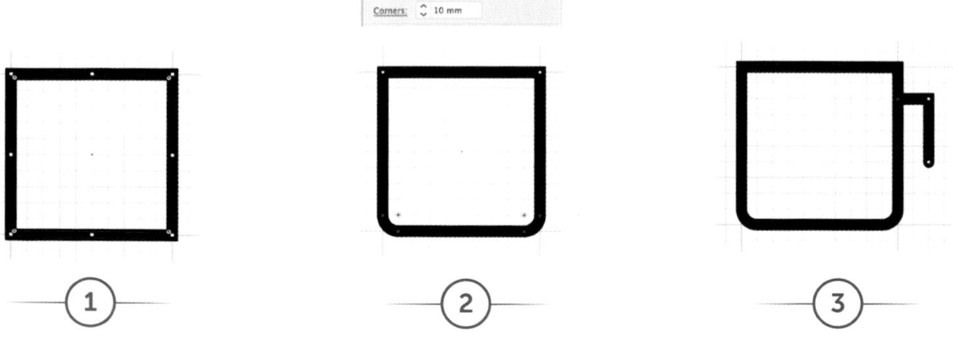

- **4** Select the inside corner of the handle with the Direct Selection tool (A) and set the corner radius to 10 mm in the Properties panel.
- **5** Select the square and then select the Pen tool and click two units down on the left side to add a point.
- **6** Select the upper-left point with the Direct Selection tool and move the point one unit left for the lip.

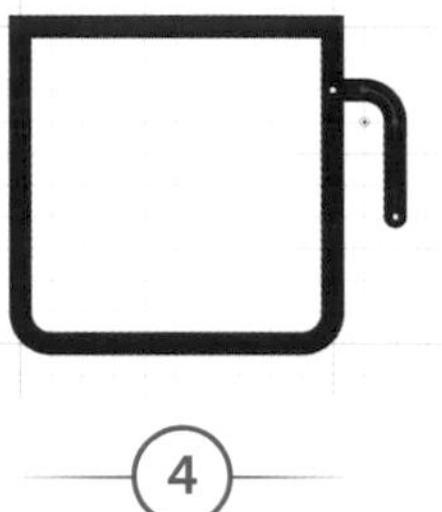

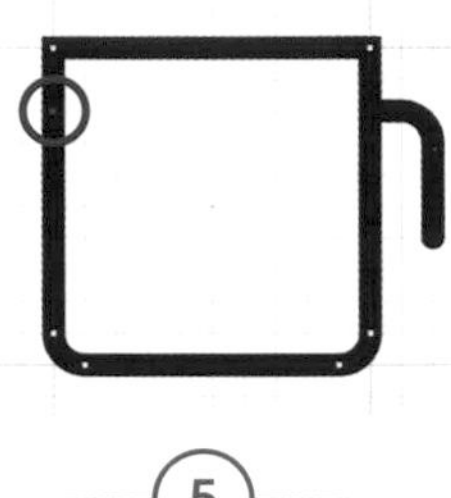

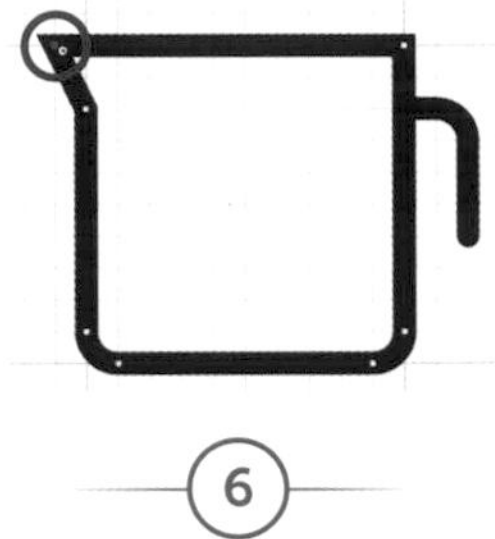

Offset Path

- **7** Draw a rectangle one unit in from the edge of the inside and fill it with a color, no stroke.
- **8** Draw a line four units long, 20 pt stroke, and rounded end caps two units from the top. Draw a line two units long, with a 20 pt stroke, and rounded end caps four units from the top. Repeat these two- and four-point long strokes for six and eight units
- **9** Turn off the grid to view your work. View > Hide Grid. One important note: when you turn off grid visibility, the Snap to Grid will still be active.

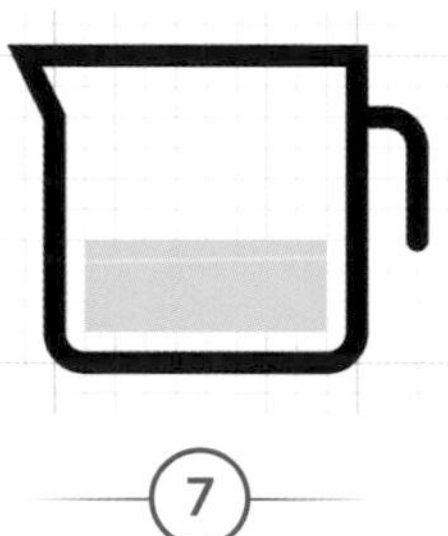

Lesson B: Coffee Pot

- 1 Let's take that measuring cup you just created and make it into a coffee pot by deleting the measuring marks and changing the fill of the inside to be a coffee-colored brown.
- 2 To take the sharp tip off the spout, select the shape and click on the rounded corner stroke in the Stroke panel. This creates a round stroke edge.
- 3 Draw a line, 20 pt weight, the width of the coffee pot. Set the ends to be a rounded cap.

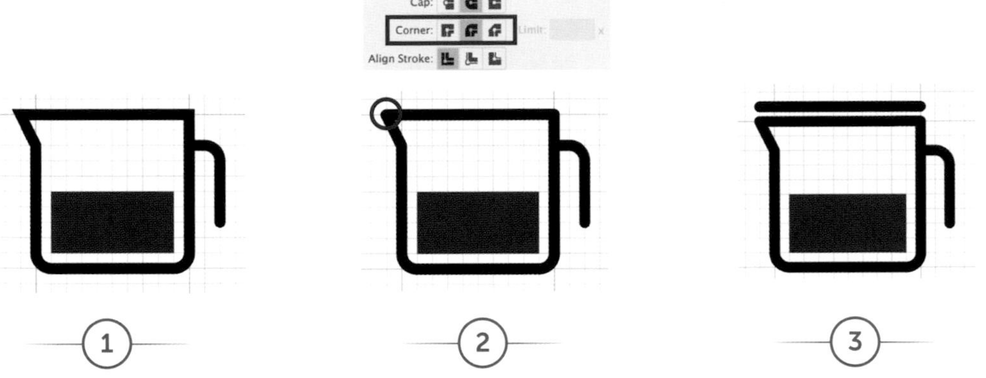

- 4 Create a circle two units by two units.
- 5 Select the bottom point of the circle with the Direct Selection tool (A) and delete the section.
- 6 Move the half circle to the top of the lid to complete this shape.

Lesson C: Spatula and Spoon

- 1 Create a rectangle eight units wide, ten units high.
- 2 With the Direct Selection tool, select the two upper corners and set the corner widgets to 10 mm in the Properties panel.
- 3 Draw a line, 20 pt weight, six units high, and apply the rounded caps in the Stroke panel. Position the line two units over and two units down from the upper-left corner. Repeat the line two units over from the first line, then two units over again to form the opening slits.

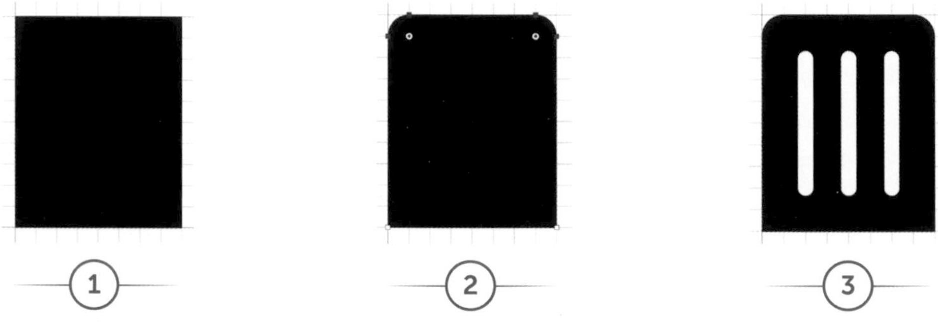

- 4 Create a vertical line with the Line tool (\), 20 pt weight, 11 units high. Position the line at the base of the spatula head.
- 5 Create a rectangle at the base of the line, ten units down from the head. This will cover one unit of the line you just created.
- 6 Select the rectangle with the Selection tool (V) and pull the corner widgets into the center to round the ends, creating a handle.
- 7 The spatula is now complete.

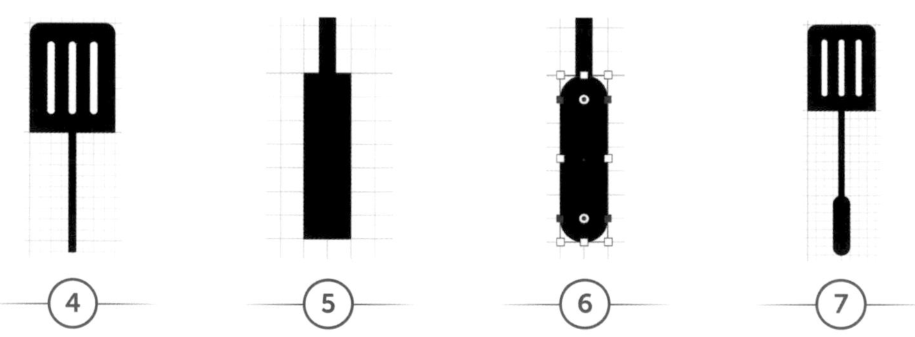

1 Create a rectangle eight units wide, ten units high.
2 Select the rectangle with the Selection tool (V) and pull the corner widgets into the center to round the ends to the maximum amount.
3 Draw a rectangle, two units wide, three units high, and position it one unit into the rounded rectangle.

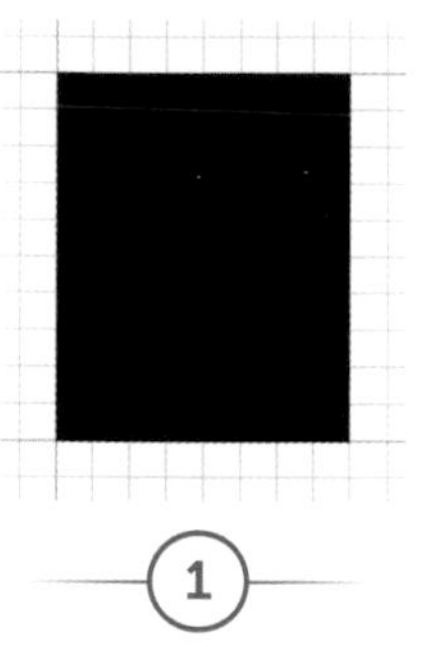
1

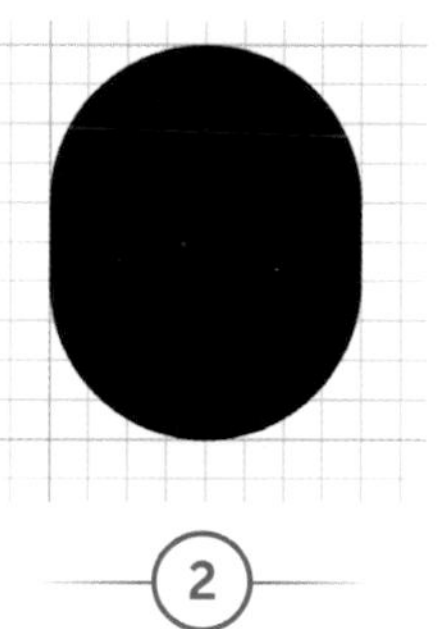
2

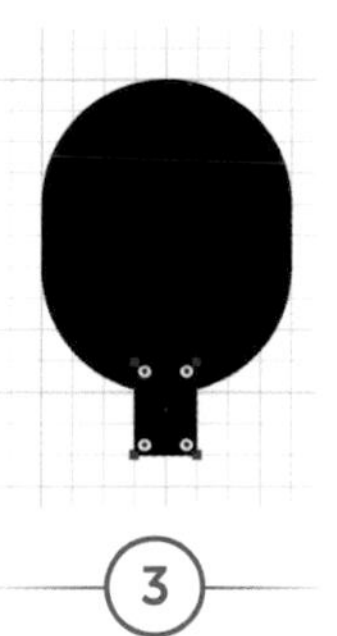
3

4 Select both shapes and use the Pathfinder Unite mode to make them one shape.
5 Select the two points where the small rectangle joins the large rounded shape with the Direct Selection tool.
6 Pull the corner widgets out from the center to the maximum amount to round the inner corners.

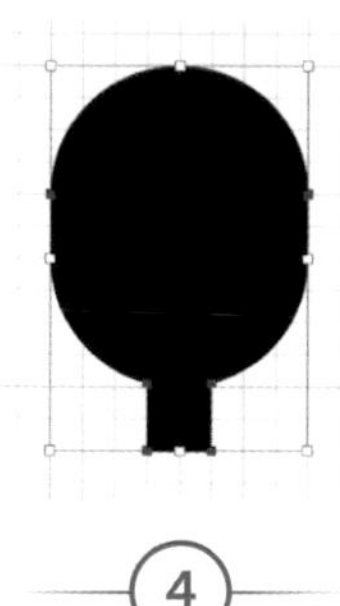
4

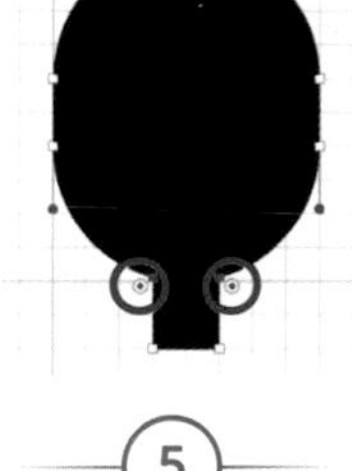
5

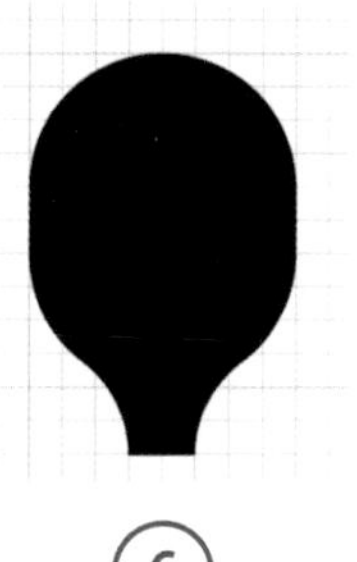
6

- 7 Create a vertical line with the Line tool (\) 20 pt weight, 11 units high. Position the line at the base of the spatula head.
- 8 Create a rectangle at the base of the line, ten units down from the head. This will cover one unit of the line just you created.
- 9 Select the rectangle with the Selection tool (V) and pull the corner widgets into the center to round the ends, creating a handle.

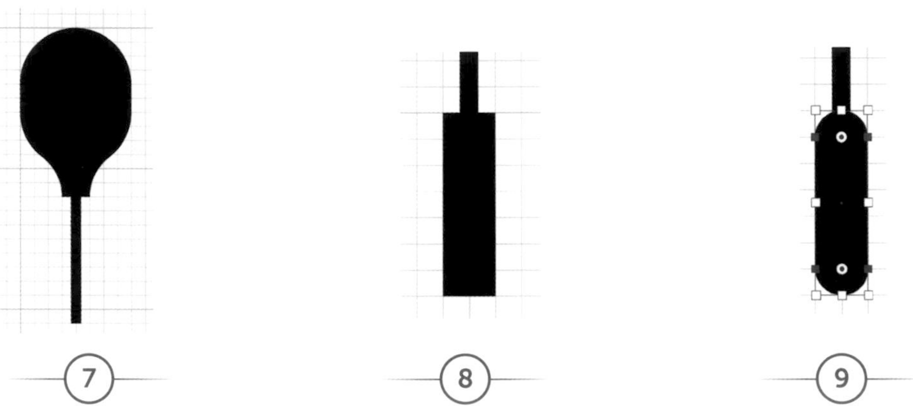

- 10 The spoon is nearly complete, it just needs a highlight noodle.
- 11 Select the Arc tool, hold Shift and draw an arc. Set the weight to 20 pts, the stroke to white, and don't use any fill. Round the end caps in the Stroke panel. Position the highlight noodle on the spoon and it is finished!

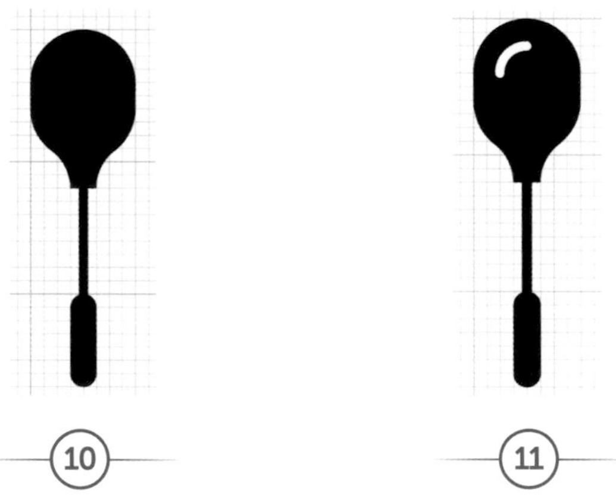

Lesson D: Boiling Water

1 Create a rectangle ten units wide, eight units high with a 20 pt stroke.

2 With the Direct Selection tool, select the upper line of the rectangle and delete it.

3 With the Direct Selection tool, select the lower corners and set the corner widgets to 20 mm in the Properties panel.

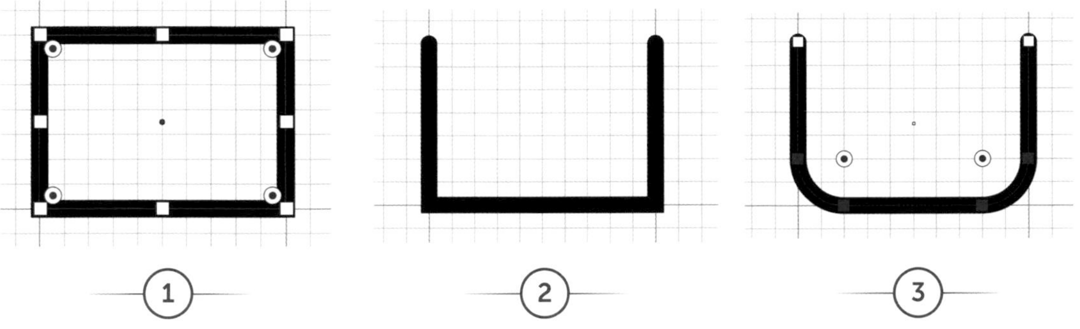

4 With the Line tool (\), draw a line seven units long. Round cap the ends in the Stroke panel.

5 Select the line and choose Effects > Distort & Transform > Zig Zag. Apply a size of 2 pts, with four ridges per segment. Select the Smooth point button and click OK.

6 Draw a circle measuring two units by two units, with a stroke of 5 pts. Draw another circle, one unit by one unit. With the Type tool (T), click and enter in an asterisk (*) to make it look like the bubbles pop.

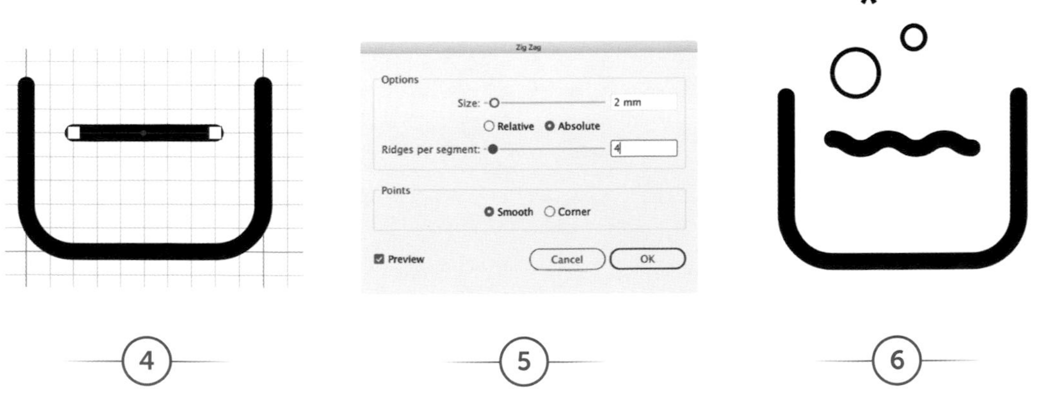

- 7 With the Ellipse tool (L), draw two circles two units by two units, with two units of space between them.
- 8 Select the top point of each circle with the Direct Selection tool (A) and move the points up one unit.
- 9 Convert the top two points from smooth points to corner points in the Properties panel to create flames.

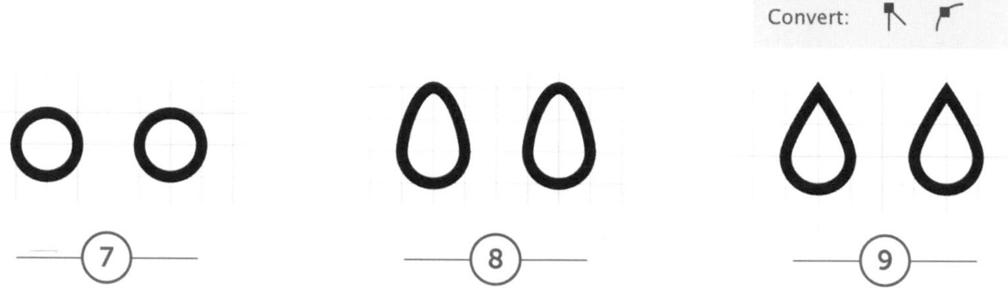

- 10 Select both flames and choose Object > Path > Offset Path. Apply a -5 mm offset and click OK.
- 11 Select the smaller inner flames and set the stroke to zero points, and fill the shape with a gray color.
- 12 Move the flames under the pot to create boiling water.

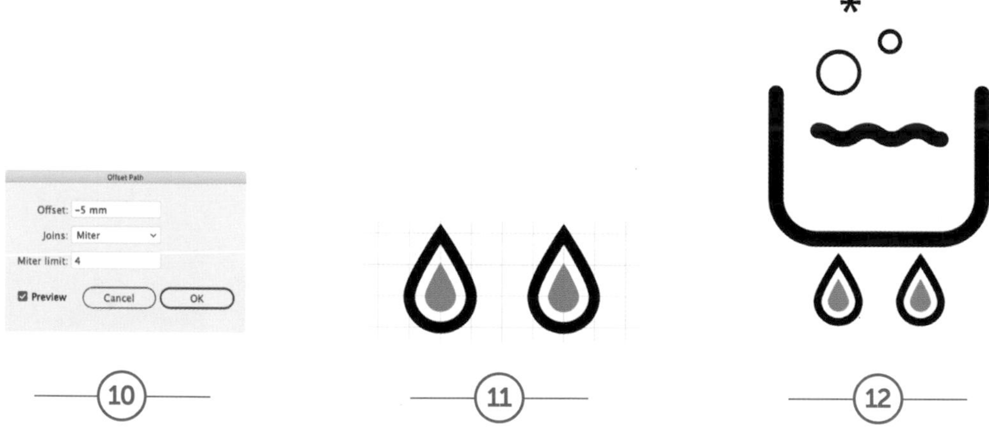

Lesson E: Toaster with Toast

1 Create a rectangle ten units wide and seven units high with a 20 pt stroke.

2 With the Direct Selection tool, select the upper two corners and set the corner widgets to 20 mm in the Properties panel.

3 Draw a line on each side of the toaster with the Line tool (\), one unit long for the handles.

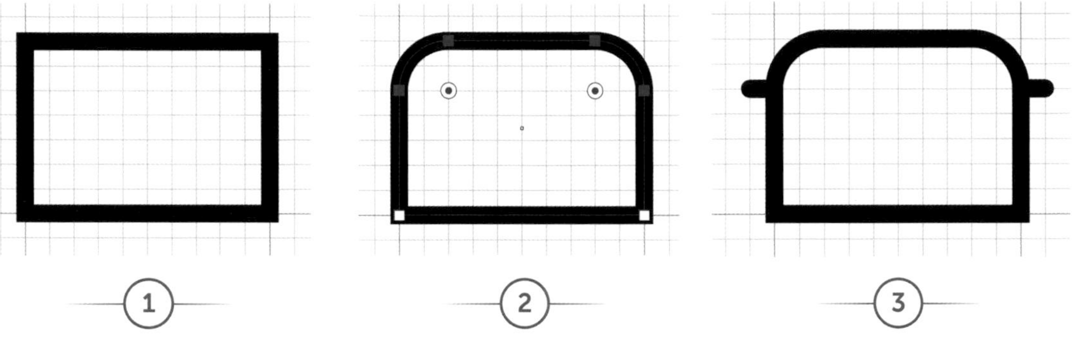

4 With the Line tool (\), draw a line seven units long and round cap the ends in the Stroke panel. Select the line and choose Effects > Distort & Transform > Zig Zag. Apply a size of 2 pts with four ridges per segment. Click the Smooth Point button and click OK.

5 Create a square measuring four units by four units with a 20 pt stroke.

6 Create an oval six units wide and two units high, and move it halfway over the square.

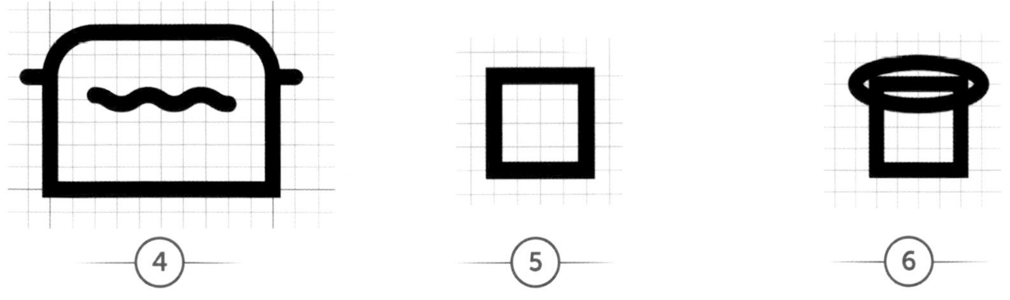

- 7 Select the square and oval with the Selection tool and choose Unite in the Pathfinder panel to make them one shape. Move the bread slice to overlap the toaster by one unit.
- 8 Select the bread and choose Object > Arrange > Send to Back. Select the toaster and fill the toaster with white to hide the bottom of the bread slice.

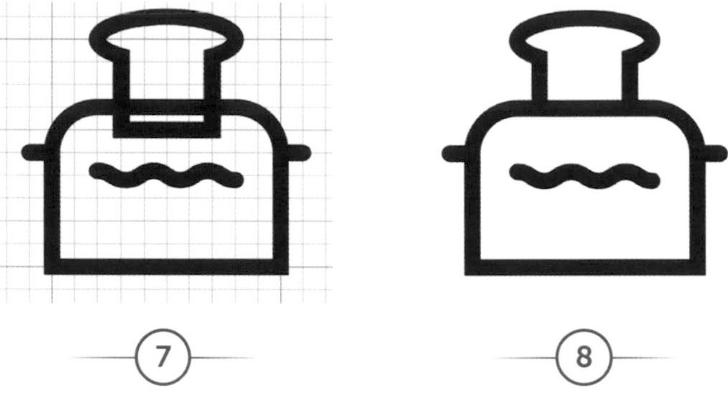

7 8

Lesson F: Egg and Avocado

1 Create a circle measuring ten units by ten units with a 20 pt stroke.
2 With the Direct Selection tool, select the upper point of the circle and move it up four units.
3 Add a fill color and a stroke color to the resulting shape to create an egg.

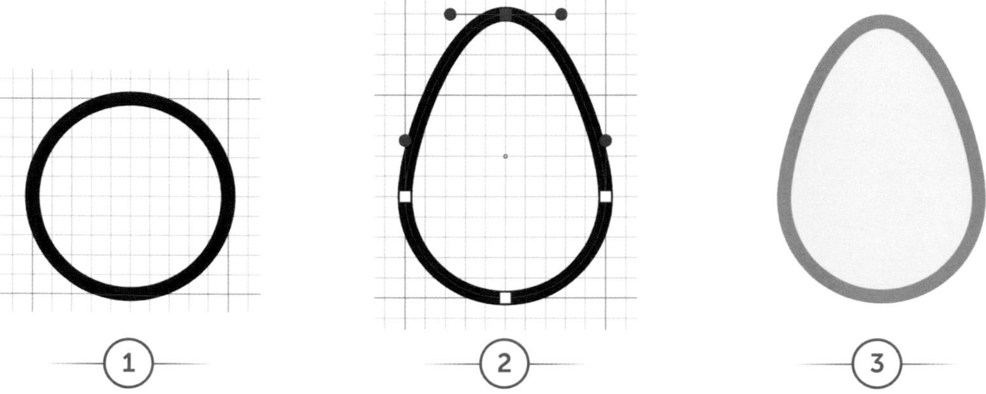

4 To turn the egg into an avocado, fill the shape with yellow green and add a dark brown stroke.
5 Draw a circle six units by six units and fill it with the same color as was applied to the stroke. Move the circle so it is centered in the bottom of the shape.

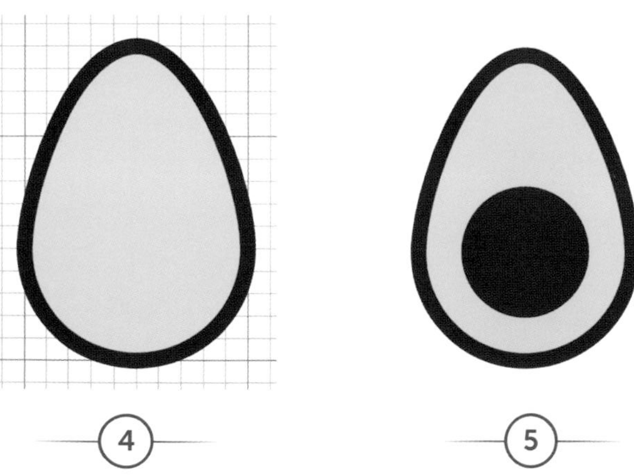

Project: Creating Plaid Fabric

In creating repeating patterns like a plaid, the grid structure is helpful for consistent spacing. With the Snap to Grid feature active, you can easily space sections of the pattern to explore your creativite ideas. Using the Transform Effect makes spacing and duplicating items much easier and cleaner than manually copying and pasting, or duplicating. In this project, you will easily explore different color variations with Recolor Artwork and learn to create different color harmonies in your artwork without having to change each object individually.

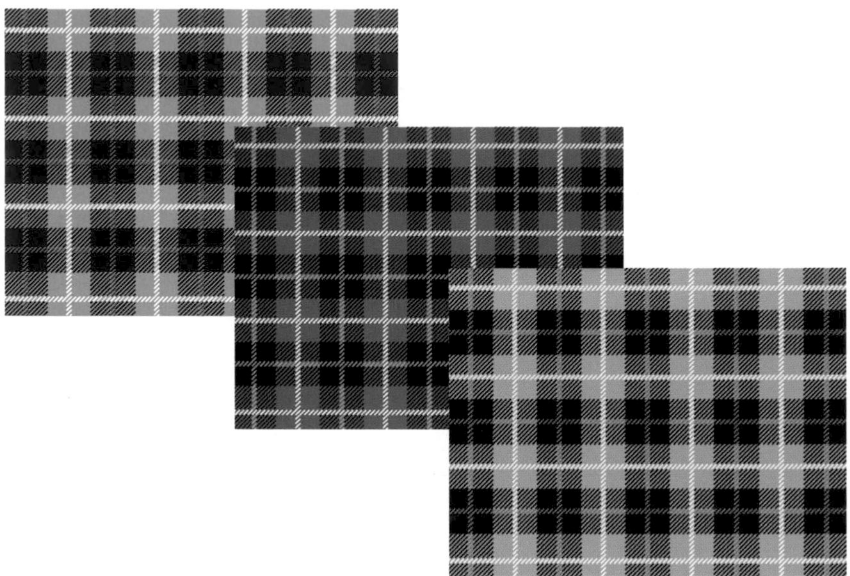

Lesson A: Create Plaid Fabric

Grid Setup

1 Set the General units in your preferences to millimeters.

2 Set the grid layout in preferences to Gridline every: 100 mm and Subdivisions: 10. This creates heavier grid lines with lighter grid lines for the subdivisions. You can uncheck the Grids in Back preference if you want to see the grid in front of your artwork.

3 Select Show Grid to see the grid. Turn on the Snap to Grid function in the View menu.

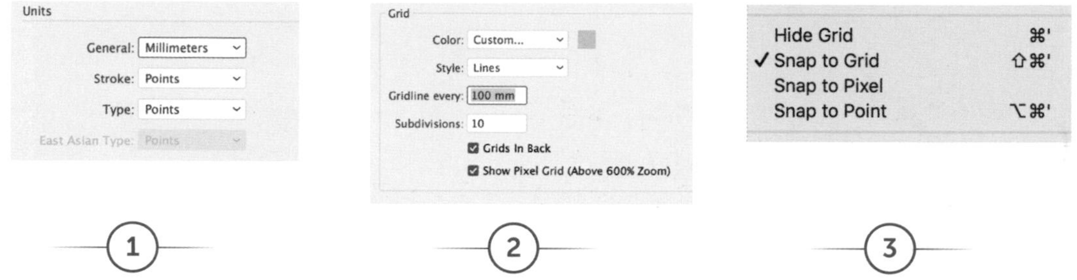

4 Each grid line will be a darker gray and the subdivisions will be a lighter gray. I refer to each darker gray section as a unit (the ten units by ten units) and each individual square a subunit.

5 Draw a rectangle nine units wide and seven units high and fill it with a color. Start the rectangle at a unit corner (darker gray grid line). This rectangle will snap to the grid if you have the Snap to Grid function turned on.

6 On the next row of units below, draw a one unit by one unit sqaure filled with black. Repeat this black square every other unit on the row, and repeat the pattern every other row below.

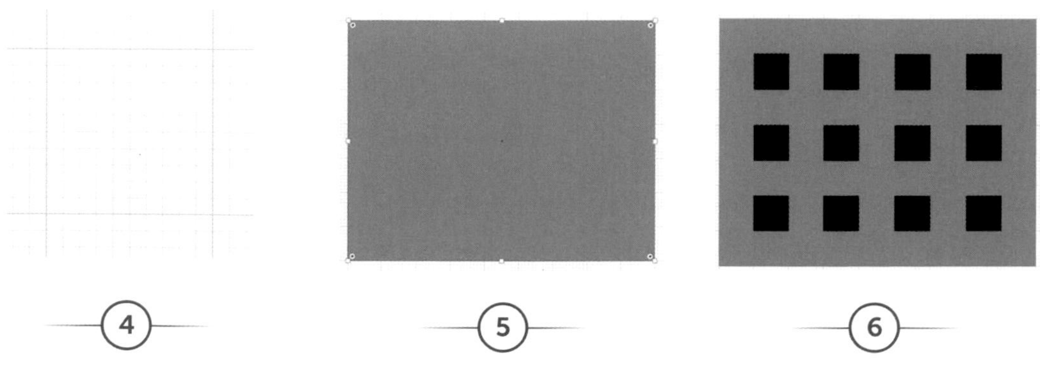

Transform

- **7** Draw a 10 pt line on the grid from corner to corner. Round cap the ends in the Stroke panel. Duplicate the line and keep it for the vertical stripe we are going to create.
- **8** Select the line and choose Effects > Distort & Transform > Transform. Set the Horizontal offset to 10 mm. Set the number of copies to 100. Click OK. This will take the single line, duplicate it 100 times to the right, skipping every other subunit.
- **9** Move the transformed set into position over the checkerboard. This is an effect so the one line is simply transformed 100 times, 10 millimeter offset each time. This effect makes the line look like 100 lines, but is really one line that appears to be 100 lines. Using an effect to transform makes it easier to edit and select a single line, rather than 100 individual lines.

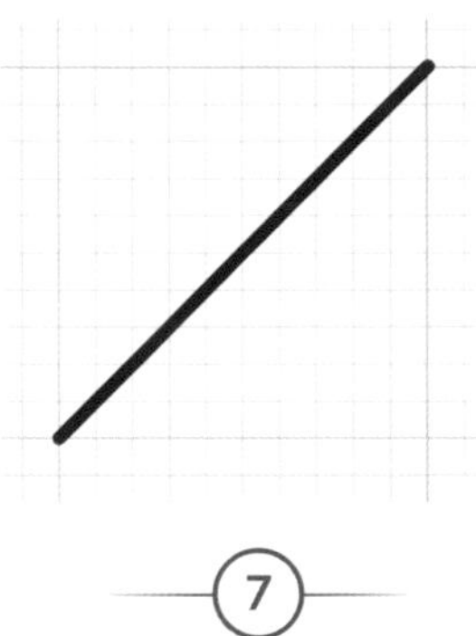

7

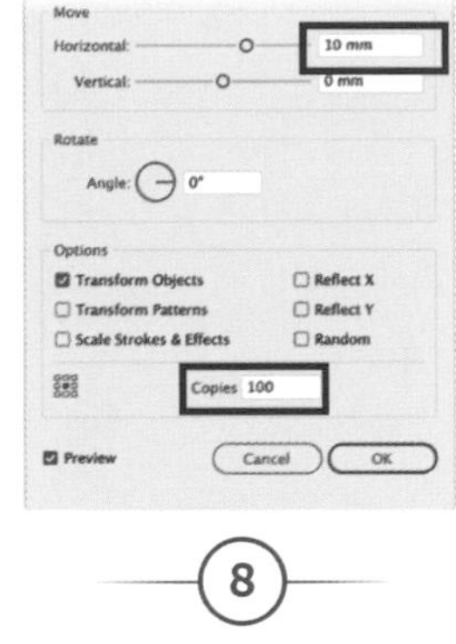

8

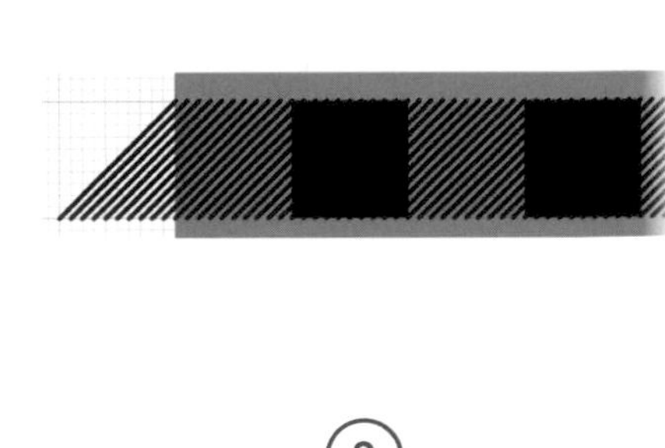

9

- **10** Select the duplicated line and move it above a row of black squares. Choose Effects > Distort & Transform > Transform. Set the Vertical offset to 10 mm. Set the number of copies to 80. Click OK. This will take the single line, duplicate it 80 times down, skipping every other subunit.
- **11** Move the transformed set into position over the checkerboard.
- **12** Repeat the transform every other row and every other column.

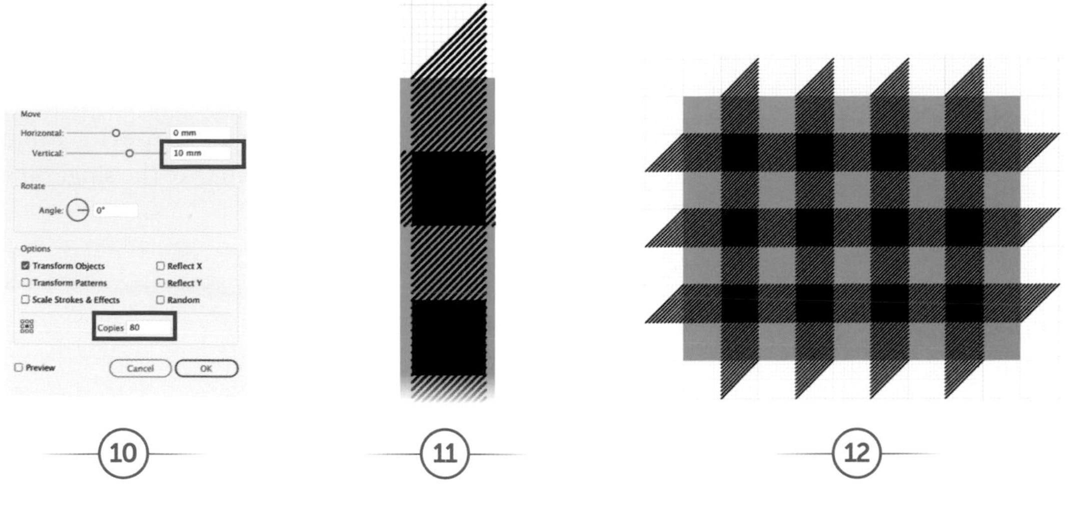

13 Create a smaller line at an angle, the size of one subunit. Add a red 10 pt stroke and round cap the ends.

14 Turn off the Snap to Grid. Move the line so the center of the line is directly in the center of one unit. *Snap to Grid will only snap objects to end and corner points, not center points. (The black lines are showing the exact center of the line aligning to the grid and are for reference only.)*

15 Select the line and choose Effects > Distort & Transform > Transform. Set the Horizontal offset to 10 mm. Set the number of copies to 100. Click OK. Move the line set to the center of a row with black squares. You may have to zoom in to get the center of the line exactly on the center of the subunits since the Snap To Grid is off.

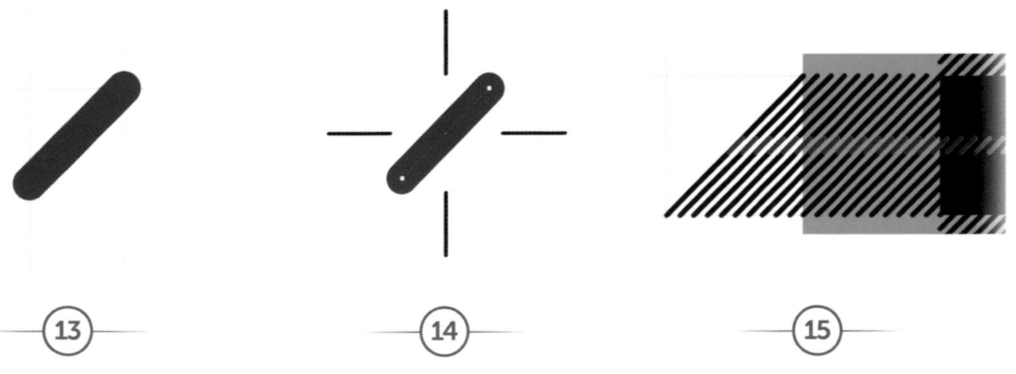

16 Duplicate the red set of lines exactly 10 subunits and change the stroke color to white.

17 Repeat the red set of lines ever other row, and the white set of lines in the remaining rows.

18 Similar to step 15, repeat the same transform with a vertical offset for the white, then red sets. Alternate the vertical sets starting with the white, then the red. Repeat this for all the vertical columns making sure that the line is centered, as in step 14.

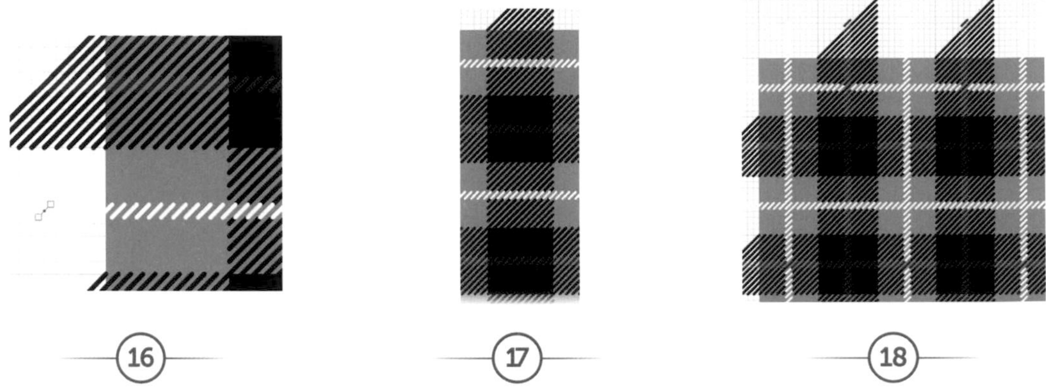

Draw Inside

19 Select all the artwork and choose Object > Group. Choose Edit > Cut to cut the artwork.

20 With the Rectangle tool (M), draw a rectangle nine units wide by seven units high. With the rectangle selected, click on the Draw Inside icon at the bottom of the toolbar.

21 Choose Edit > Paste to insert the cut artwork inside the nine by seven rectangle. Click on the Draw Normal mode at the bottom of the toolbar. Nice work creating a tartan plaid!

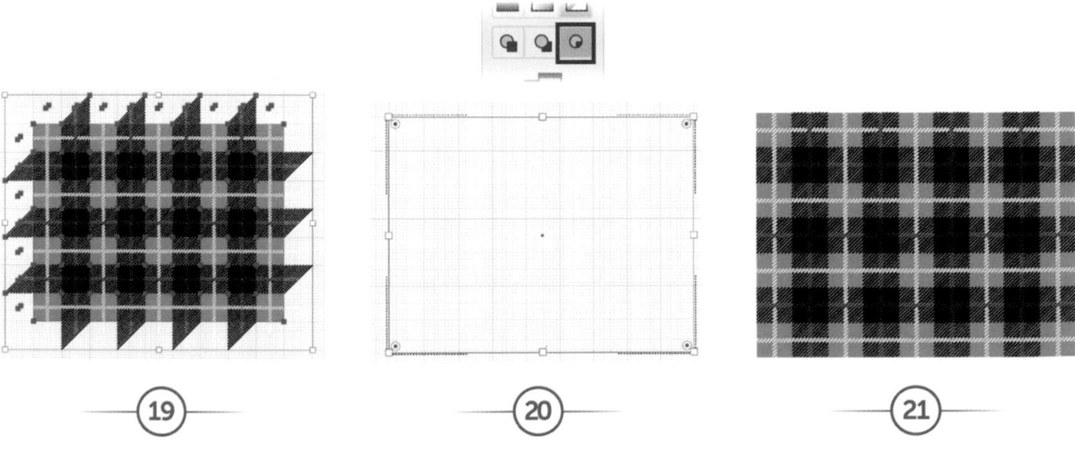

Recolor Artwork

22 Select the artwork and choose Edit > Recolor Artwork. Click on the Edit tab at the top of the Recolor Artwork panel. Click on the link icon in the lower right of the color wheel to link the colors together.

23 Move the colors' handles around the wheel to actively change them in the plaid artwork.

24 Try several different color renditions with Recolor.

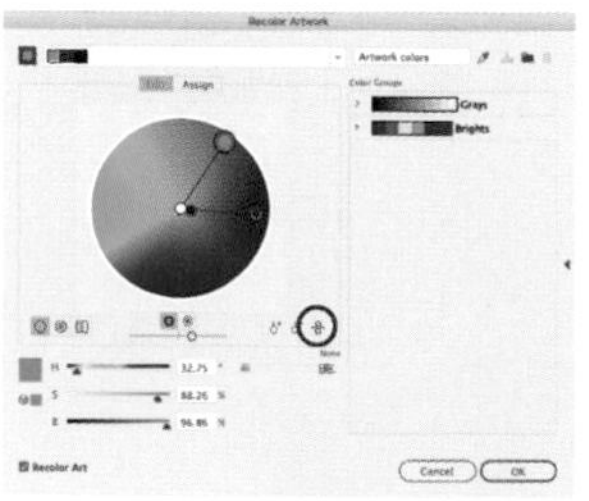

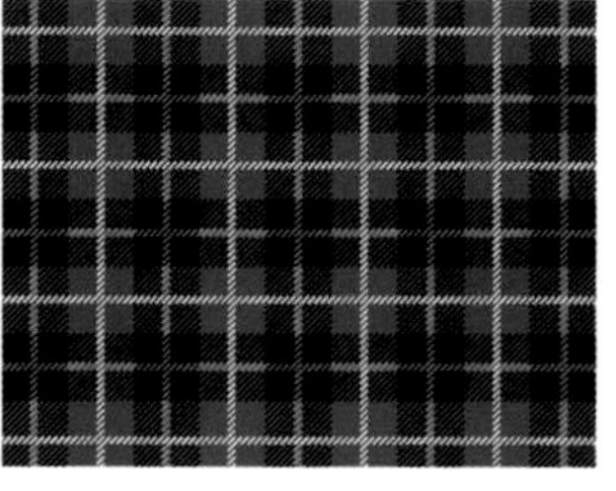

Project: Building a Sewn Patch

Build a sewn patch using the Appearance panel. The Zig Zag effect creates the sewn look. This build is a bit tricky as it requires some fancy footwork in terms of getting everything in the right order in the Appearance panel. The cool part is we are not drawing multiple shapes; this build is done with one circle and ordering the effects and strokes in the Appearance panel. The end result is very cool—give it a try.

Lesson A: Create a Sewn Patch

Appearance Panel

1 Start with the Ellipse tool (L), and click on the artboard to open the Ellipse dialog box. Set the width and height to 50 mm and click OK.

2 Open the Appearance panel from the Window menu.

3 Set the stroke on the circle to be 10 pts and set a color.

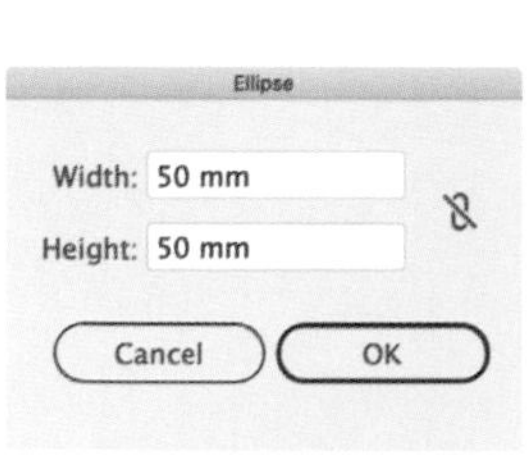

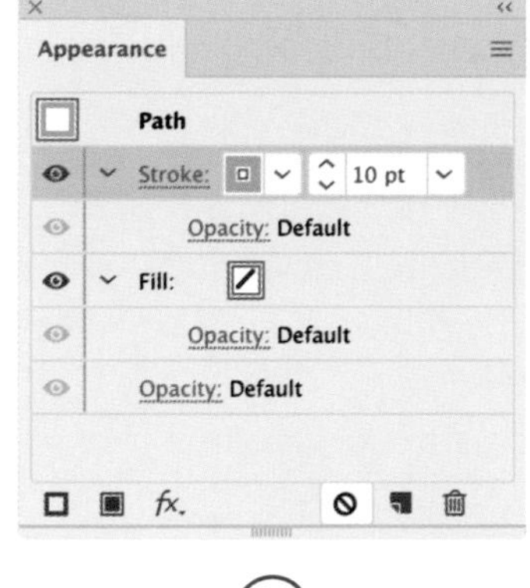

2

3

4 Select the circle and click on the Stroke section in the Appearance panel. Choose Effect > Distort & Transform > Zig Zag. Apply a size of 1 mm and click the Absolute option, 50 Ridges per segment, the Smooth points option, then click OK.

5 This will create a wavy zigzag look on the stroke of the circle. Select the circle.

6 Click on the New Stroke icon in the lower left of the Appearance panel. This will create a new stroke in the panel on top of the other stroke.

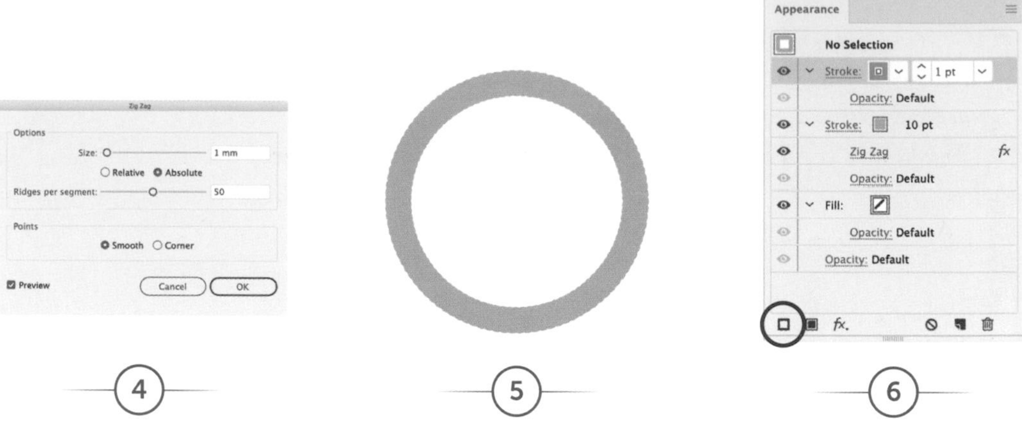

7 Click on the top-most stroke in the list, choose a color, and set the point size to 1. This creates a stroke on top of the existing stroke.

8 With the circle selected and the new 1 pt stroke selected in the Appearance panel, choose Effect > Distort & Transform > Zig Zag. Select a size of 1.5 mm, 100 Ridges per segment, and Smooth points. Click OK.

9 This creates a stitch effect on the stroke without creating a new shape.

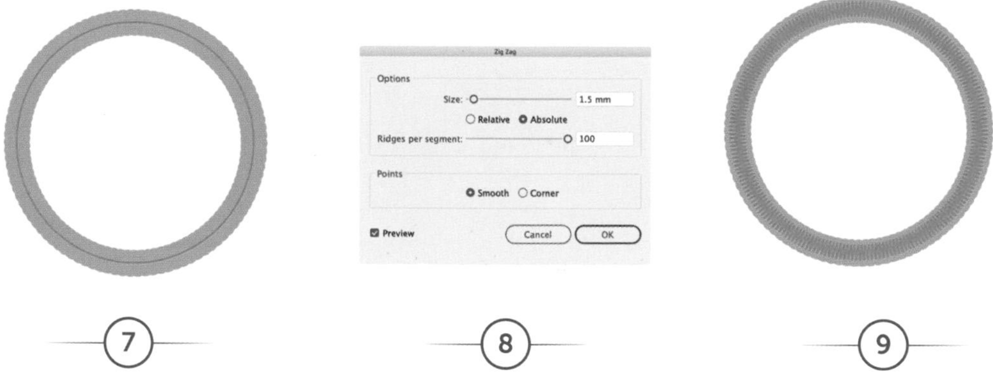

Effects

10 With the circle selected, and the stroke selected in the Appearance panel, choose Object > Effect > Distort & Transform > Transform. Set the Horizontal and Vertical scale to 97% each. Click OK.

11 This scales the stroke and the applies the effect on a smaller level.

12 Next, choose Object > Effect > Stylize > Outer Glow. Set the mode to Multiply, the color square to black, the opacity to 75%, the Blur to .35 mm, and click OK.

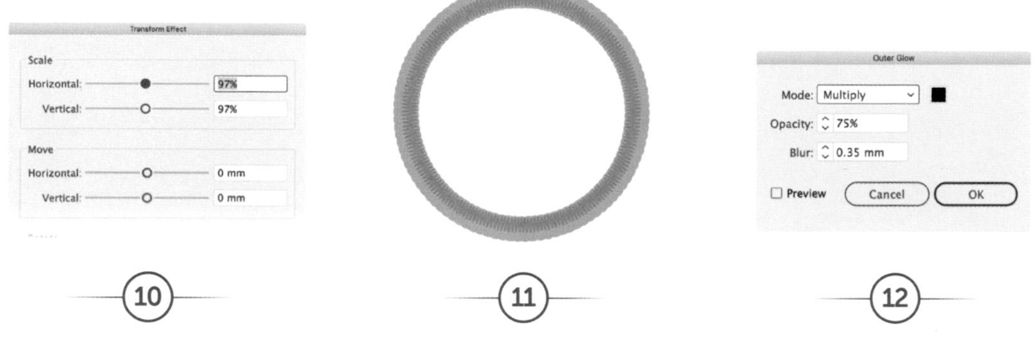

Multiple Strokes

13 The Outer Glow will create a nice shadow separating the strokes.

14 With the circle selected and the stroke selected in the Appearance panel, press Option+click (Mac) / Alt+click (PC) and drag the stroke down below to duplicate it and all its attributes.

15 In the newly duplicated stroke, change the color to a darker shade. Click on the transform link under that stroke to edit the transform effect.

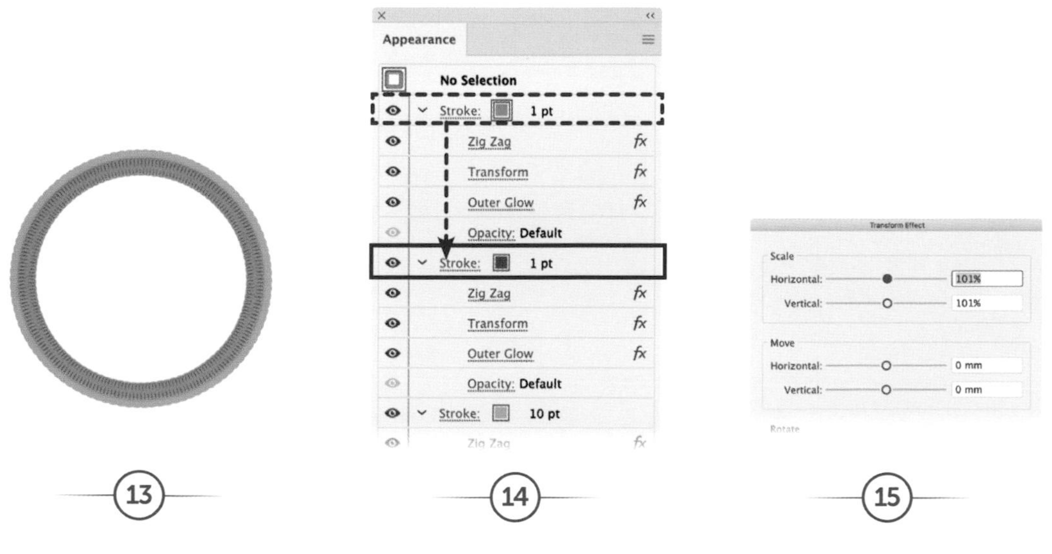

16 The third stroke is slightly larger than the top stroke to create the look of stitching beneath the other stitching.

17 With the circle and the stroke selected in the Appearance panel, press Option+click (Mac) / Alt+click (PC) and drag the stroke above to duplicate it and all its attributes.

18 In the newly duplicated stroke, change the color to white. Click on the Zig Zag effect and remove it from the list by clicking on the Trash Can icon at the lower right of the panel. Click on the stroke link to open the stroke options. Set the weight to 1.5 pts, round cap the ends, click on the dashed line checkbox. Set the dash to 5 pts and the gap to 5 pts.

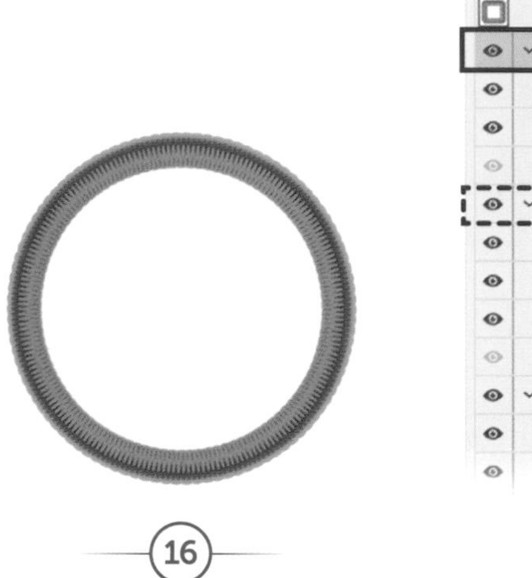

16

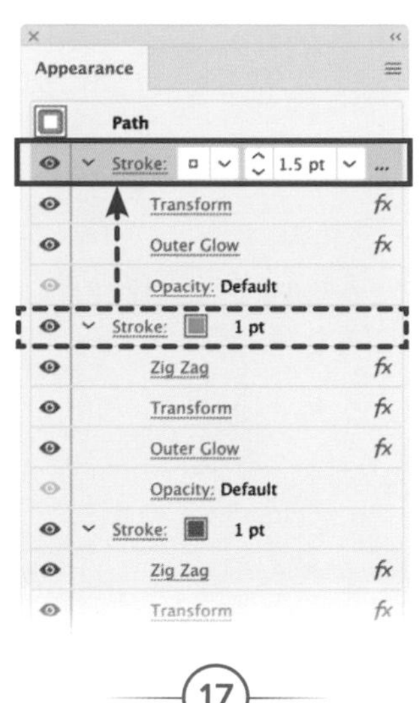

17

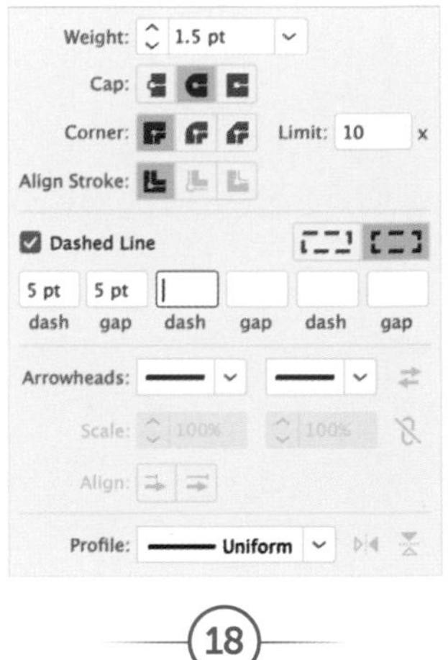

18

19 With the circle and the dashed white stroke selected in the Appearance panel, click on the transform effect and set the Horizontal and Vertical scales to 82%. This will reduce the size of the dashed stroke so it appears inside the patch.

20 With the Rectangle tool (M), click on the artboard and create a rectangle 60 mm wide, 1 mm high. Click OK.

21 Fill the rectangle with a color. Select the rectangle and use Option+click (Mac) / Alt+click (PC) and drag to duplicate the rectangle directly below the first rectangle. Fill the second rectangle with a slightly lighter color than the first. Duplicate the second rectangle and fill it with an even lighter color.

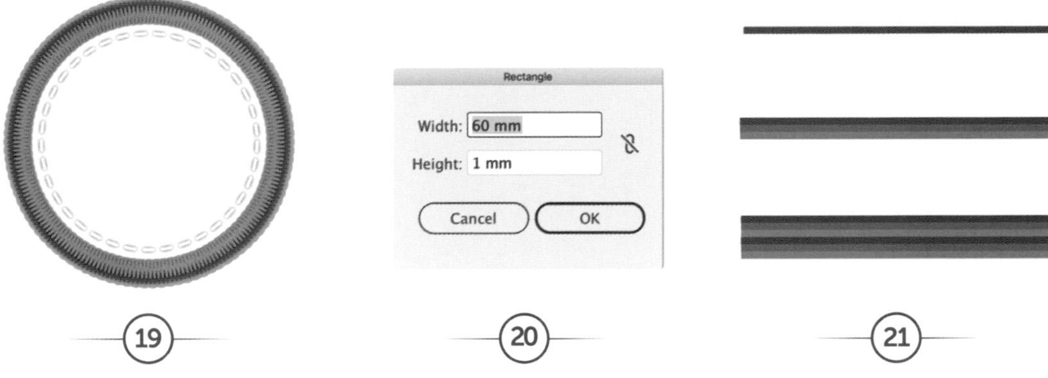

19 20 21

Draw Inside

22 Select all three rectangles and duplicate them below the first set. Use Command+D (Mac) / Ctrl+D (PC) to duplicate the sets for a total of 18 sets of three. Select all the sets and choose Object > Group. Choose Edit > Cut.

23 Create a new circle, 50 mm x 50 mm, as in step 1. Select the circle and click on the Draw Inside mode at the bottom of the toolbar. Dotted lines will appear around the shape.

24 Choose Edit > Paste to paste the grouped set of lines into the circle. Then click on the Draw Normal mode at the bottom of the tool.

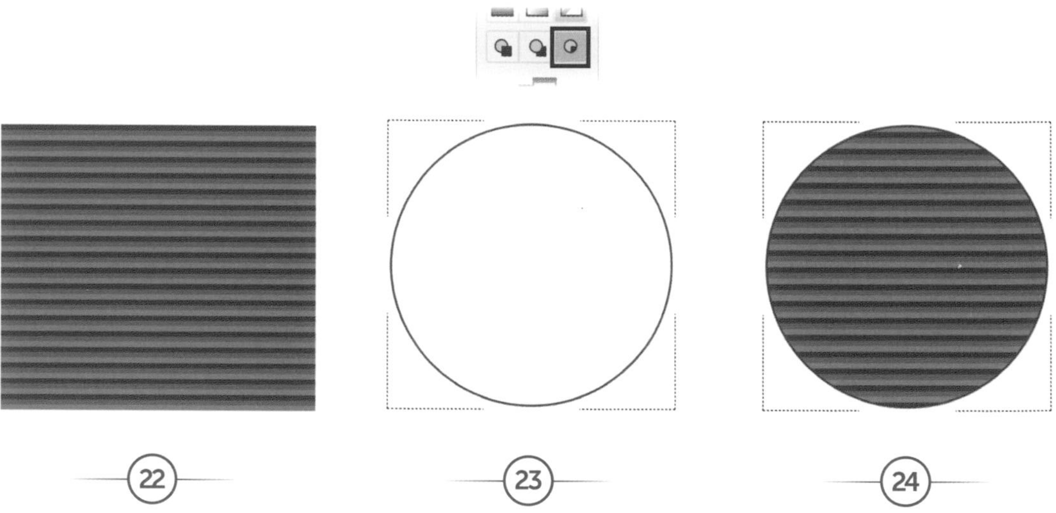

Recolor Artwork

25 Select the circle with the lines, move it into position over the circle with the strokes, and choose Object > Arrange > Send to Back. This will move the lines to the back of the shape.

26 Select the filled shape behind it and choose Edit > Recolor Artwork. Click on the Edit tab near the top of the panel. Click on the link at the lower right of the color wheel to lock the colors in harmony. Move the paddles around the wheel to change the color. Use the slider at the bottom of the color wheel to change the brightness/darkness of the colors.

27 This is an easy way to change the color and brightness of any selected object.

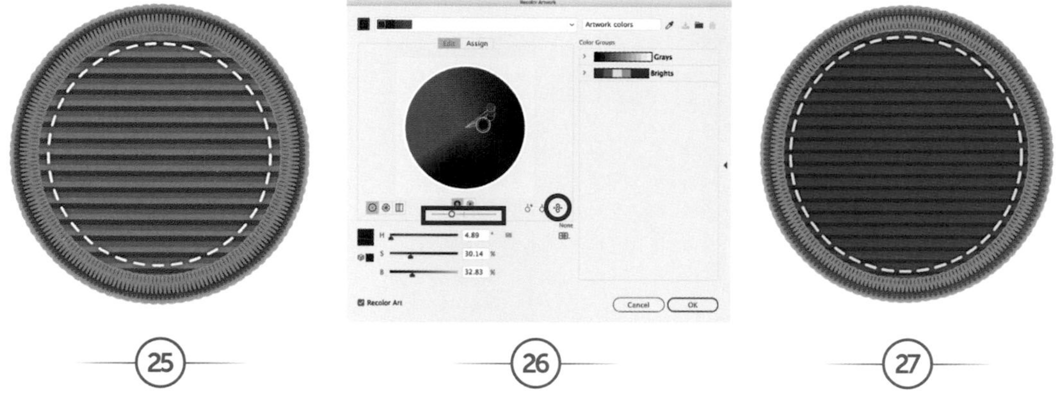

28 Click on the artboard with the Star tool. Set Radius 1 to 15 mm, Radius 2 to 5 mm, 5pts. Click OK.

29 Choose Effects > Distort & Transform > Zig Zag. Set the size to 1 mm, Ridges to 35 per segment, and Points to Smooth. Click OK.

30 Set a stroke and fill color for the star. Notice the ends of the star create weird tangents where the Zig Zag effect goes around the points of the star.

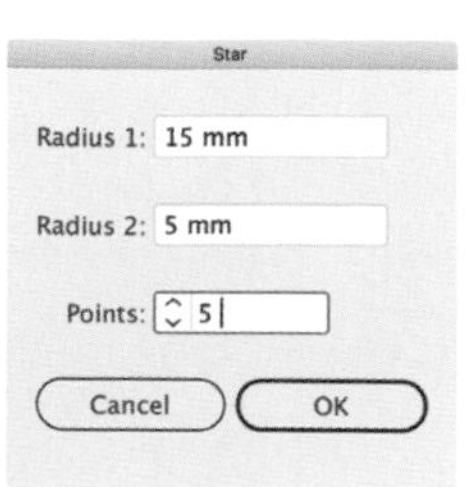

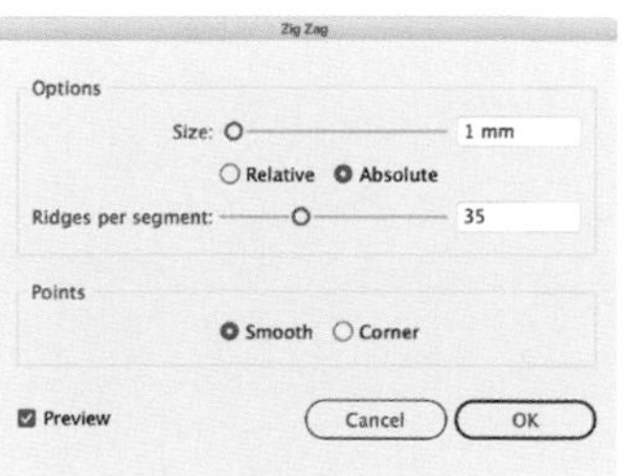

- **31** Select the star with the Direct Selection tool and pull the corner widgets in toward the center slightly to round the tips and remove the end tangents.
- **32** Enjoy your sewn patch! It looks embroidered.

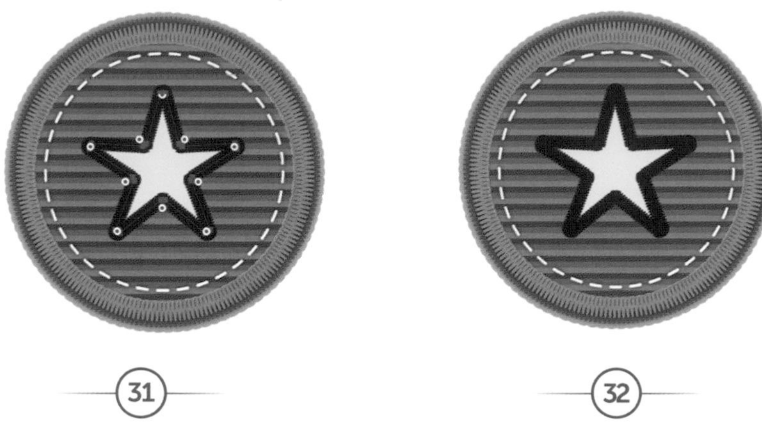

Project:
Big Build—Camping Gear in an Outdoor Scene

Lesson A: Create a Hot Air Balloon

Hot Air Balloon

- 1 Start with a circle. Add a tall rectangle to the bottom of the circle.
- 2 Select both shapes and use the Pathfinder Unite mode to make them into one shape.
- 3 Use the Direct Selection tool to choose the two corner widgets on the inside corners. Pull the corner widgets out.

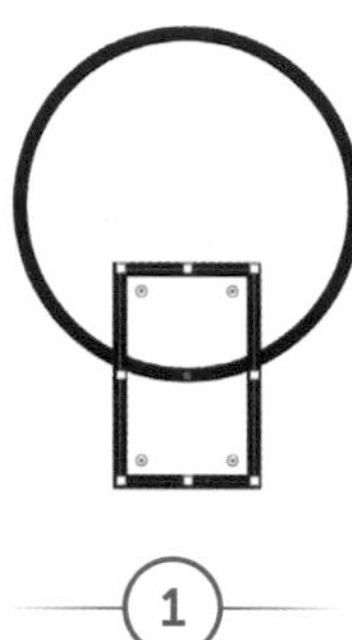

1

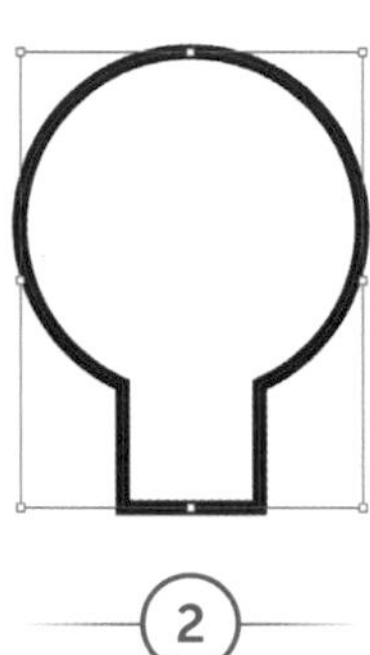

2

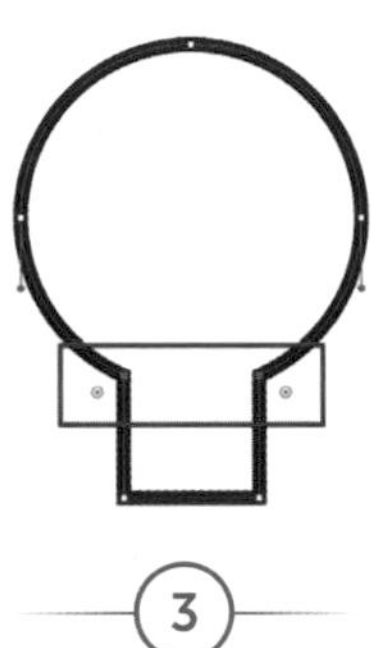

3

Copy and Flip

- 4 This is the result of the corner widget pull. This is also a great way to create a light bulb!
- 5 Use the Direct Selection tool to select the right half of the balloon. Copy and paste the shape.
- 6 Paste the shape again and flip it over to create two halves. Place them together so they touch at the top. Select the halves.

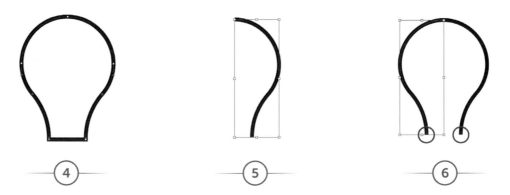

4

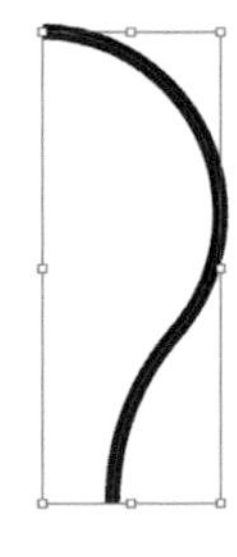

5

6

Blend Options

- **7** Choose Object > Blend > Blend Options. Set the Spacing to Specified Steps and enter in 5 steps. Click OK.
- **8** Choose Object > Blend > Make to create the curved lines in between the end lines.
- **9** The Blend Shape needs to be expanded to be able to be converted into a Live Paint Object. Choose Object > Expand. Click OK.

7

8

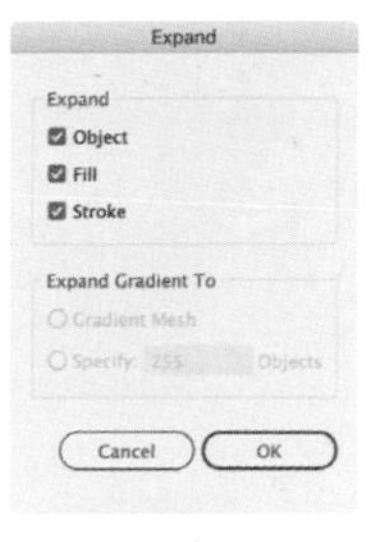

9

Live Paint Mode

- **10** With the shapes expanded (it will not look any different), add a rectangle to the bottom of the balloon.
- **11** Select all the shapes and choose Object > Live Paint > Make. Live Paint objects will have snowflake-looking pull handles.
- **12** Select the Live Paint Bucket tool (K). On the upper part of the Live Paint Bucket cursor there are three boxes; this is your color palette. Use the left or right arrows to scroll through the color palette indicator. The center color is your choice. Click in the area where you want to add that chosen color.

10

11

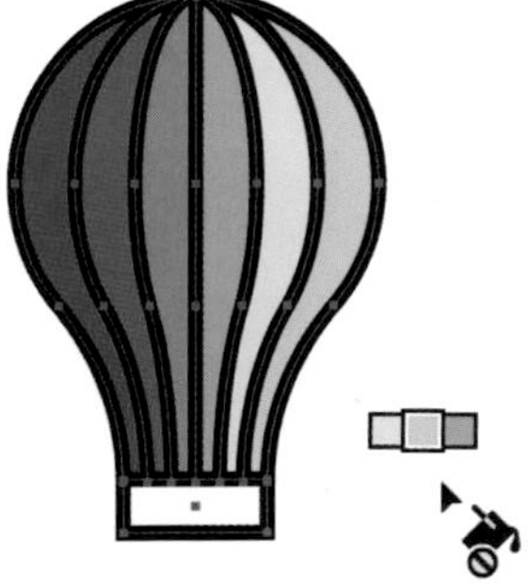

12

- 13 Add a rectangle underneath to create the basket.
- 14 Use the Direct Selection tool to select the lower two points and pull in the corner widgets slightly.
- 15 Add a top to the basket with a rounded-corner rectangle.

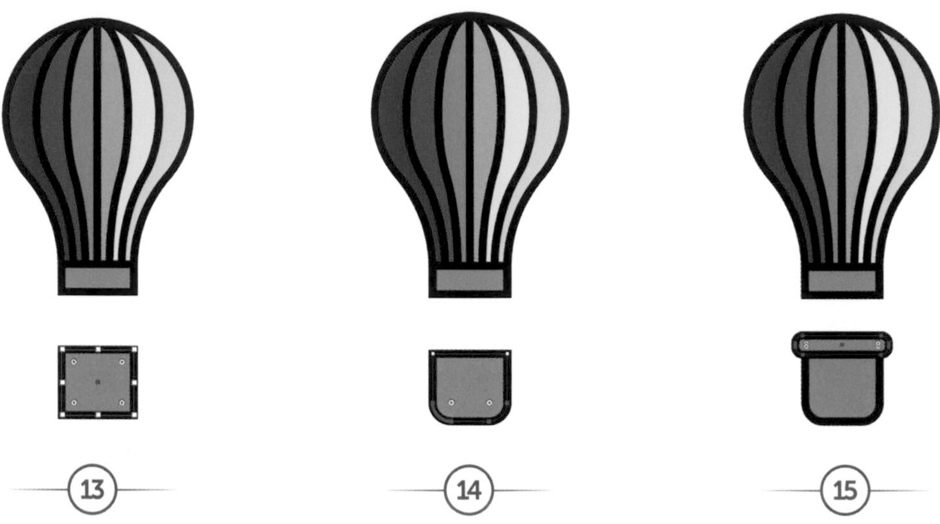

- 16 Change the color of the top of the basket to a darker color to add contrast.
- 17 Add attaching lines between the balloon and the basket with the Line tool (\). To change the color of all the strokes at one time, select one of the attaching lines, then choose Select > Same > Stroke Color.
- 18 With all the same stroke color selected, change the color of the strokes in the Swatches panel.

Duplicate and Repeat

- **19** To create the cloud, start with a circle.
- **20** Duplicate the circle and place it halfway over the original.
- **21** Use Command + D (Mac) / Ctrl + D (PC) to duplicate the circle several times.

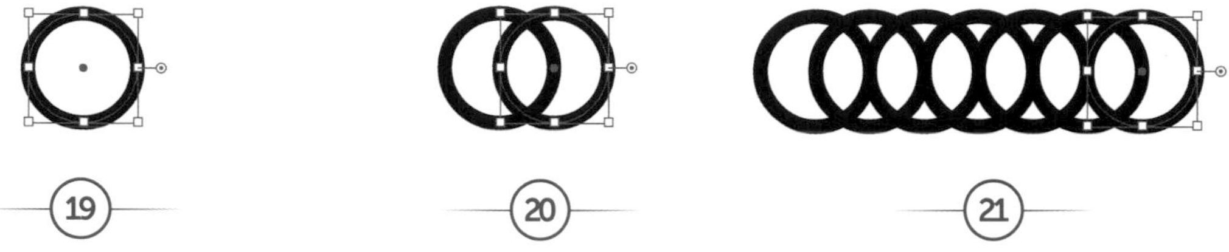

Pathfinder

- **22** Select a set of circles, copy it, and move it halfway up the group. Repeat the step again, with a smaller group, stacking sets on top of each other.
- **23** Select all the shapes and choose the Pathfinder Unite mode to make them all one cloud.
- **24** Use the Direct Selection tool to select and then delete the lower section of the cloud.

25 With the open cloud, choose Object > Path > Join to join the shapes together.

26 Add a fill to the cloud and place it in front of the balloon using Object > Arrange > Bring to Front. Duplicate the cloud, fill it with a darker shade of the same color, and choose Object > Arrange > Send to Back.

27 Add lines with rounded end caps in front of and behind the clouds and the balloon. Create shorter and longer lines to give the balloon the look of moving through the sky.

Lesson B: Create a Tent

Camping Tent

1 Start the tent off with the Polygon tool. While drawing the Polygon, use the down arrow to decrease the number of sides to three, creating a triangle. Add a fill color and an 8 pt stroke weight. Choose a darker color for the stroke.

2 Draw a line in the middle of the triangle.

3 Create a square with an 8 pt stroke weight.

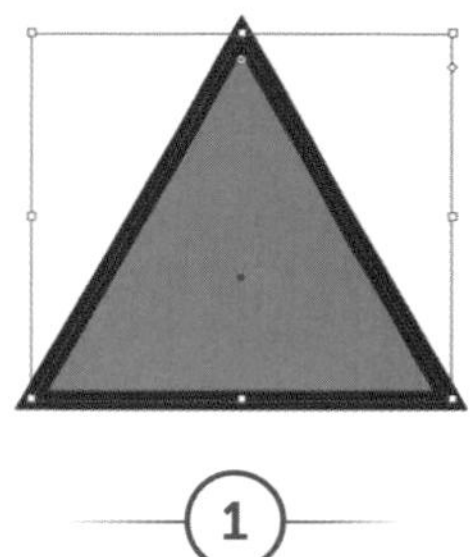

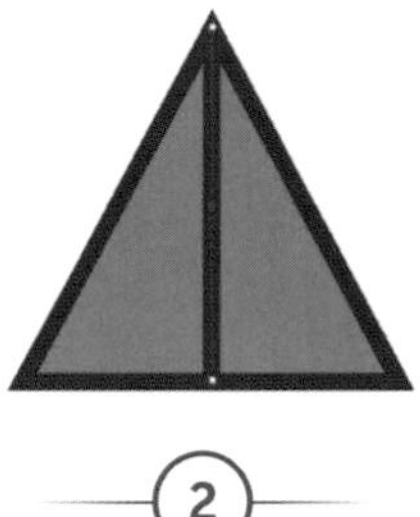

Join

4 With the Direct Selection tool (A), select the lower-right corner point and delete it.

5 Choose Object > Path > Join to close the shape.

6 In the Stroke panel, set the Stroke Corners to rounded.

7 Duplicate the triangle and rotate it 180 degrees. Fill the shape with a darker shade of the fill color.

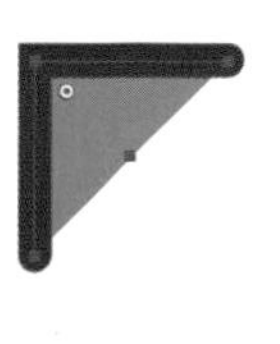

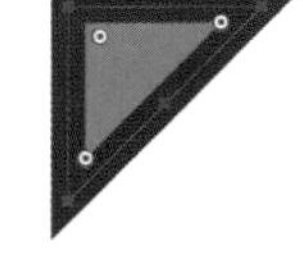

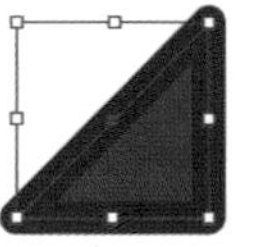

- **8** Place the two triangles at the center of the tent to create the look of an open flap.
- **9** Draw a line, 16 pt weight, with rounded end caps. Choose Object > Arrange > Send to Back and move the line behind the tent to anchor it on the ground.
- **10** Create two circles with an 8 pt stroke weight and overlap them slightly.
- **11** Select both circle and use the Pathfinder Minus Front mode to leave a crescent shape.

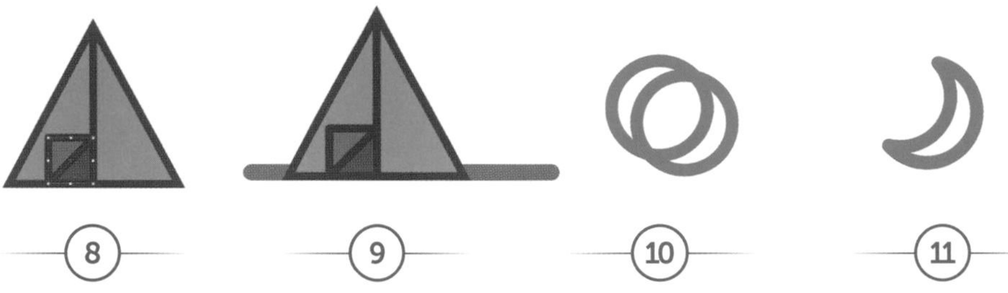

Pathfinder

- **12** Create a rectangle over the crescent shape. Select both shapes and use the Pathfinder Minus Front mode to remove the lower portion of the shape.
- **13** This will create a blade of grass or green plant.
- **14** Duplicate the shape and use the flip horizontal feature in the Properties panel. Scale the duplicate shape larger and move it to the right.
- **15** Select both shapes and use the Pathfinder Unite mode to make them one shape.

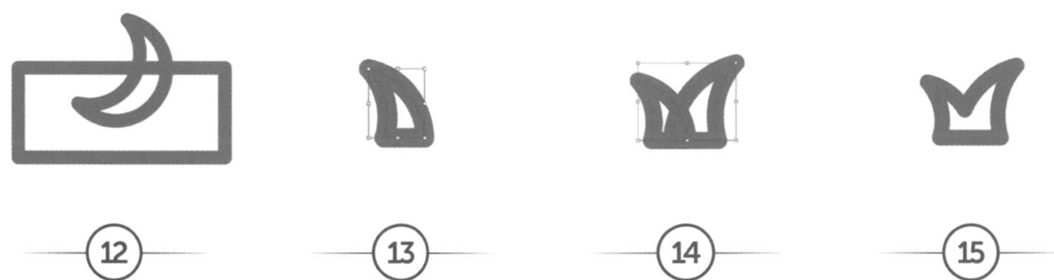

16 Fill the grass with a darker shade of the stroke color. Move the foliage into position next to the tent.

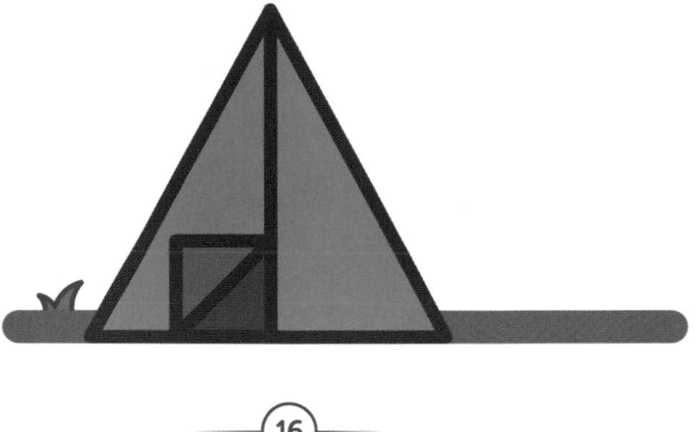

16

Lesson C: Create a Campfire

Campfire

1 Create a rectangle with a fill color.

2 Use the Direct Selection tool (A) to select the lower-right corner and pull the corner widget in slightly.

3 Select the Skew tool and click on the lower-right corner of the rectangle. Hold Shift and pull the top of the rectangle to the right to skew the shape.

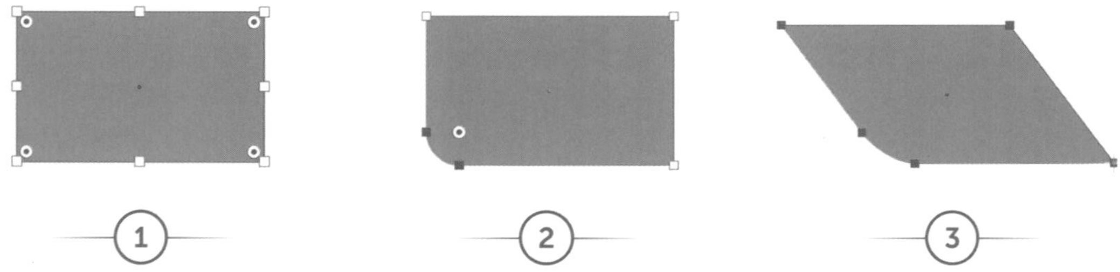

Pathfinder

4 Duplicate the skewed shape and overlap the original shape.

5 Select both shapes and use the Pathfinder Minus Front mode to remove the front shape.

6 Duplicate the shape and rotate it 180 degrees and position it over the original shape. Use the Pathfinder Unite Front mode to make it into one shape.

7 Select the shape and make it narrower by dragging the pull handles closer together. Rotate the shape to make the tips point vertically, like a flame.

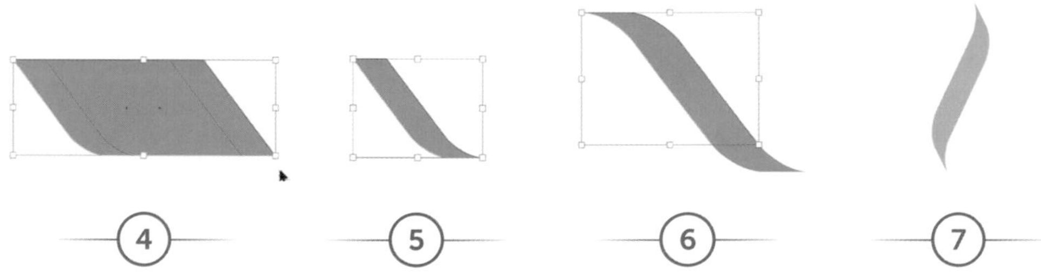

Duplicate and Draw Inside

8 Duplicate the flame shape. Scale it smaller, and repeat the duplicate and scale.

9 Add a red fill to some flames to mix in with the orange flames. Use the Object > Bring to Front or Object > Send to Back to arrange the flames.

10 Add 8 pt strokes to the flames with a darker shade of the fill color. Select all the flames and group them. Cut the flames using Edit > Cut.

11 Draw a rectangle with no fill or stroke over the flames. Click on the Draw Inside mode at the bottom of the toolbar. Paste the cut flames into the rectangle and move them so their bottom tips are hidden by the rectangle. Click on the Draw Normal mode at the bottom of the toolbar when you're done.

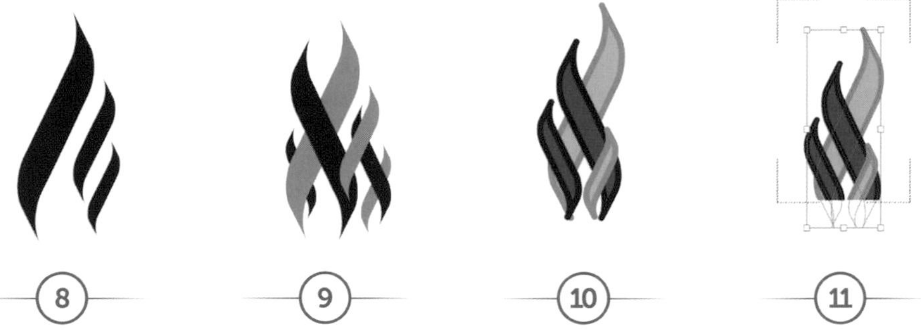

12 Add some wood to the fire by creating a rectangle, wide and short.

13 Use the Direct Selection tool to select the right end and corner widget the ends fully. Move the wood under the fire.

14 Use the Direct Selection tool to select and remove one end of the rectangle. In the Stroke panel, set the end caps to be rounded to indicate a log.

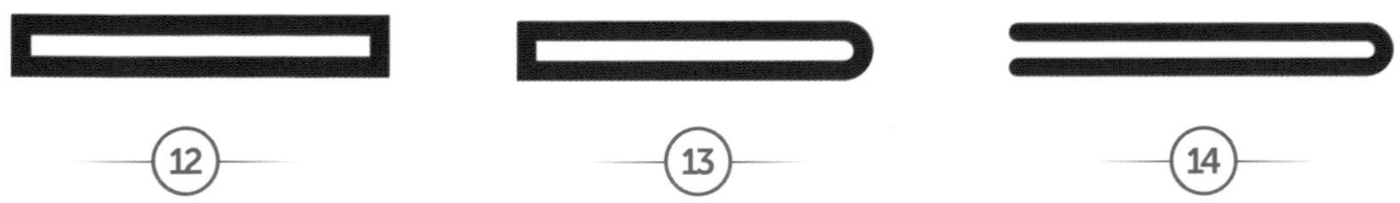

Snap, Crackle, and Pop

- **15** Duplicate the log and rotate it 180 degrees. Apply a lighter color stroke. Move the logs under the fire.
- **16** Add some snap and crackle to the fire by creating small circles with orange, red, and yellow fills. Add lines to the flames for texture.
- **17** Add some pop to the snap and crackle by creating lines to form an X and placing them above the fire.

Lesson D: Create Pine Trees

Blend Options

1 Create a triangle by using the Polygon tool. Draw the polygon and while drawing, hold the down arrow to reduce the number of sides. Fill it with a lighter green and add a heavy stroke of a darker green. Use the Stroke panel to round the corners of the stroke.

2 Duplicate the first triangle, then increase the size and place it below the first triangle. Do not scale the stroke and effects while sizing; open the Transform panel and make sure Scale Stroke and Effects is off before you scale.

3 Choose Object > Blend > Blend Options. Set the spacing to Specified Steps and choose 3 steps. Click OK.

4 Choose Object > Blend > Make to blend the shapes together.

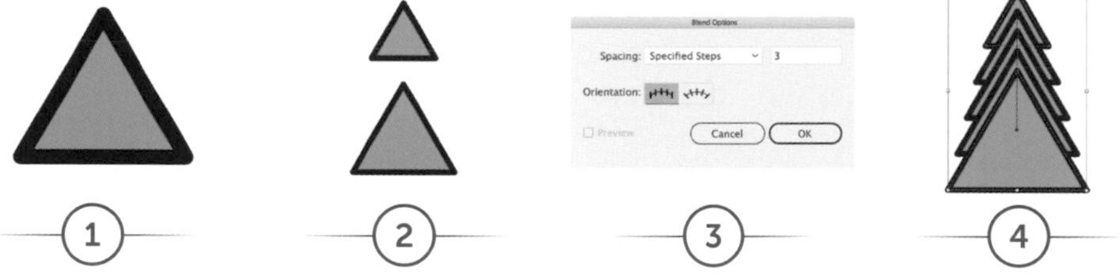

Expand

5 Select the blended shapes and choose Object > Expand.

6 This creates shapes for each step of the blend. Select all the shapes.

7 Choose the Pathfinder Unite mode to make all the shapes into one shape.

8 Duplicate the shape and set the stroke weight to zero. Duplicate the tree and position it up and to the right slightly.

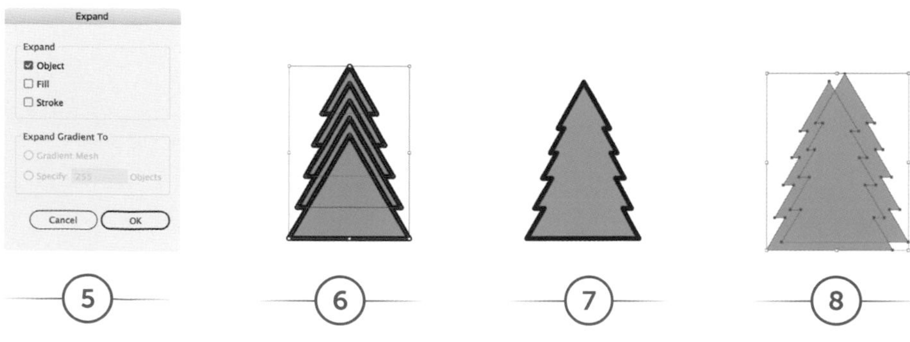

Draw Inside

- **9** Choose the Pathfinder Minus Front mode to leave a small section. Fill the small section with a darker shade of the fill color. Cut the shape using Edit > Cut.
- **10** Select the tree shape and choose the Draw Inside mode at the bottom of the toolbar.
- **11** Paste the darker section in and move it to the left, making a slip shadow. Click on the Draw Normal mode.
- **12** Add lines in the same color and weight as the stroke of the tree. Add a circle at the end of the rounded lines for visual texture. Select all the objects and group them.

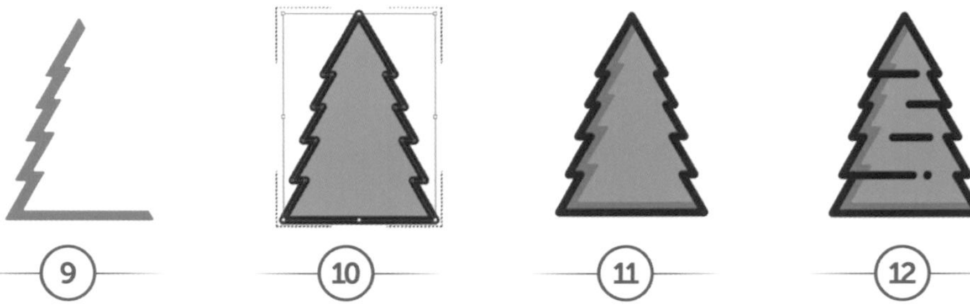

Lesson E: Create a Backpack

Backpack

1 Create a rectangle to start the backpack.
2 Pull the corner widgets in toward the center of the rectangle to round the edges.
3 Add a stroke to the shape. Duplicate this shape twice for later use.
4 Create a rectangle for the lower section of the backpack. Add a fill color. Cut the rectangle with Edit > Cut.

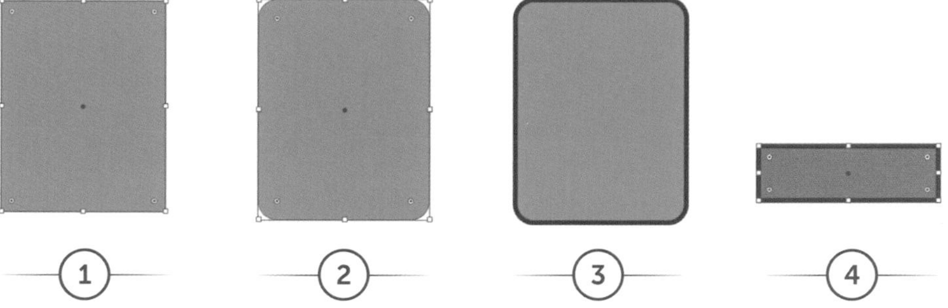

Draw Inside

5 Select the main rectangle and click on Draw Inside mode at the bottom of the toolbar.
6 Paste the shape into the rectangle and move it to the bottom to create the lower section of the backpack. Click on the Draw Normal mode.
7 Create a rectangle for the front pouch. Choose a lighter fill color. Pull the corner widgets in toward the center of the shape to round the corners.
8 Copy the pouch, reduce the height, and with the Direct Selection tool (A), select the lower corners. Pull the corner widgets out so the corners are not rounded. Move the shape to the top of the pouch as the cover.

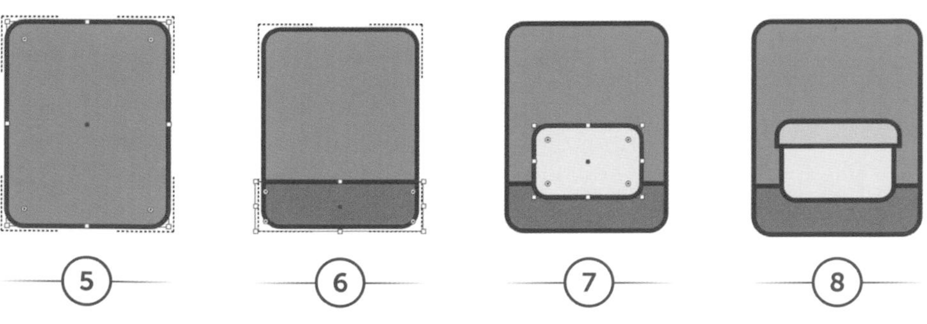

- 9 Take the duplicate of the main rounded rectangle in step 4, and reduce the height. With the Direct Selection tool (A), select the upper corners and pull the corner widgets out so the corners are not rounded. Move the shape to the top of the backpack as the cover and scale it to be larger than the backpack.
- 10 Create a strap with a buckle by creating a vertical rectangle with no stroke. Draw a rectangle over the strap for the buckle. Select the strap and the buckle and group them using Object > Group.
- 11 Duplicate the strap and buckle to the other side of the backpack.

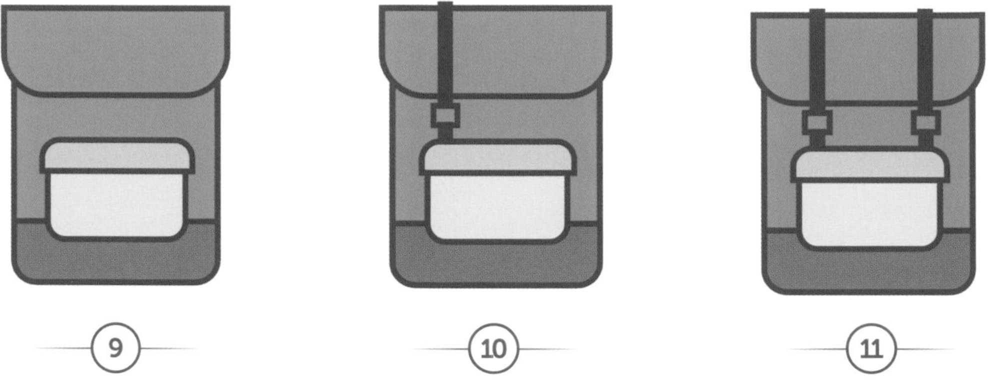

Arrange Objects

- 12 Create a zipper on the pouch with two lines, one vertical and one horizontal. Round cap the ends in the Stroke panel.
- 13 Create an oval and move it behind the backpack with Object > Arrange > Send to Back.
- 14 Take the duplicate shape from step 4 and remove the fill. Make the stroke a darker color.

Opacity

15 With the Direct Selection tool (A), click on the upper-right corner and delete it, leaving just the lower corner.

16 Set the Opacity of the line to 60% (or less, depending on the color) and set the Blend mode to Multiply.

17 Move the line over the lower-left area of the backpack to create a shadow line. Create a line near the top of the pouch. Set the Opacity and Blend modes as you did in step 16. Create a lighter line at the top of the backpack for a highlight and repeat the Opacity and Blend mode settings. You may want to adjust the opacity to make the lines look like they work well with the colors. Add a small rectangle for the logo on the backpack pouch flap.

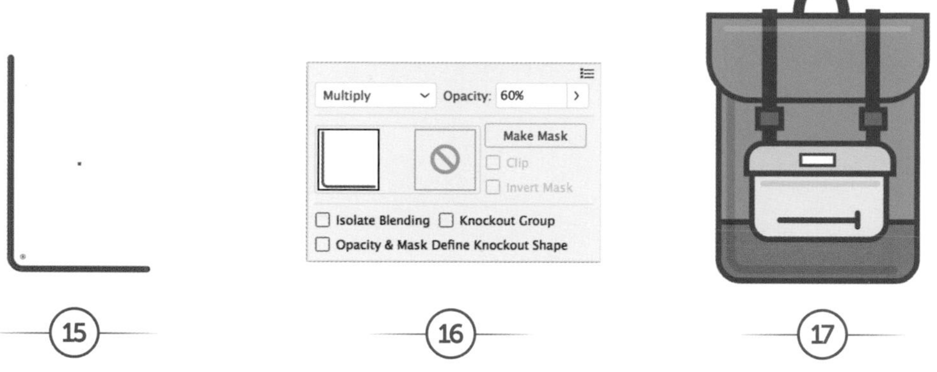

Recolor Artwork

18 To create multiple backpacks of different colors, select the backpack, group it, and duplicate it. Choose Edit > Colors > Recolor Artwork. Click on the Edit tab at the top of the Recolor Artwork panel. Click on the link icon to the lower right of the color wheel, linking the colors together.

19 Move the color paddles around the wheel to change the color harmonies, or use the sliders under the color wheel to change the brightness or colors.

Lesson F: Assemble the Scene

Camping Day and Night

Let's assemble the scene with all your creations from the lessons. Group each set of creations together and then copy and paste from one artboard into this artboard. Scale the objects to their appropriate sizes. Remember to open the Transform panel and choose the Scale Stroke & Effects button if you want the stroke to increase and decrease with the scale. Or uncheck it to have the stroke not scale with the size.

Useful Tip for Scaling Strokes: If an object has a stroke that is too large for the scene, uncheck the Scale Stroke & Effects, and scale the object larger. Click on the Scale Stroke & Effects box, then scale it smaller. The stroke will scale smaller along with it.

To create an evening camping scene, make a dark-colored background and add stars and a moon from the Weather icons creations. Clouds from the Hot Air Balloon project can be used, or just use the lines as wisps of clouds.

Project: Output

Asset Export

1 Open the Asset Export panel from the Window menu.

2 Drag your grouped artwork into the Asset Export window. If the artwork is not grouped, each piece will become a unique asset. Or, you can hold Option (Mac) / Alt (PC) while dragging the creation and it will be a single asset.

3 Select the asset to export in the upper section. Choose the size of the asset, the file format, and click Export.

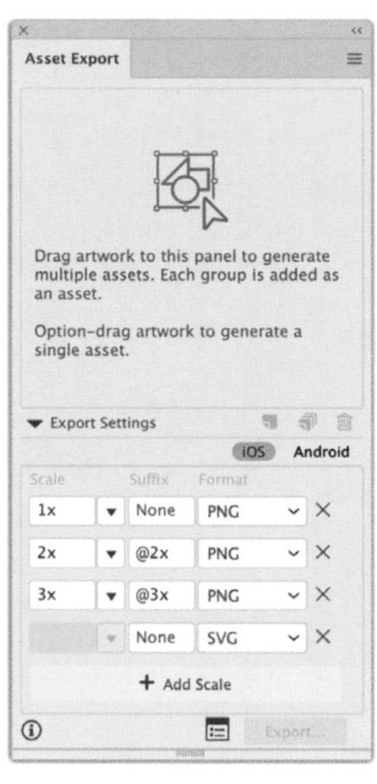

1

2

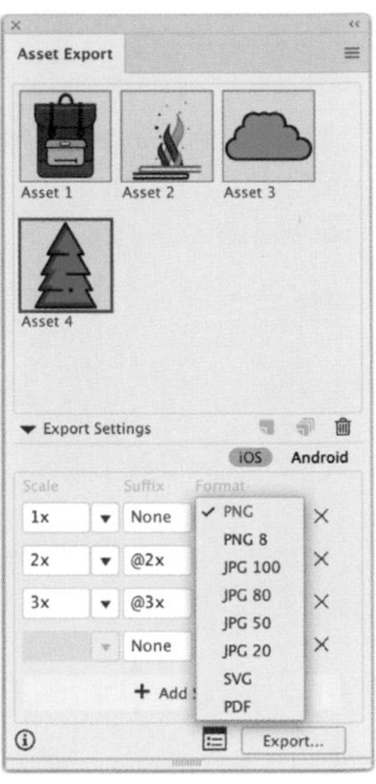

3

Saving and Printing Files

1 To save a file as an Illustrator document, choose File > Save. When saving as an Illustrator format, users who have Illustrator can open and edit the files. Many other applications will accept native Illustrator files (.AI files) being imported directly into the application without the need for opening the Illustrator file.

2 To save as an Illustrator PDF, choose File > Save or Save As and choose Adobe PDF from the options list. An Adobe PDF file can be edited using Illustrator. If the end user does not have Illustrator, the file can be opened in Adobe reader, printed, or placed in many other applications.

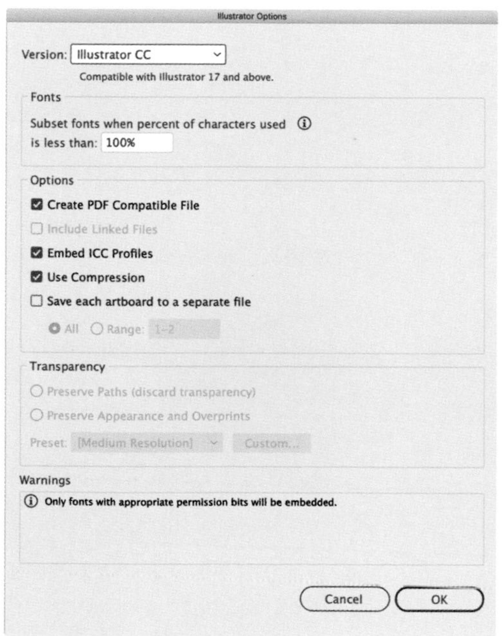

1

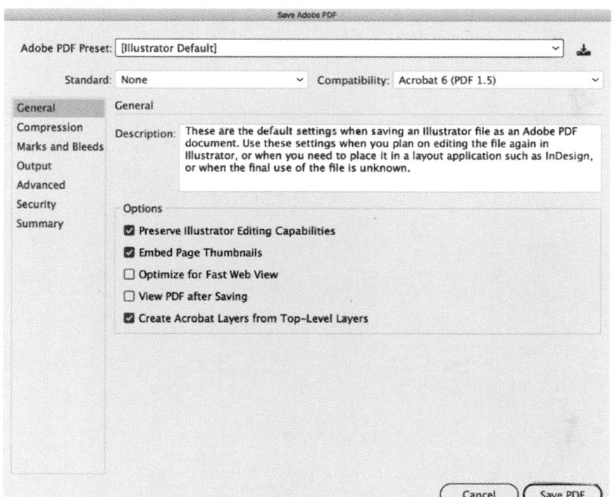

2

Printing

1 Printing an Illustrator file can be done by choosing File > Print. The print dialog box may look slightly different depending on your printing device. Here you can choose which artboard to print. You can choose the paper size and scale the artwork to fit the paper. You can move the image around in the Preview panel to position it on the printed page.

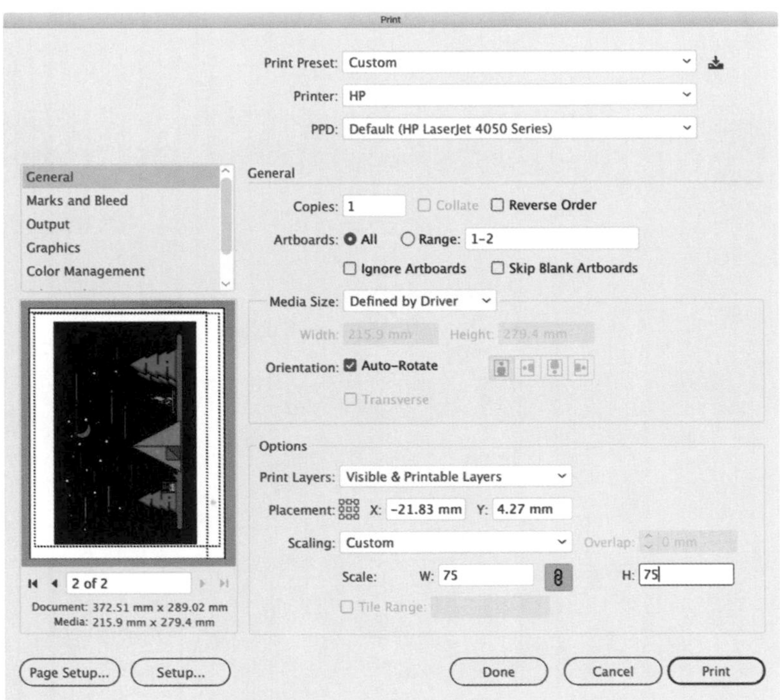

1

THE COMPENDIUM

1 Preferences and Workspaces

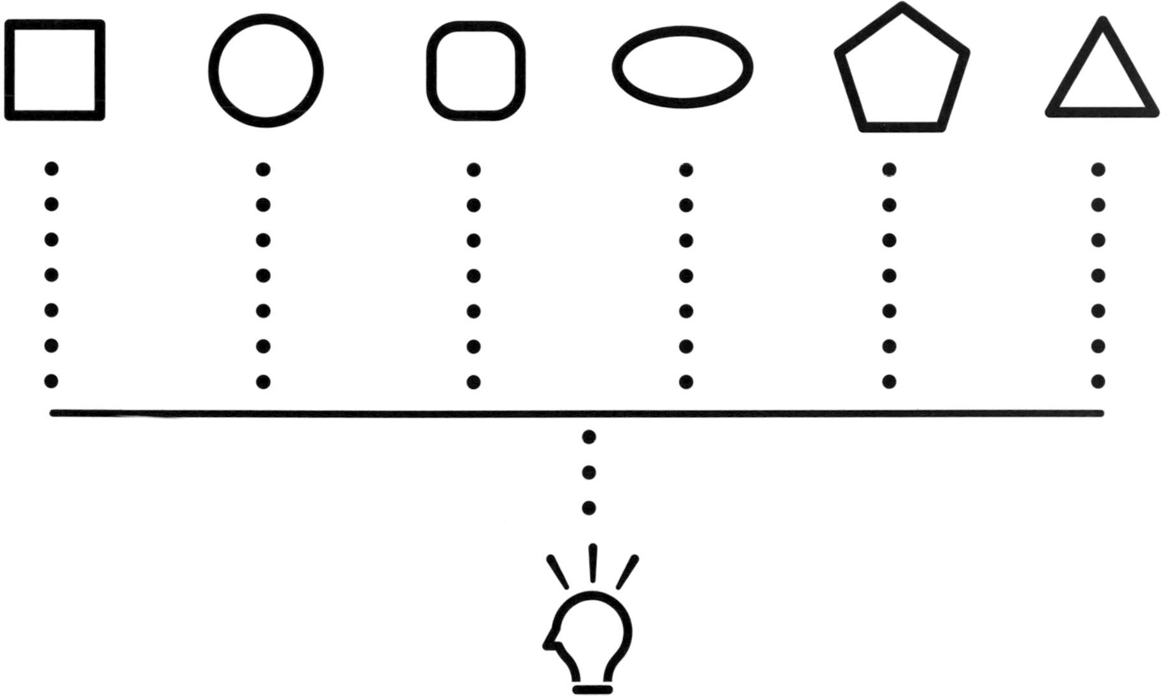

Preferences

Document-Specific and Global

For Illustrator to run as smoothly as possible within your workflow, you will need to set up your preferences to your liking. Preferences on a Mac are under the Illustrator menu > Preferences. On a PC they are under the Edit Menu > Preferences.

VERY Important Note: I strongly recommend that you change the preferences with no document open because this will set the preferences for the application. Any preferences set with a document open will *only* be applied to that document.

At times, preferences may become damaged and may need to be reset to their default values. To restore preferences, press and hold Option+Command+Shift (Mac) or Alt+Control+Shift (PC) as you start Illustrator. The new preferences files are created the next time you launch Illustrator.

General

The General section contains the preferences that don't fit neatly in the other categories. These are all global preferences.

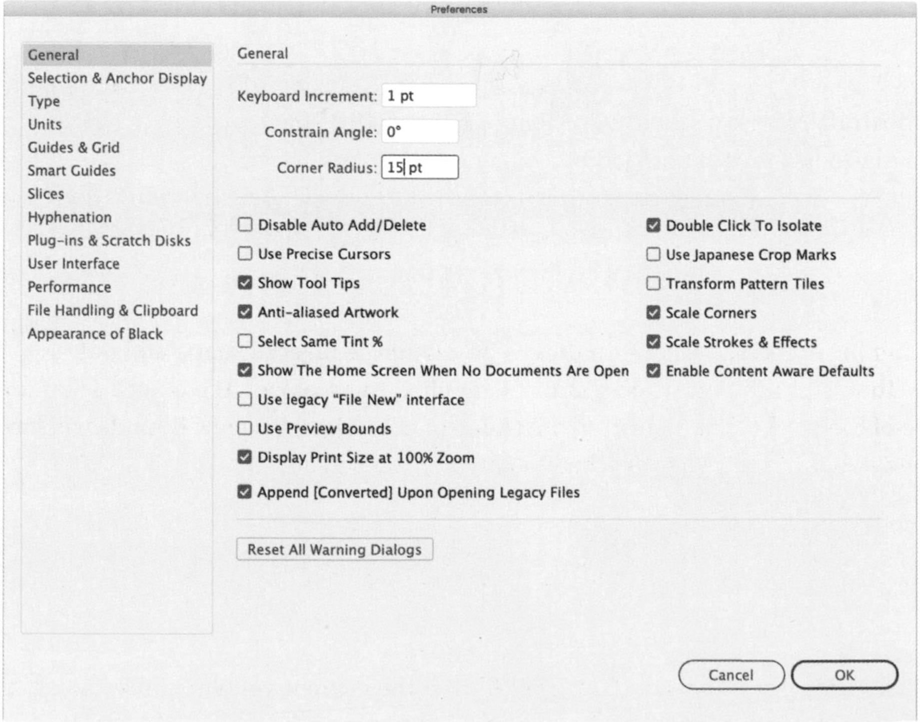

Keyboard Increment

This value represents how much an object will move when you use the arrow keys on the keyboard.

Constrain Angle

This the degree to which an object will rotate when you hold down the Shift key while turning the object.

Corner Radius

This describes the default rounded corner size when drawing a rounded rectangle.

Show Tool Tips

The Tool Tips become visible when you hover the cursor over a tool. This feature is useful for new users but these impact the performance of Illustrator.

Anti-aliased Artwork

Anti-aliasing is the smoothing of jagged edges when the artwork is being viewed in Pixel Preview.

Select Same Tint %

This allows the selection of a color that has a tint applied to it and other colors with that same tint percentage.

Show The Home Screen When No Documents Are Open

The Home Screen shows recent documents, new document presets, and Adobe stock templates.

Use Legacy "File New" Interface

The legacy File New interface is an older version of creating a New File.

Use Preview Bounds

When aligning objects, the coordination is based on the object's shape and not its boundary or edge. That is to say that when a stroke is applied to an object, the object will not align to the edge of the stroke, but instead to the shape of the object. Preview Bounds will have them align to the edge of the stroke or what you preview on screen.

Display Print Size at 100% Zoom

Shows the actual size of the printed piece when the zoom size is at 100%.

Append [Converted] Upon Opening Legacy Files

This setting will convert and update older files to the current version of Illustrator. A Dialog box appears informing you of how the file is being converted.

Double Click To Isolate

This function is used isolate a layer, sublayer, path, or group of objects without having to ungroup the file or find the objects in the layers panel. If you find that you are a frequent "double-clicker" and you are in Isolation mode inadvertently, turn this option off.

Use Japanese Crop Marks

This creates crop marks for Japanese printing.

Transform Pattern Tiles

This transforms the pattern in an object when the object is scaled. This can also be set in the Transform panel.

Scale Corners

When scaling or transforming an object with rounded corners, this option scales the corners in proportion with the object. This can also be turned on and off in the Transform panel.

Scale Stroke & Effects

When scaling or transforming an object with a stroke and/or effect, this option scales the stroke/effect in proportion with the object. This can also be turned on and off in the Transform panel.

Enable Content Aware Defaults

This feature works with the Puppet Warp tool to add pins where it identifies the best location for warping.

Reset All Warning Dialogs

From time to time, dialog boxes will pop up warnings informing you of important items. Those dialog boxes may have a checkbox that says "Don't Show Again." Once that box is checked and you click OK, the dialog will not appear again. Click the reset button to see the dialog boxes again.

Selection & Anchor Display

This is where you adjust how lines and anchor points act and appear.

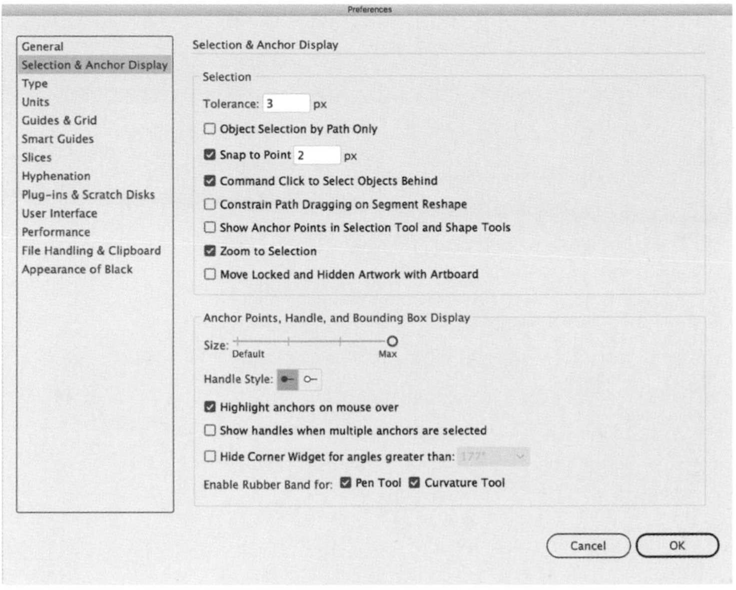

Selection

This defines how items can be selected, and how they will look once selected.

Tolerance

This sets the pixel range for anchor points. A lower values means you have to be closer to an anchor point to be able to select it.

Object Selection by Path Only

This limits selection of an object to clicking on just the path—not inside the shape. I would not use this option because it makes it more work selecting objects.

Snap to Point

This snaps objects to anchor points and guides. You can set the distance between the object and anchor point or guide; larger number means it will snap from farther away.

Command Click to Select Objects Behind

When objects are stacked on each other, use Command+Click (Mac) or Control-Click (PC) to select objects on the bottom of the stack without moving the objects on top out of the way.

Constrain Path Dragging on Segment Reshape

This feature allows you to reshape a path in a relative transformation. This keeps the control handles in alignment on their original angles when transforming.

Show Anchor Points in Selection Tool and Shape Tools

When this is enabled, anchor points will show on an object when using the Selection or Shape tools.

Zoom to Selection

When this feature is enabled, the point of the zoom will be centered on a selected object.

Move Locked and Hidden Artwork with Artboard

Moving artboards around will move all the visible and unlocked content. To have everything move with the artboard, check this box to enable this feature.

Anchor Points, Handle, and Bounding Box Display

Size

This sets the size of the handles and anchor points.

Handle Style

This sets the look of the handles but does not change their functionality.

Highlight Anchors on Mouse Over

With this function enabled, anchor points on an object will be indicated when you mouse over them.

Show Handles When Multiple Anchors Are Selected

By default, when an anchor is selected, it will only show the handles associated with that point. Click this box to have it show all the handles on the shape when a point is selected.

Hide Corner Widget for Angles Greater Than:

This hides the corner widget indicator when angles are set to a specified value.

Enable Rubber Band for:

When you enable the Rubber Band for any lines made with either the Pen tool or the Curvature tool, it will show both the curve of a path and how the points on the line connect.

Type

This section controls text behavior and display.

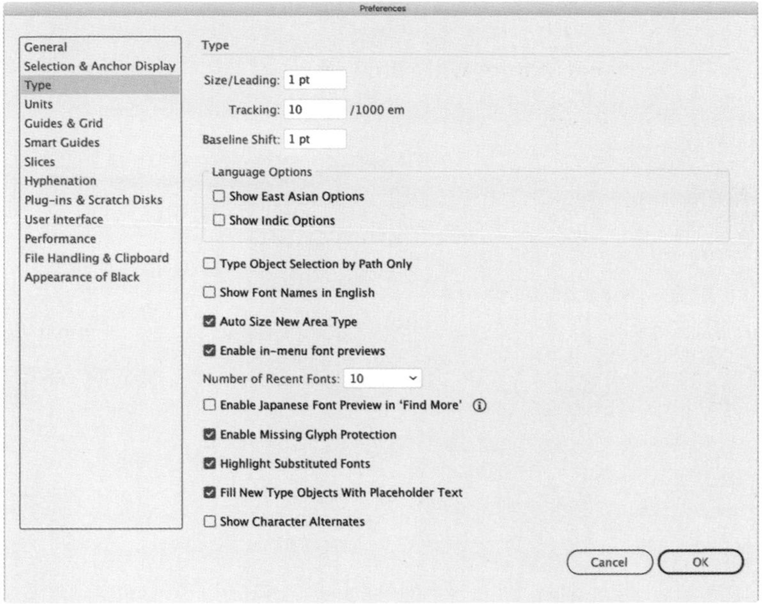

Size/Leading:

This controls the increments in which the size/leading will change using the keyboard shortcuts. The default setting for the size of type is 2 pt; I suggest 1 pt for finer adjustments. Press ⌘+Shift+</Ctrl+Shift+< to decrease the size or ⌘+Shift+>/Ctrl+Shift+> to increase the size (respectively). To adjust the leading, press ⌥+↑ / Alt-↑ to decrease leading or ⌥+↓ / Alt-↓ to increase the leading.

Tracking:

This controls the increments in which the kerning will change using the keyboard shortcuts. The default is 20/1000; I suggest 10 pt for finer kerning adjustments. To kern the space between two character, place the type cursor between the characters and use ⌥+← / Alt-← to tighten the letter spacing or ⌥+→ / Alt-→ to loosen the letter spacing. Tracking works the same as kerning; the only difference is tracking is spacing that is applied to multiple characters at one time. The shortcuts for tracking and kerning are the same.

Baseline Shift:

This controls the increments in which the baseline will change using the keyboard shortcuts. The default is 2 pt; I suggest 1 pt for finer adjustments.

Language Options

When creating Illustrator files in different languages, check the applicable language box.

Type Object Selection by Path Only

This limits the selection of text to clicking on just the path, not on the text areas. This is not recommended because it makes for more work selecting text.

Show Font Names in English

This displays font names in English if the fonts are not in English originally.

Auto Size New Area Type

This function automatically resizes the Area Type container to fit the amount of text entered. It's a helpful feature because it saves a step in closing and opening up Area Type containers.

Enable In-Menu Font Previews

This displays the name of the font and the selected text in that font.

Number of Recent Fonts

This displays the listed number of recently used fonts in the font menu.

Enable Japanese Font Preview in 'Find More'

Click this box to display Japanese fonts when you click on the Find More (fonts) item in the font menu.

Enable Missing Glyph Protection

Choose this item to automatically select incorrect, unreadable characters between roman and non-roman fonts.

Highlight Substituted Fonts

When opening an Illustrator file, there may be fonts that are missing from your system and don't display correctly. This feature highlights the incorrect font to make you aware that it is substituted.

Fill New Type Objects with Placeholder Text

When drawing a new area text container, this feature autofills the container with placeholder text.

Show Character Alternates

Character Alternates are different styles of the same letter. With this option enabled, a small menu will appear below a letter with alternate options available in that typeface.

Units

Units are basic settings for how things are measured.

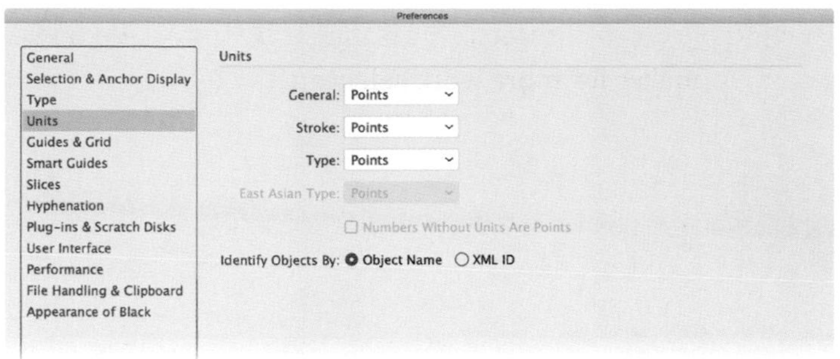

General:

These are the increments for measuring objects, position, spacing, and size. The available choices are points, inches, millimeters, centimeters, and pixels.

Stroke:

Your choices for units to measure the weight of the stroke are points, inches, millimeters, centimeters, and pixels.

Type:

Here you can chose a unit to measure the size of the type. Points are the standard for measuring type.

Identify Objects By:

This feature identifies the object by its description. XML identifies each object by a unique tag.

Guides & Grid

For alignment and control of artwork, guides and grids are helpful.

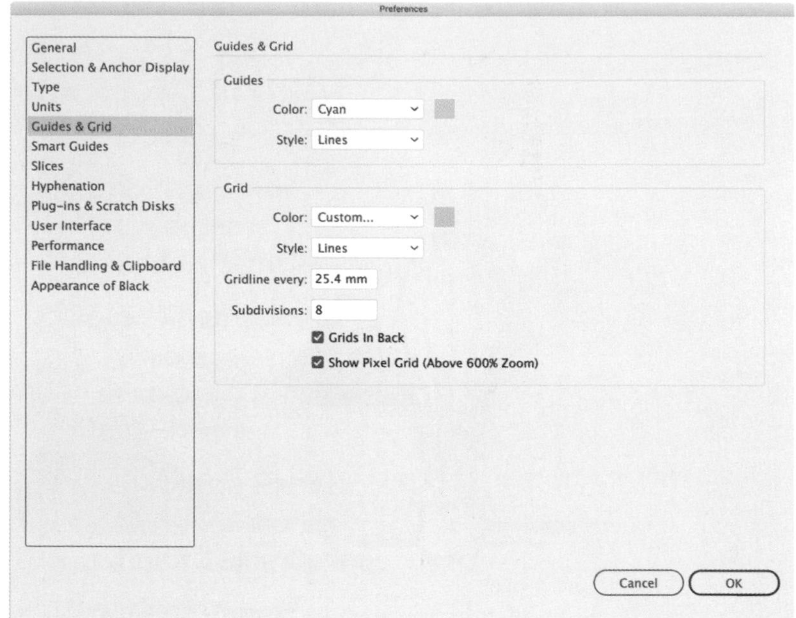

Guides

Color:

Set the color of the guides by clicking on the color square.

Style:

Guides can be displayed as colored lines or dots.

Grid

Color:

Set the color of the guides by clicking on the color square.

Style:

Guides can be displayed as colored lines or dots.

Gridline Every:

This draws major grid lines based on this value.

Subdivisions:

This divides the major grid lines into subdivisions based on the value.

Grids in Back:

This creates the grid to appear behind the artwork.

Show Pixel Grid

When working in Pixel Preview mode, this option will show the grid only when the file is zoomed into 600% or greater.

Smart Guides

These options provide you with more ways to give structure to your layout. Here you set the appearance and action of Smart Guides.

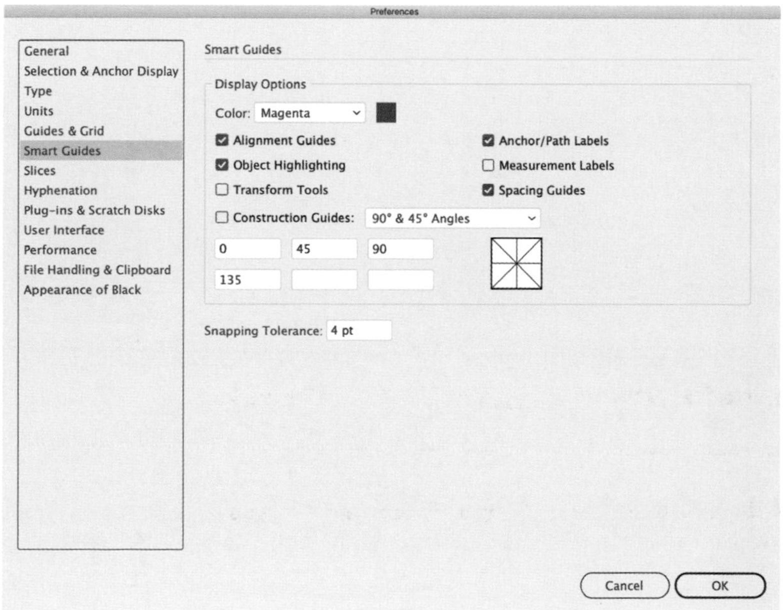

Display Options

Here you can choose how close the Smart Guides appear and function.

Color:

Here you can choose the color of the Smart Guides from a full spectrum of color.

Alignment Guides

This feature shows alignment to edges and centers of other objects and the artboard when Smart Guides are active.

Anchor/Paths Labels

This features shows tool hints identifying an anchor, point, path, or center on an object when these items intersect or align.

Object Highlighting

With this function enabled, objects will be highlighted as you drag them around the workspace. The highlight color matches the object's layer color.

Measurement Labels

This function displays size information for many drawing tools and text tools. Labels show the position of the object on the artboard using X and Y coordinates. While you create, select, move, or transform objects, these labels appear.

Transform Tools

This displays size and angle values when you scale, rotate, and shear objects.

Spacing Guides

These display guides show the spacing between two objects when a third object is being placed to match the spacing of the first two objects.

Construction Guides:

These guidelines appear as you draw new objects. The values represent the angles at which the guidelines will be drawn from the anchor points of a nearby object.

Snapping Tolerance:

This value specifies the number of points the object must be from another object for Smart Guides to take effect.

Slices

This section is for slicing a file for web use.

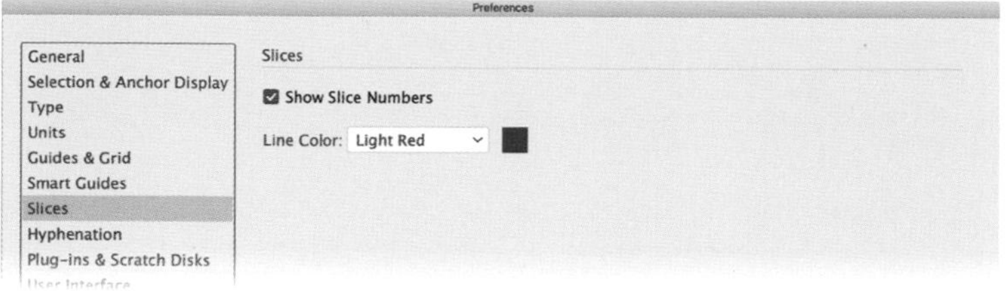

Show Slice Numbers

This labels each slice with a number for easier identification.

Line Color:

Here you can choose the color of the divider lines that indicate the slices.

Hyphenation

This section allows you to set the language for the hyphenation regulations. These corrections, (or exceptions) are typically stored in the User Dictionary.

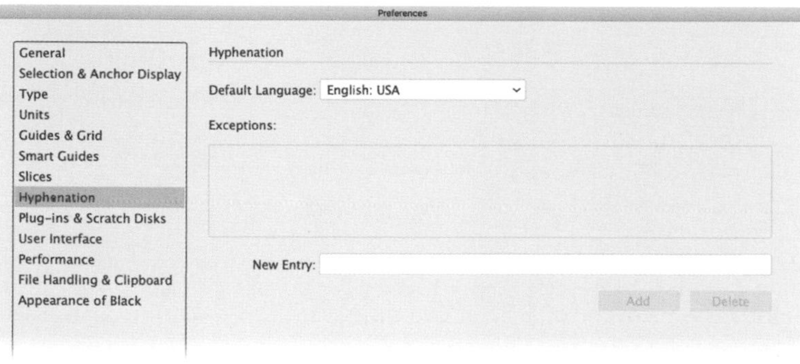

Plug-ins & Scratch Disks

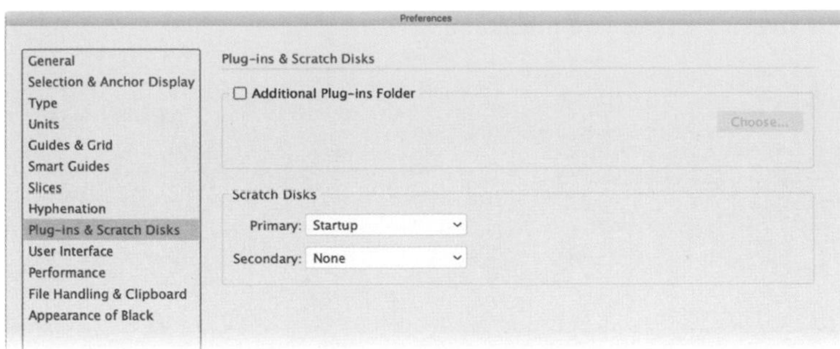

Additional Plug-ins Folder

Plug-ins add more features and capabilities and are generated by third parties outside of Adobe. Check the box to choose the appropriate folder to locate the plug-ins.

Scratch Disks

Scratch space on the hard disk drive is dedicated to temporary storage while working in Illustrator. A disk that is near full or full can slow down the performance of the program.

Primary

This indicates the main hard drive in your device.

Secondary

This indicates a backup device to allow for more room as needed if the primary become full.

User Interface

These options allow you to configure how light or dark the interface appears, how the panels function, and the overall size of the interface.

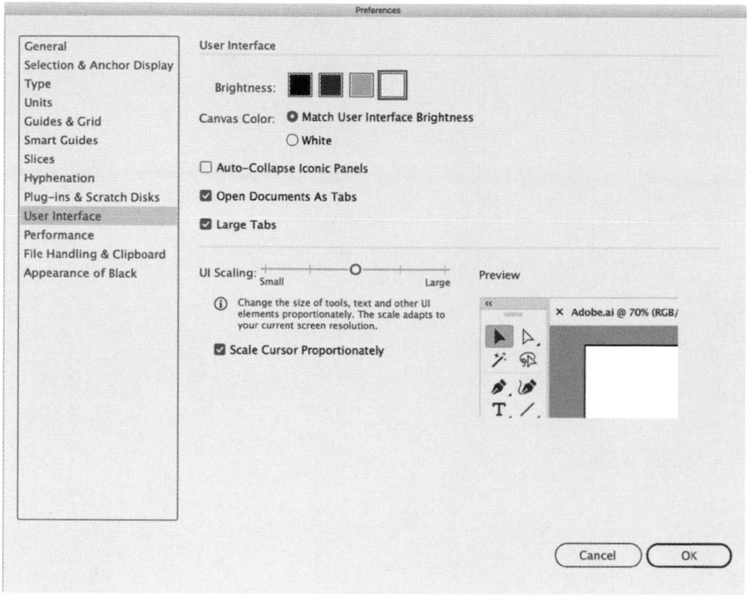

Brightness:

This allows you to choose either a dark interface with light callouts or a light interface with dark callouts.

Canvas Color:

This allows you to set the canvas area outside the artboard to your chosen interface color. Select white if you don't want the canvas to match the interface.

Auto-Collapse Iconic Panels

When using the panels, this option allows you to expand and auto-collapse a panel when done. If you need the space on your screen, this helps. If you don't need space, this adds the extra step of opening the panel each time you need to use it.

Open Documents as Tabs

This opens all documents in one window with each document identified as a tab instead of an individual floating window.

Large Tabs

This creates larger tabs that identify each open document window.

UI Scaling

This scales the user interface up so it's larger and easier to see. As indicated, it changes the size of tools, text, and other UI elements proportionally. The scale adapts to your current screen resolution.

Scale Cursor Proportionately

This scales the cursor size when the user interface is scaled.

Performance

Not your performance—the performance of Illustrator!

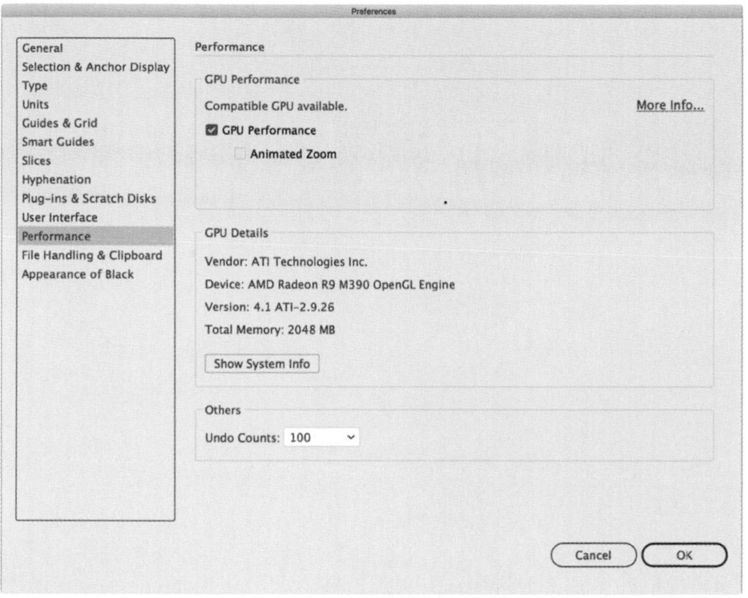

GPU Performance

The Graphics Processing Unit (GPU) is part of the video card/display and is designed to rapidly execute commands for manipulating and displaying images. There are known issues that may be encountered when GPU Performance is enabled. They are: Rendered artwork appears jagged, black, or white; or Illustrator crashes. Documents appear in black-and-white. You may receive an error message that GPU Performance Features are not available in the document you are currently working on, or the width lines may look tapered, or the ends vanish.

Turn off GPU Performance to see if these problems are part of the GPU.

Animated Zoom

This makes zooming actions smooth and animated.

GPU Details

This displays an overview of your system.

Others

Undo Counts:

This value controls the number of undos available. The default is 100. More undos may slow performance.

File Handling & Clipboard

The options in this section control file saving and the copying of data.

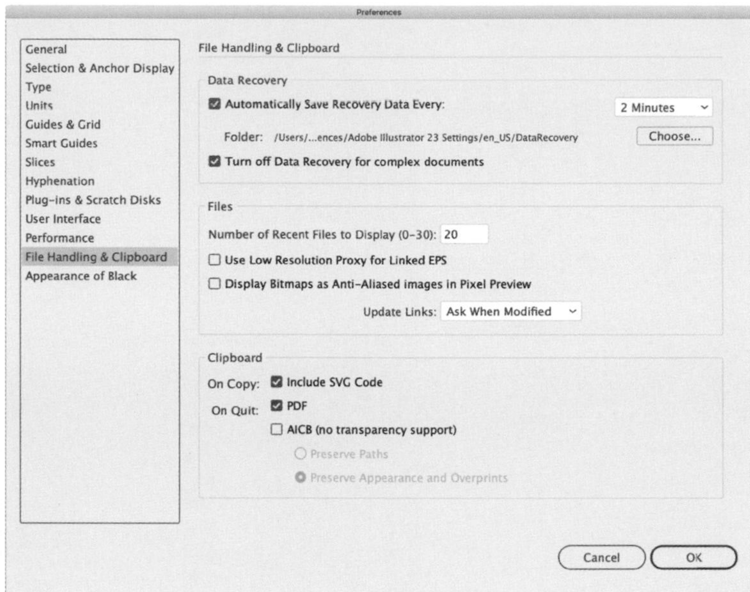

Data Recovery

Automatically Save Recovery Data Every:

Yes, you do want to automatically recover data in case Illustrator crashes (and it will crash). This options sets up an "auto save" every two minutes as a default. This is not a substitute for frequent saving of the file but it is a worthwhile safety net.

Turn off Data Recovery for Complex Documents

When data recovery is turned on, Illustrator may pause while backing up large or complex files. If this slows down or interrupts your workflow, select the checkbox to turn off data recovery.

Files

Number of Recent Files to Display (0-30):

This sets the number of recent files that will appear in the File menu or the Startup window.

Use Low Resolution Proxy for Linked EPS

In this mode, linked EPS files will display as low resolution to save memory and increase program performance.

Display Bitmaps as Anti-Aliased Images in Pixel Preview

This smooths the appearance of a Bitmap image when in Pixel Preview mode.

Updates Links

This determines how links will be updated. The options are Automatically, Manually, or Ask When Modified.

Clipboard

On Copy:

When copying content, check this option to include SVG data.

On Quit:

If you have copied content and quit Illustrator, a dialog box will appear and ask what you want to do with the clipboard contents. Content can be saved as a PDF, as paths, or as paths that preserve the Appearance and Overprints as they were created in Illustrator.

Appearance of Black

Just when you thought black was black! These options allow you to set the appearance of black elements in your workspace.

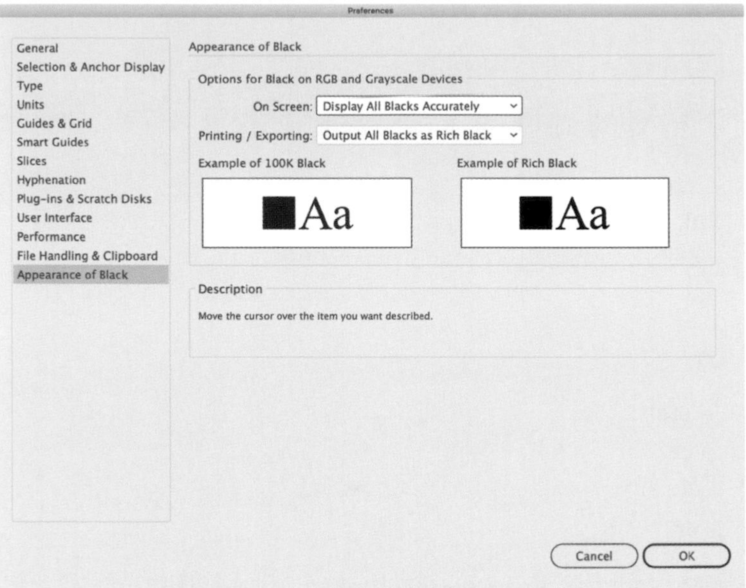

On Screen:
I choose to display all blacks accurately on-screen. This means that items that use the pure black (100% K) swatch will look slightly lighter or more gray than a rich black, which is made from black ink plus other inks to make it appear more saturated.

Printing/Exporting:
This is the confusing one. Choose Output All Blacks as Rich Black, then hover your cursor over that choice and note the description below that attempts—but fails—to clarify what that means. What it means is that on a device that has only black ink or toner, both rich blacks and pure blacks will print as dark as possible. RGB devices refers to common desktop inkjet printers. However, with this choice, color laser printers and printing presses (true CMYK devices) will still produce blacks accurately: pure black will use black ink or toner only, and rich black will be output as designed. Since I output to all of these devices, I choose the confusing Output All Blacks as Rich Black setting. It works!

Workspaces

Panel Locations

Illustrator has lots of panels that can crowd your work area. Each panel can be moved, minimized, attached to other panels, or made free-floating. You can create a combination of panels and designate them as a workspace. Setting up workspaces based on your workflow can be very helpful. I use certain panels for creating logos and other panels when creating infographics, and I set a workspace based on the panels I use for that task.

Choose A More Useful Initial Workspace

When you open or create a document, select Window > Workspaces and choose from a preset list of workspaces depending on the type of creation you intend to work on.

In each of these preset workspaces, Illustrator has included a set of panels it recommends for that type of work. Once you choose a workspace, the panels will appear on your screen. This can be a great way to get started; as you progress you may want to change the arrangement or location of the panels to best suit your needs.

Automation
✓ Essentials
Essentials Classic
Layout
Painting
Printing and Proofing
Tracing
Typography
Web

Reset Essentials
New Workspace...
Manage Workspaces...

Creating a New Column of Panels

To create a new dock of panels alongside the first, drag a panel by its name toward the original dock until a dense blue vertical "drop zone" appears. Collapse the new column with the small double arrow (>>) in its upper-right corner. Drag in more panels under the first. I suggest adding the Properties panel to this second dock to easily access (and collapse) it.

Adjust the width of this new dock by dragging its left edge. You can shrink it until the names are gone and only icons remain if you like.

Individual panels have drop zones, too. If you drag one panel to the bottom edge of a free-floating panel, you will create a free-floating dock. Or, if you drag a panel's name next to another panel's name, the drop zone is the small window itself; that is, both panels' tabs will be side by side in the same window.

Once your panels are located where you want them, return to the workspace menu and choose New Workspace.... We can name this "Real Essentials." Later, if this workspace becomes untidy, we can once again use the workspace menu and choose Reset Real Essentials.

Customizing Menus and Keyboard Shortcuts

Keyboard shortcuts are fantastic. When you are starting to learn a new application there may be a bit of time hunting and pecking through the menus trying to find commands, but these tips will help for both menus and tools shortcuts.

What's on the Menu?

Go to Edit > Keyboard Shortcuts to be presented with a large dialog box with lots of options. There is a dropdown list for setting shortcuts for the tools and for menu commands.

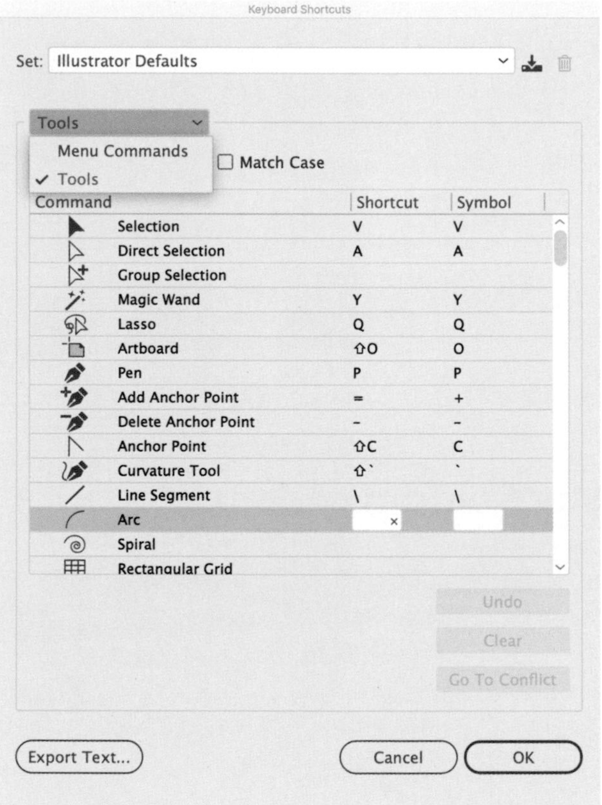

The Keyboard Shortcuts dialog box will list all the tools and any shortcuts associated with those tools. Not all tool will have shortcuts, but the most-used tools will. If your workflow requires the use of tools that do not have shortcuts applied as a default, click on the tool and the Shortcut and Symbol field will become active. Enter in a shortcut that best fits your workflow. If you enter a shortcut that is already in use (that is, it's applied to another tool), there will be a conflict. You can override that conflict and use your shortcut as the default or choose different shortcut command, keeping the original shortcut intact.

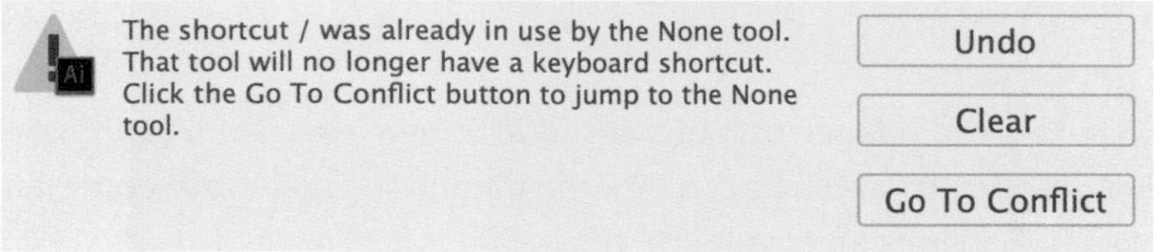

The shortcut / was already in use by the None tool. That tool will no longer have a keyboard shortcut. Click the Go To Conflict button to jump to the None tool.

Undo

Clear

Go To Conflict

Undo will return the original shortcut to the default. Go To Conflict will go to the tool that has that keyboard shortcut you want to use. Clear will clear the default shortcut on that tools and allow you to enter that shortcut for another tool.

Once a new default shortcut is created, the Tool Hint will reflect that new shortcut when you hover your cursor over the tool.

This process is the same when you choose Menu Items from the list. Any new shortcuts will be reflected next to the Menu command.

Any changes to the default shortcuts will require you save the shortcuts as a new set. The Illustrator defaults cannot be overwritten so any changes will require a newly named set of shortcuts when you click OK. A dialog box will appear asking for a new name for your custom presets. Enter in a name and click OK.

Keys to Success

You have seen those posters online that can show you all the shortcuts—and you can purchase it for $25.00. Instead, you can make your own for free! Click on the Export Text... button and you will get a text file that reflects all the default and customized shortcuts.

Creating a New Document

Let's gets started. You will need to create a new document so you can begin creating. Choose File>New... to begin. The New Document window displays preset sizes, any recent files, Adobe Stock Templates, presets for the type of file you will be creating, as well as sizes and artboard orientation.

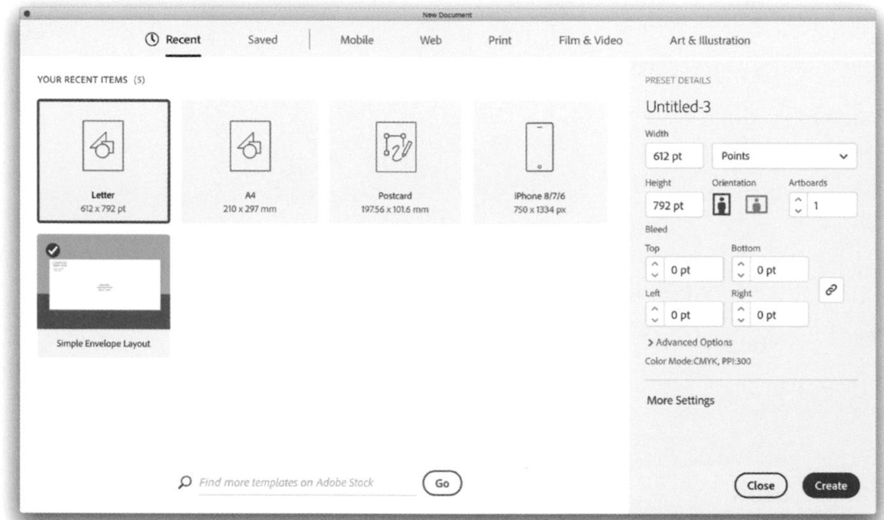

Presets

At the top of the New Document window there are presets for Mobile, Web, Print, Film & Video, and Art & Illustration. The presets will give you a list of common document sizes to choose from to begin creating. Or you can create a custom-size document from the settings on the right side of the dialog box.

Mobile, Web, and Film & Video: This preset displays common phone and tablet sizes (Mobile), commonly used web sizes, or common film and video formats. Below the presets you have access to Adobe Stock Templates, some of which are free. These presets use pixels as the unit of measure, the document is in RGB mode, and the resolution is set to 72 ppi.

Print: This preset displays paper sizes and Adobe Stock Templates for print work. Print presets use points as the default unit of measure, which can be changed via the Unit dropdown menu and the document is in CMYK mode. The resolution is set to 300 ppi.

Art & Illustration: This preset displays paper sizes and Adobe Stock Templates for creating artwork or Illustrations work. Art & Illustrations presets use points as the default unit of measure (which can be changed) and the document is in RGB mode. The resolution is set to 72 ppi.

Document Sizes

Creating a new custom document can be done easily if the presets are not what you are looking for. On the right side of the New Document window you can set the parameters for your document.

Set the width and height of the document. The unit of measure can be changed from the dropdown menu. Choices include Points, Picas, Millimeters, Centimeters, Inches, and Pixels.

Set the orientation of the document to be portrait or landscape.

Set the number of pages... umm, artboards. Illustrator does not use pages since it is not a page layout application—that is what Adobe InDesign is for. Illustrator uses artboards yet you may hear them refereed to as pages as a generic descriptor.

Bleed can be set on all sides of the document. This does not change the size of the document, it creates guides outside the artboard for extending artwork for bleed purposes.

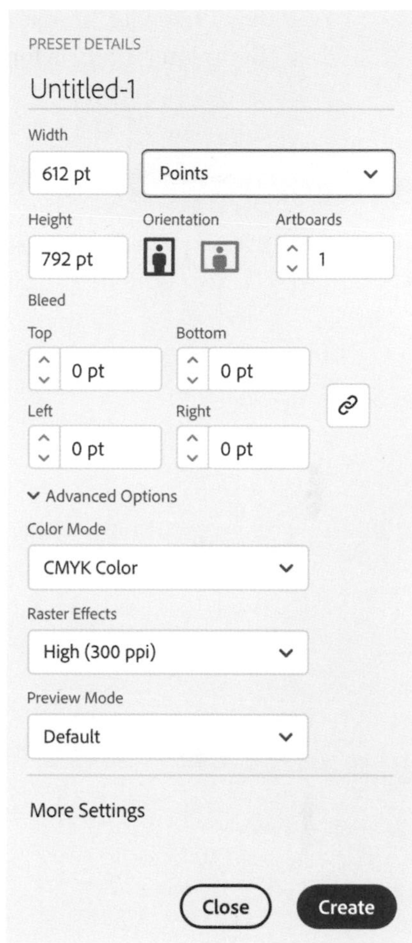

Advanced Options

To set more options for the new document, click on the Advanced Options arrow.

Color Mode: Choose RGB for work that will be displayed on phones, tablets, computers, or TV screens. Choose CMYK if the creations will be printed.

Raster Effects: This setting allows the rendering of any raster effects to be high resolution (300 ppi), medium resolution (150 ppi), or low resolution (72 ppi). This may not seem relevant since Illustrator is a vector-based application but there are some effects that can be applied that render as raster (pixel-based) output.

Preview Mode: Choose from Default preview, Pixel preview, or Overprint preview. I leave it as Default and change it in the document when I need to preview artwork in the other modes.

More Settings: Click to find a few extra settings for setting up artboards.

Artboards

Illustrator has artboards instead of pages. Since Illustrator is more about creation and than it is about formatting books, we use artboards instead of page layouts. When creating a new document, you can specify how many artboards you would like, or you can easily add them later.

Artboard Panel

The Artboard panel is located under Window>Artboards. This displays the artboards in order and by name, and offers options to edit, add, or delete artboards.

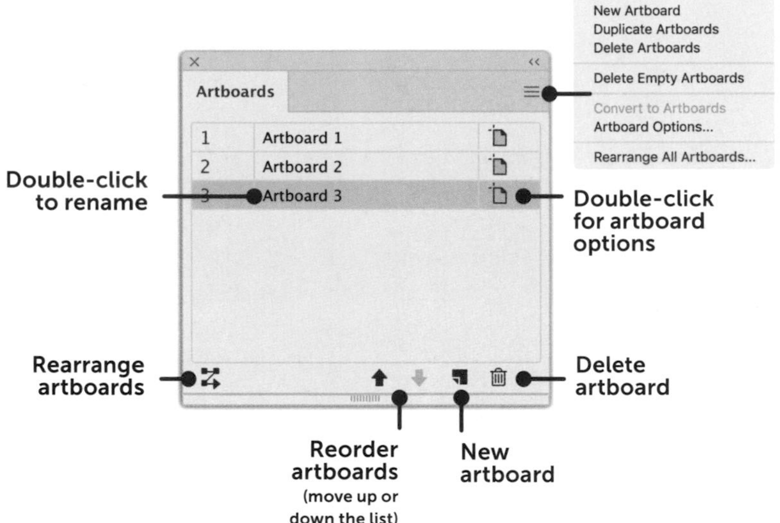

Naming the artboard can be done by double-clicking on the name to open the edit field. Type in the new name and click Return. Options can be accessed by double-clicking on the artboard icon. Use the dropdown menu to access other options such as add, delete, or duplicate artboards.

Artboard Tool

The Artboard tool (Shift + O) is an easy way to edit, add, or modify artboards. When you click on the Artboard tool, the artboard(s) in your document will be directly editable. The background around the artboards will darken and one of the artboards will have a blue dashed line around it idicating it is ready for editing.

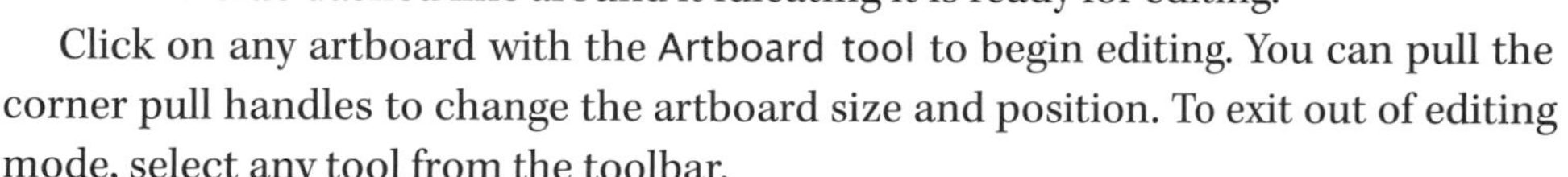

Click on any artboard with the Artboard tool to begin editing. You can pull the corner pull handles to change the artboard size and position. To exit out of editing mode, select any tool from the toolbar.

Artboard Options

To get more Artboard Options, double-click on the Artboard tool to open the options panel.

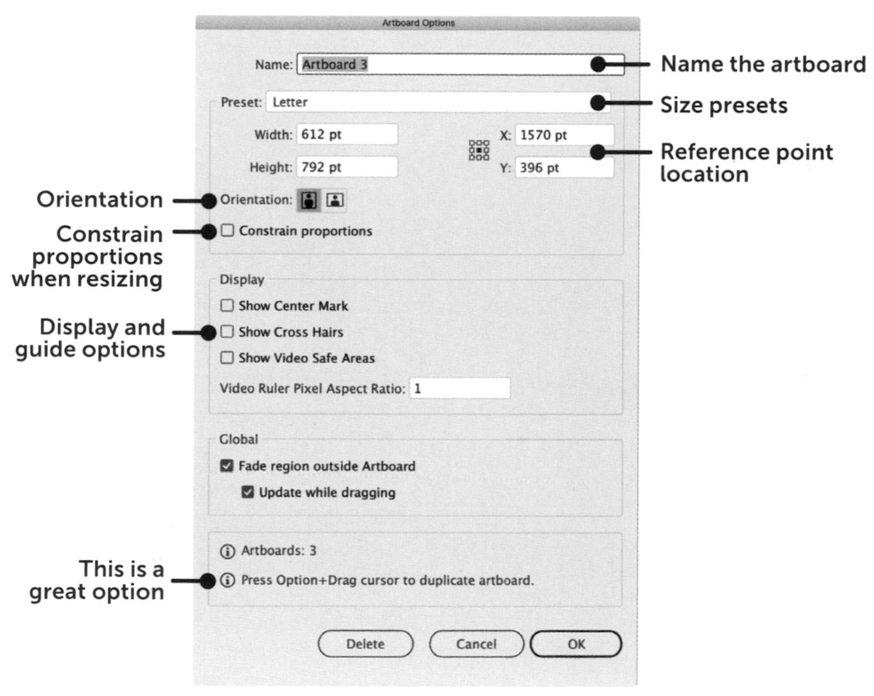

Layout Artboards

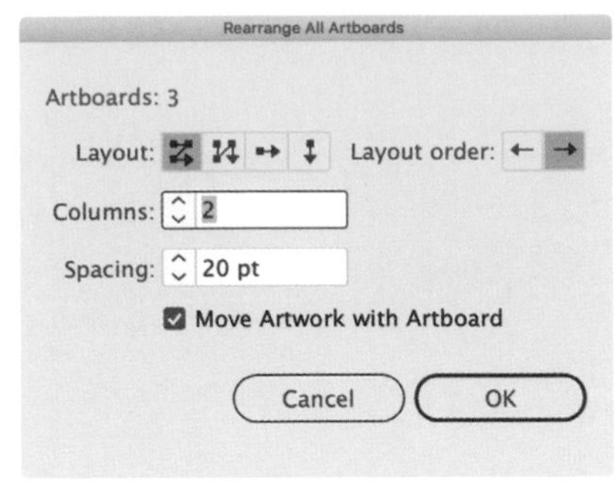

To layout and arrange the artboards, double-click on the zigzag icon at the bottom left of the Artboard panel. You can also choose Rearrange All Artboards... from the Artboard panel drop- down menu.

The default layout is left to right. Change the layout order by clicking the icon from up to down or in a line vertically or horizontally. Layout order can read left to right or right to left.

If the column layout is chosen and there are enough artboards, you can set the number of columns. Spacing between the columns is set here as well.

Move Artwork with Artboard is a good option that will keep track of artwork if the artboards move.

Convert to Artboards

If using the Artboard tool to create artboards is not working, or the add artboard command is to hard to get to, you can draw a rectangle or square using the Rectangle tool. Select the rectangle and choose Convert to Artboard from the Artboard panel dropdown menu. This does not work with any other shape tool but the Rectangle tool.

Duplicate Artboards

I love this feature. When I am working on a logo for a client and they want several different versions, I will make one logo per artboard. Then, for different versions I will duplicate the artboard and begin a new version. Why create multiple artboards instead of just one artboard with several logos? Good question! It is easier to reference the designs if they are on separate artboards and when I save them as a PDF, each artboard will be a page in the PDF document. I can also print any artboard or artboards I want instead of one artboard with all the logos on it.

My favorite method of duplicating an artboard is to select the Artboard tool and Option + click (Mac) and drag the artboard to duplicate it (use Alt + click and drag on a PC). You can select the artboard and choose Duplicate artboard from the Artboard panel dropdown menu as well.

Printing Artboards

When it comes time to print, the artboards can be printed as separate pages. Any artwork that is not on an artboard will not print so make sure all your work is done on an artboard and nothing is off the edge. We will discuss this is greater detail in the section on printing.

Layers Panel

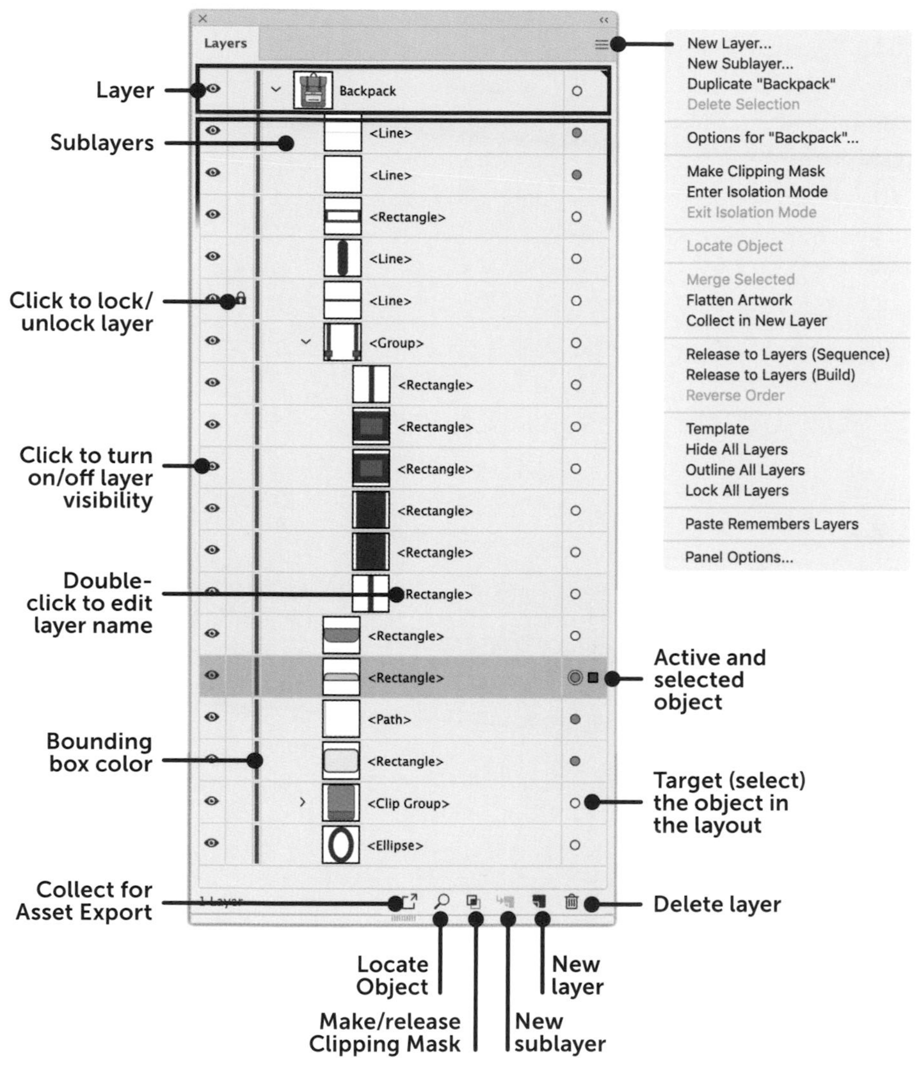

About Layers

When creating artwork in Illustrator things can get complicated. It can be challenging to keep track of, select, group, edit, move, and find all the items of your artwork. Objects can be inside groups or in a clipping mask, behind other items, or locked, and selecting that object becomes difficult or frustrating. Layers provide a way to manage all the items that you have created in your artwork. One way of understanding Layers it to think of them as clear folders that contain artwork. You can put items into the folder (those are sublayers) and easily move the layers above or below to bring them in front of or behind other layers. These layers can be grouped to move or transform as one unit and can be locked or hidden so they are protected.

The Layers panel (Window > Layers) will help organize, list, and edit the objects in a document. The blue bounding box around each shape is the color indicator of the layer. The same color shows the bounding boxes, paths, anchor points, and then center point of the object. This color is a reference to quickly locate what layer the object is on in the Layers panel. You can change the layer color to suit your needs.

Every new Illustrator document contains one layer, and each object you create is listed under that layer as a sublayer. You don't need the layers panel for simple artwork, but it will make navigating, editing, working with, and creating so much easier and efficient that once you get the swing of it, it will become a must-have in all your Illustrator creating.

Layer Panel Overview

The Layers panel has one layer by default. Under this layer is where all the content you create in the artwork will be—these are sublayers.

When a layer in the Layers panel contains other sublayers, a triangle shows in the upper left of the layer's name. To see all the sublayers click on the arrow to the left of the layer thumbnail to open it.

Visibility Column

An eye icon shows the visibility of the layer or sublayer. Click the eye to turn on or off the visibility. Keep in mind that when you print a file, that layer will not print unless you either turn the visibility back on or you choose Print All Layers from the Print layers section of the Print dialog box.

Edit Column

Second in from the eye icon shows whether the layer or sublayer is locked or unlocked. Click the area to show the lock and that layer will not be moveable or editable. Click the lock to unlock the layer for editing.

Target Column

Target seems like the wrong word to use here if you think of a target as something that is being selected. Illustrator uses the word target to indicate whether the item has an effect or attribute applied from the Appearance panel. When the target button appears as a double ring icon, the item is targeted (it has an effect or attribute applied); a single ring icon indicates that the item is not targeted. The effect or attribute could be a drop shadow or transform effect or an opacity on the object, and the target informs you that this object has something extra.

Selection Column

This column indicates whether or not items are selected via a small box (the same color as the one on the far right of the layer, next to the target icon). When an item is selected within a group, a smaller color box appears next to the parent item (the group that it is in) as well as the top-level layer of sublayers. If all of the objects within the parent item are selected, the selection color boxes are the same size as the marks that appear next to selected objects. You can dim any linked images and bitmap objects to make it easier to edit or trace over artwork on top of the image.

Layer Management Best Practices

There is no ideal way of managing layers, layer order, and naming in Illustrator; yet there are some very helpful tips to make your life, and your Illustrator files, easier to manage and navigate.

Let's look at the artwork as though you are building a house. Start building the house (your new document) by choosing the size and color modes.

The build process may begin with an image placed in a file on a layer so you can trace over it; or the sketch may be in your mind ready to go. Build your foundation, then build the house on the foundation. Think about what goes into all the behind-the-scenes content. Naming the layer Background or Base may be a good start.

Once the base is done, create new layers for the more detailed levels of the "house": first floor, second floor, roof layer. Now you can go back and turn on and off the visibility of the layers and begin to add details. You may have a layer right above the first floor that is all the rooms. You may want to break out each room on that floor as a layer—name the rooms something that makes sense and is easy to understand.

How far do you go with layers? Each room as a layer may be just right. Is there furniture in each room? Is there a layer for each piece? Possibly: it can be very easy to select the "chair" and edit it as one layer. I would not, however, break down the chair into further folders beyond that one item.

Naming

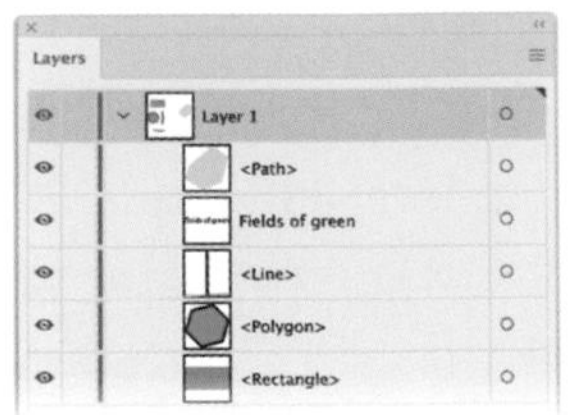

An important aspect of working with layers is naming everything in a way that helps you stay organized. The default layer-naming convention is Layer 1, Layer, 2, Layer 3…, yet when you create artwork, Illustrator assigns names to each sublayer (indented below the layer) based on what the shape, type, or image is. For instance, if you draw a rectangle, that sublayer will be labeled sublayer <Rectangle>. You can edit this title by typing in a line of text that describes the contents of the layer. This is very helpful for identifying each object that makes up your artwork.

When I create artwork I begin with the base artwork (the content in the background). What goes on the artboard first? It may be a shape filled with color, a pattern, or an image. Since layers build on top of each other, any layer that is added to the Layers panel is added on top of the previous layer by default. Layer order can be changed, but I like to think of a process of building this list like I think about the house. Start with the foundation and frame and build the house one floor at a time. Put the roof on it, then start adding the colors and details.

Be consistent when naming layers: be careful of uppercase and lowercase letters, and be brief and descriptive for optimum efficiency. The Layers panel doesn't take up much screen space, so using long layer names may require you to open the panel up wider—taking up more working space. Be especially careful about naming the top layer, because that's the title heading by which you'll search for filed objects.

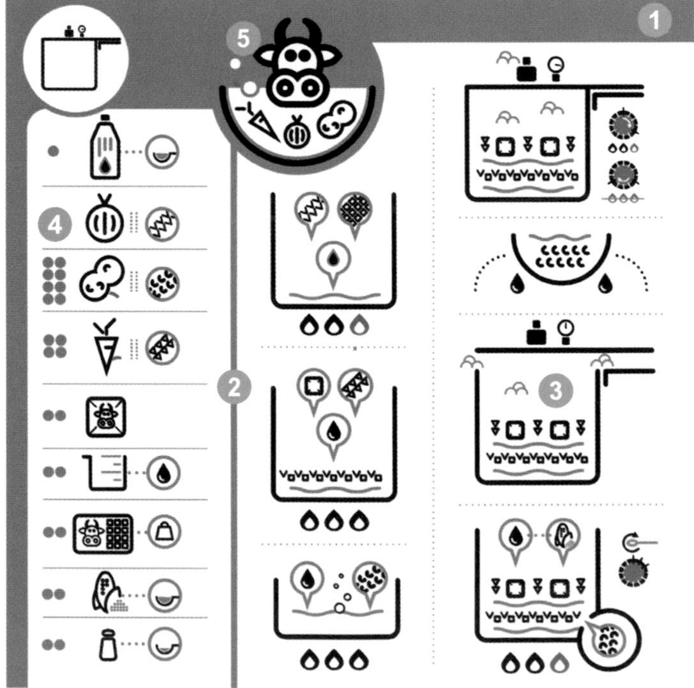

Think of layer order and naming like building a house.

Start with the foundation, build each floor, add the roof, and then add the details.

In this example, I use a recipe for beef stew. In this infographic I break down the main components into folders, then each complete step (or ingredient) gets a folder within it.

For Segregating Content

With the example of the beef stew recipe, the main components of the recipe are the Background, Framework, Ingredients, Steps, and Hero image. Segregating the content into these top-level folders makes it much easier to find and isolate each component within the artwork.

For Protecting Content

When your file has many components, trying to select, edit or move, modify or remove a shape, path or fill can be harder than you think. Some may be grouped together, behind another object, or inside a Clipping mask. If I wanted to select the background shape, I may end up selecting the content above and not be able to get to the background shape without moving my upper level content.

Layers allow you to segregate the content into its main components, which can be locked or have their visibility turned off so that other content can be accessed without modifying or moving anything to get to the underlying content. This makes the workflow very efficient and much less frustrating.

Creating New Layers

In the Layers panel, click the name of the layer above which (or in which) you want to add the new layer. Layers are added on top of the layer you have selected

To add a new layer above the selected layer, click the Create New Layer button in lower-right side of the Layers panel. To create a new sublayer inside the selected layer, click the Create New Sublayer button in the Layers panel.

Layer Options

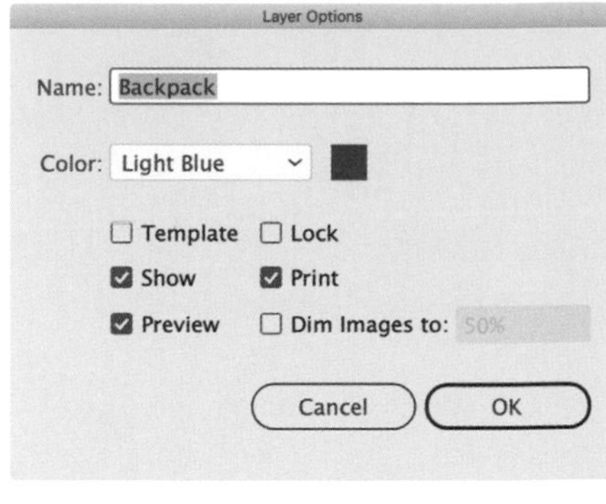

To set layer and sublayer options, double-click the item name in the Layers panel or select the layer name and choose Options For <item name> from the Layers panel dropdown menu.

Name: Add a name for the layer or sublayer.

Color: Select a color that identifies that layer. The layer color shows on the selected items' bounding box to indicate the layer the object is on.

Template: This makes the layer a template layer.

Lock: This prevents changes to the item.

Show: This turns on or off the visibility of all artwork contained in the layer on the artboard.

Print: This makes the artwork contained in the layer printable or not printable.

Preview: This displays the artwork contained in the layer in color instead of an outline.

Dim Images: This reduces the intensity of linked and bitmap images contained in the layer to a specified percentage.

Panel Options

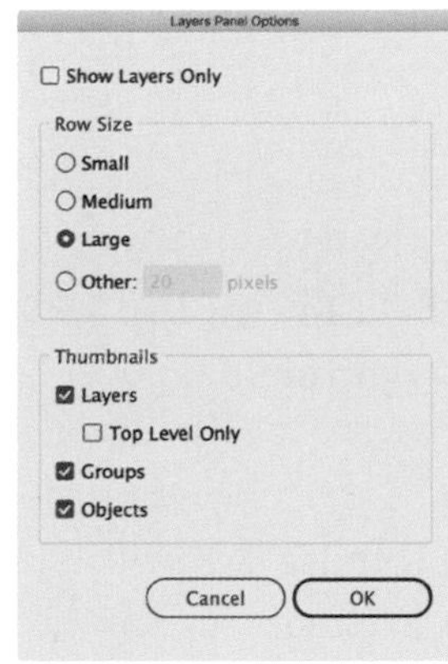

Row Size

In these options you can set the size of the layer thumbnails. By default they are small, so you have your choice of a preset size or to enter in a pixel preview size in the Other field.

Thumbnails

You can set the visibility of thumbnails for all the layers and sublayers (they are all on by default). Unchecking the boxes will show the layer with no corresponding thumbnail.

Layer Order

Moving Objects to Layers

You can move an object from one layer to another in a few different ways. Make sure the layer with the object and the destination layer are both unlocked.

Select the object you want to move with the Selection tool. Click the name of the destination layer in the Layers panel, then choose Object > Arrange > Send To Current Layer from the Layers panel dropdown menu.

Drag the selected-art indicator (small color square), located at the right of the layer in the Layers panel, to the layer you want.

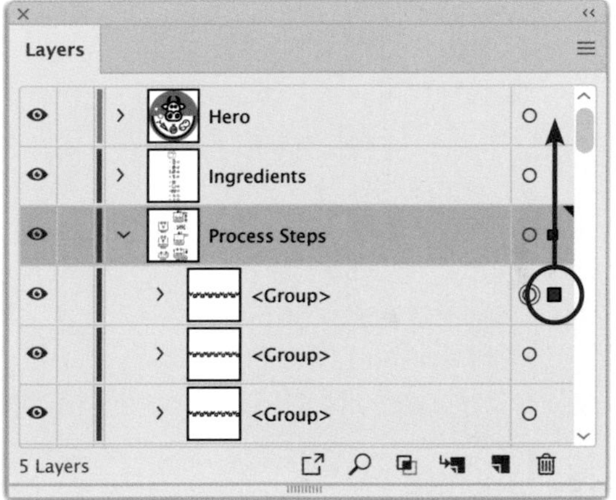

Create a New Layer While Moving

You can move objects or layers into a new layer by selecting the items and choosing Collect in New Layer from the Layers panel dropdown menu. Hold down Command (Mac) or Control (PC) to select nonconsecutive items; hold down Shift to select consecutive items.

Reordering Layers

This is helpful for moving items into and out of groups, as well as reordering layers and their content. To move or reorder layers or sublayers, click on the name of the layer or sublayer and drag it up or down the Layer panel to reorder the content.

Bring to Front/Send to Back

You can use the Object>Arrange> (Bring Forward, Bring to Front, Send Backward, Send to Back) to reorder the layers as well.

- **Bring Forward**: Moves the object forward one layer at a time.
- **Bring to Front**: Moves the object forward to the top of the group or to the top of the layer.
- **Bring Backward**: Moves the object backward one layer at a time.
- **Bring to Back**: Moves the object backward to the bottom of the group or to the bottom of the layer.

Selecting Objects

When your files get complicated, trying to select or locate them can be a monumental task if you're trying to click on it with the selection tool. In the Layers panel you can select the object by clicking on the layer, then clicking to the right of the target icon. A color box will appear showing that the object is selected. That object will now be selected in the artwork.

Locating Objects

You may be working on your artwork and have an item selected, and now you want to find where it is in the Layers panel. Use the Locate Object command from the Layers panel drop-down menu. This command is especially helpful for locating items in collapsed layers.

Editing Groups

To edit a group of items, you do not need to ungroup the objects, make an edit, and then try to select all the parts and regroup them. Keep everything grouped and locate the group in the Layers panel.

Open the group by using the arrow to the left of the group name. Inside the group, select the layer you want to edit, then click on the circle (target) at the right of the layer. A color square will show next to the target, and the object will be selected on the artboard. This isolates the object so you can make edits to the selected object artwork without having to ungroup it.

Add Objects to a Group

To add objects to a group without ungrouping the content, select the layer that is outside the group and drag the layer into the group in the stacking order you want.

Remove Objects From a Group

To remove objects from a group without selecting the layer that is in the group, drag the layer outside of the group or into another group.

Editing Clipping Masks

There are several ways to work on Clipping masks, but the Layers panel makes it easy because it works in the same way as editing groups. The one difference is the mask itself acts as a "window" where the artwork appears.

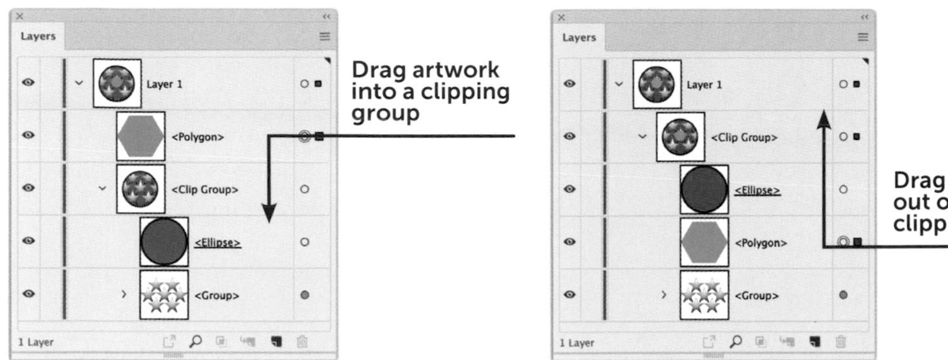

Adding to an existing mask requires releasing the mask, adding or editing the content, and remasking. In the Layers panel, you keep the mask and clipping intact and add, edit, or delete content like you would a layer group.

Consolidate Layers and Groups

Merging and flattening layers are similar in that they both let you consolidate objects, groups, and sublayers into a single layer or group. Merging layers allows you to select which items you want to consolidate into one layer. Flattening takes all the visible items in the artwork and consolidates them into a single layer. Both merging and flattening retains the stacking order of the artwork, but other layer-level attributes, such as clipping masks, aren't preserved.

Merge Selected

To merge items into a single layer or group, hold Command (Mac) or Ctrl (PC) and click the names of the layers or groups that you want to merge. This allows nonconsecutive selection of the layers. To select a set of layers that are consecutive, click on the first layer then hold Shift and click on the last layer. This selects everything between the two layers or groups you clicked on. Then, select Merge Selected from the Layers panel dropdown menu. All items will be merged into the layer or group you selected last.

Layers can only merge with other layers that are on the same hierarchical level in the Layers panel. Likewise, sublayers can only merge with other sublayers that are within the same layer and at the same hierarchical level. Objects can't be merged with other objects.

To flatten layers, click the name of the layer into which you want to consolidate the artwork. Then select Flatten Artwork from the Layers panel dropdown menu.

Release Items to Separate Layers

Release To Layers is used to prep files for web animation in which each object needs to be on a layer. The Release To Layers command redistributes all of the items in a layer and creates individual layers for each object.

- **Release each item to a new layer**: Choose Release To Layers (Sequence) from the Layers panel menu after you click on the layer.
- **Release To Layers (Build)**: To release items into layers and duplicate objects to create a cumulative sequence, choose Release To Layers (Build) from the Layers panel dropdown menu. The bottommost object appears in each of the new layers, and the topmost object only appears in the topmost layer. This is how cumulative animation sequences are created using this feature.

2 Shape Creation

Drawing Vector Shapes

Vector shapes and objects are infinitely scalable lines and shapes that maintain their crisp details when resized. They are resolution-independent because they are defined by mathematical objects called vectors, which define a shape using its geometric characteristics. It's not made up of pixels of a finite size.

Logos, type faces and fonts, illustrations, and icons are several examples of why vector graphics are the best choice for artwork. Vector creations can be used at various sizes in different graphics applications, and can be saved in a PDF format. When printed, vector graphics will print clear and sharp at any size.

Rectangle

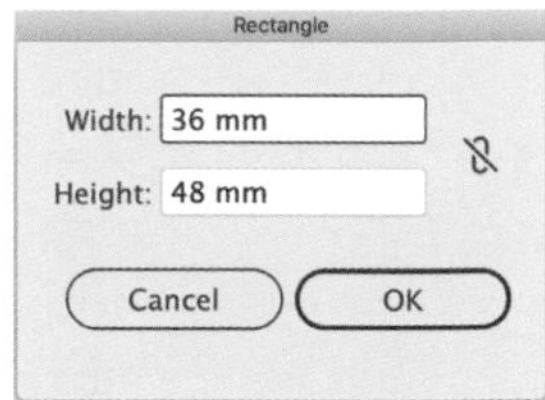

Select the Rectangle tool and click and drag on the artboard to draw a shape. While drawing the rectangle, a measurement label will show outside the lower-right corner of the shape. This will show the size of the shape being drawn. These shapes will not snap to an exact measurement, so sizing them using these labels is not recommended. Measurement labels can be turned on or off in the Preferences > Smart Guides > Hide Measurement Labels.

To create a rectangle at a preset size, select the Rectangle tool from the toolbar and click on the artboard. This will open a dialog box for the parameters of the shape. To make the width and height the same, click on the link icon to the right of the entry fields.

Holding the Shift key while drawing a rectangle will create a perfect square. Shift can be held before or during drawing; however, remember to release the mouse first before letting go of the Shift key to keep the rectangle constrained. If the Smart Guides are turned on (View > Smart Guides), a magenta line will show from upper-left to lower-right corners indicating the shape is constrained as a perfect square.

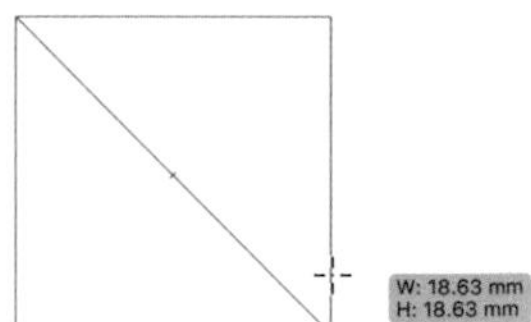

SMART SHORTCUT

To draw a rectangle from the center of the shape, hold Option (Mac) / Alt (PC). This can also be used with Shift to draw a square.

Ellipse

Select the Ellipse tool and click and drag on the artboard to draw a shape.

To create an ellipse or circle at a preset size, select the Ellipse tool from the toolbar and click on the artboard. This will open a dialog box for the parameters of the shape. To make the width and height the same, click on the link icon to the right of the entry fields.

Ellipse
Width: 15 mm
Height: 15 mm
Cancel OK

Holding Shift while drawing an ellipse will create a perfect circle. Shift can be held before or during drawing; however, remember to release the mouse first before letting go of the Shift key to keep the circle constrained. If the Smart Guides are turned on (View > Smart Guides), a magenta line will show from top to bottom and left to right indicating the shape is constrained as a perfect circle.

> **SMART SHORTCUT**
> To draw from the center of the ellipse or circle, hold the Option key while drawing. Shift can also be used in combination with Option to draw a perfect circle from the center.

Polygon

Select the Polygon tool and click and drag on the artboard to draw the polygon.

The default for the Polygon tool is six sides. If you alter the number of sides in the imput field, that becomes the new default for subsequent polygons.

To create a polygon at a preset size and number of sides, select the Polygon tool from the toolbar and click on the artboard. This will open a dialog box for the parameters of the Polygon tool. Polygons draw from the center of the shape so the measurement is the radius (or half the overall width) of the shape.

> **SMART SHORTCUT**
> To change sides while drawing a polygon, the up and down arrows will increase or decrease the number of sides while the mouse is drawing. If the up and down arrow is used after the polygon is drawn, it will move the shape around on the artboard.
>
> To keep the polygon positioned so that a side is horizontal when drawing, hold the Shift key down.

Once the polygon is drawn, move the cursor over the diamond shape on the right side. The cursor shows a +/- that can change the number of sides on the shape by dragging it up or down. The maximum number of sides that can be created this way is 11; the minimum is three.

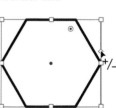

You may wonder why there is no Triangle tool in the toolbar. Set the number of sides on the polygon to three to create a triangle.

Star

Select the Star tool and click and drag on the artboard to draw a shape. Radius 1 and Radius 2 can be a bit confusing. Stars, like polygons, draw from the center of the shape. Radius 1 is the distance from the center of the star to the outside point. Radius 2 is the distance from the center of

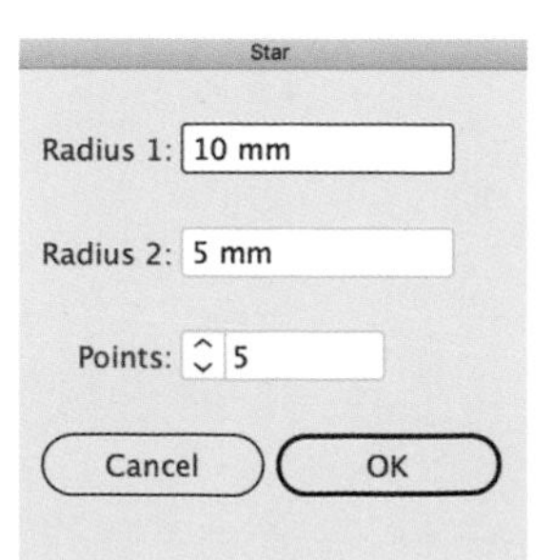

the star to the inside point. The larger the difference between Radius 1 and Radius 2, the longer the points on the star will be.

These dialog boxes do not offer a preview so you have to guess what it will look like. If it is not what you expected, you will have to draw the shape again because there is no way to edit the points of the radius after the shape is created. Once the new star is drawn, the number of points and the radius will be the new default for the star until they are changed.

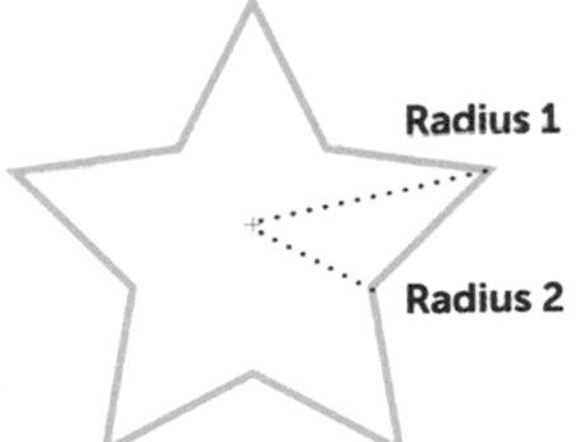

SMART SHORTCUT

To make up for the lack of a preview feature in the Star dialog box, you can actively change the points and radius while drawing by using the up arrow to add more points or down arrow to reduce the number of points.

To change the difference between Radius 1 and Radius 2 use the Command key and pull the cursor out from the center while drawing a star, this will make the star points longer. Command and pull in to the center while drawing a star to make the star points shorter.

Editing Corners/Corner Widgets

Once a rectangle has been drawn, the corners can be edited using the corner widgets. Actually, any shape with corners has corner widgets, however on a star they are not visible by default but they can be accessed. Corner widgets are the small targets that appear in the corners of a selected shape. Use corner widgets to create rounded, inverse rounded, or flat beveled corners.

Corner widgets appear when the shape is drawn or selected with the Selection tool. Click on the corner widget target and pull in toward the center to round the corners more; pull out from the center to round the corners less. The cursor will show what shape the corners will be with the corner icon (in this example, the rounded corner).

SMART SHORTCUT

To change the corner style, Option + click on the target to change it from rounded, to flat, to inverse rounded.

To change just one corner independently of the other corners, click on the target (to turn it into a donut with no center dot). Once the corner is changed, the donut will turn back into the target icon and that corner will no longer be independent of the others.

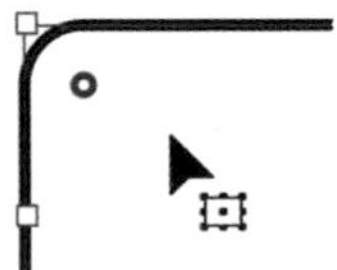

A single corner can also be selected with the Direct Selection tool. Click on the corner to be edited with the Direct Selection tool and change the attributes of the corner widgets.

The Transform panel and the Properties panel allows for the editing of each corner size and style together or independently. The dropdown menus next to the corner values allow for the three different corner styles: rounded, inverse rounded, and bevel. The link between all the corner values will keep them the same. Unlink the icon to set each corner value and style independently.

Polygons show one corner widget to adjust all the corners together. Select the polygon with the Selection tool and the corner widgets will appear at the top of the shape.

Stars have corner widgets too, but they don't show when the shape is selected with the Selection tool. To see the corner widgets on a star, select the shape with the Direct Selection tool and all the corner widgets will appear for editing. Editing the star's corner widgets can only be done on the shape with the Direct Selection tool because the corner options are not available in the Transform panel or Properties panel.

Drawing Lines

The Line section of the toolbar contains the Line Segment tool, Arc tool, Spiral tool, Rectangle Grid tool, and Polar Grid tool. Only the Line Segment tool has a shortcut assigned to it. If the toolbar doesn't contain the Arc tool, click on the three dots at the bottom of the toolbar to add the Arc tool to the set of drawing tools.

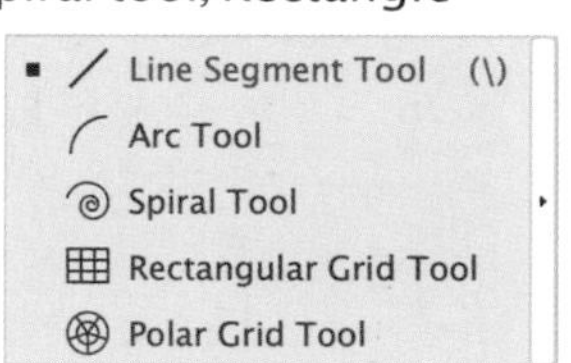

Line Segment

Select the Line Segment tool and click and drag on the artboard to draw a line. To create a line at a preset length, select the Line tool from the toolbar and click on the artboard. This will open a dialog box for the parameters of the line. The length of the line and the angle can be set. The Fill Line checkbox is somewhat of an oddity here, as a line normally has a Stroke Color and not a Fill Color. See the next section on drawing arcs to learn more about the concept of a Fill Color applied to a line.

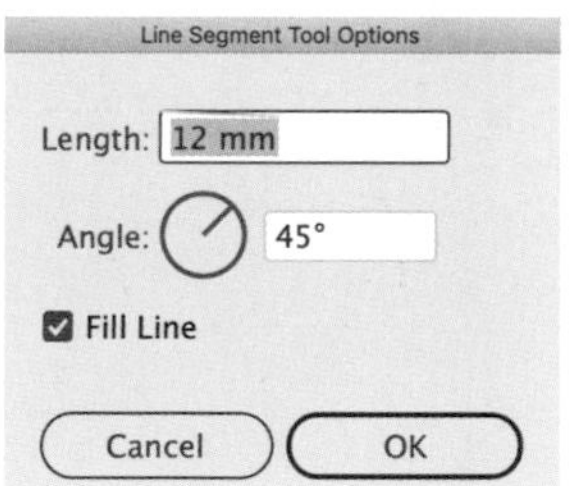

SMART SHORTCUT

Holding Shift while drawing a line will keep it horizontal, vertical, or at a 45-degree angle depending on which direction you draw the line. Holding the Option key while drawing will start the line from the middle instead of from the end.

Arc Segment

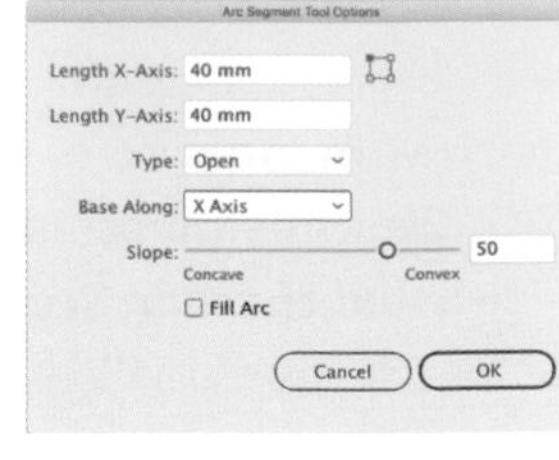

Select the Arc tool and click and drag on the artboard to draw an arc. Click on the artboard with the Arc tool to open the dialog box.

- **Length X-Axis** controls the width of the arc.
- **Length Y-Axis** controls the height of the arc.
- **Type** controls creating a 1/4 path that is open, or a closed path, creating a 1/4 segment of a circle.
- **Base Along** controls the direction of the arc that is created.
- **Slope** controls the slop of the arc, that is, how flat or steep the slope is. A negative value will create a concave (inward) slope and a positive value will create a convex (outward) slope. A slope of 0 creates a straight line, but we have the Line Segment tool for that.
- **Fill Arc** fills the arc with the current fill color. The Fill Color will fill from open end to open end of the arc even though the lines are not closed. If the Fill were added to a Line Segment, it would not be noticed.

> **SMART SHORTCUT**
>
> While drawing an arc, hold Shift to create a 1/4 circle arc.
>
> In the Arc Segment dialog box, there is a box icon to the right of the Length X-Axis field. This is the reference point from where the Arc Segment will start drawing from. The corner highlighted is the point of origin.

Spiral

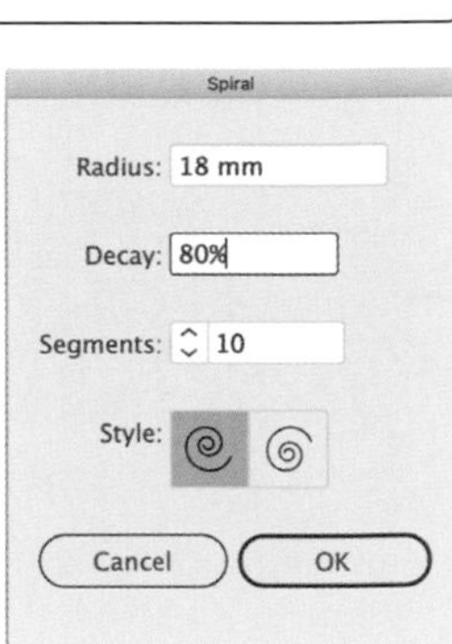

The Spiral tool creates a spiral effect that can spiral either left or right. Click on the artboard with the Spiral tool to open the dialog box.

The Spiral tool draws from the middle of the shape, and the size is measured by the radius.

- **Decay** is how tight or loose the spiral appears. A high percentage of decay will create a tight spiral; a low percentage of decay will create a very loose spiral.
- **Segments** of the spiral are defined as quarters of a turn. There are four segments for each turn of the spiral. Ten segments would indicate two and a half turns of the spiral from to end point to end point.
- **Style** affects which direction the spiral turns.

> **SMART SHORTCUT**
>
> To change the number of segments while drawing a spiral, use the up arrow or down arrow to increase or decrease the number of segments while the mouse is drawing the shape. One important note: hold the mouse still while using the up and down arrows to avoid runaway segment creation. If you move the mouse while using the up arrow or down arrow it will create hundreds of segments instantly.
>
> To change the rate of decay while drawing, use the Command key and pull out from the center to decrease the rate of decay (higher percentage) or pull into to the center to increase the rate of decay (lower percentage). Move the mouse in or out in very small increments or the decay will max out instantly. A little movement goes a very long way.

Rectangular Grid

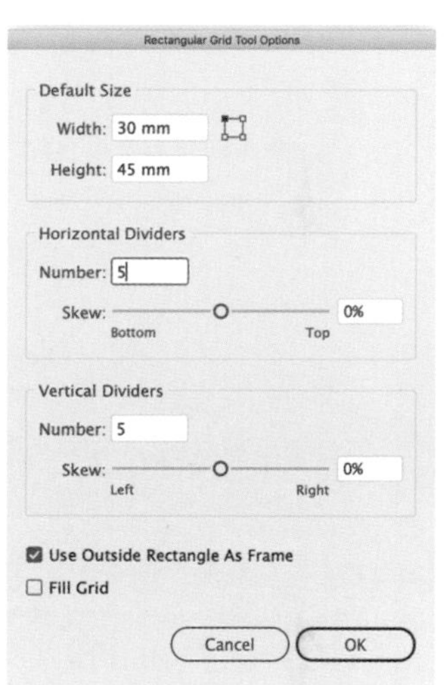

Select the Rectangle Grid tool and click on the artboard and drag until the grid is the desired size. The box icon to the right of the Width field is the reference point from where the grid will start drawing from. The corner highlighted is the point of origin.

- **Default Size:** Set the width and height of the overall grid size in these input fields.
- **Horizontal Dividers**: This specifies the number of horizontal dividers between the top and bottom of the grid. The Skew determines how the horizontal dividers are weighted (or skewed) toward the top or bottom of the grid. Changing the skew along the slider will make the dividers progressively closer together.
- **Vertical Dividers**: This specifies the number of dividers between the left and right sides of the grid. The Skew works the same way as it does for the Horizontal Dividers.
- **Use Outside Rectangle As Frame**: This replaces the top, bottom, left, and right segments with a separate rectangular object in the event you want to edit or delete the outside shape independent of the grid structure inside.
- **Fill Grid**: This fills the grid with the selected fill color. If unchecked, the fill is set to none.

SMART SHORTCUT

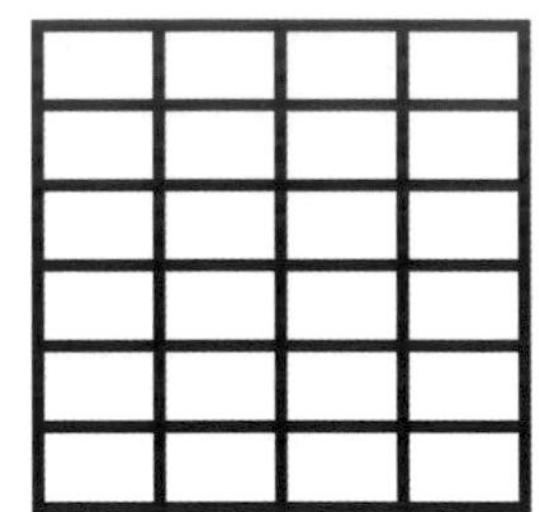

While drawing a Rectangular Grid, hold the Shift key and it will create an overall square grid. This will not make the cells inside the grid perfect squares; only the overall grid shape will be a square.

To change the number of horizontal dividers while drawing the grid, use the up arrow or down arrow to increase or decrease the number of dividers.

To change the number of vertical dividers while drawing the grid, use right arrow or left arrow to increase or decrease the number of dividers.

Polar Grid

Polar Grid Tool Options
Default Size
Width: 35 mm
Height: 35 mm
Concentric Dividers
Number: 5
Skew: 0%
In Out
Radial Dividers
Number: 5
Skew: 0%
Bottom Top
Create Compound Path From Ellipses
Fill Grid
Cancel OK

Select the Polar Grid tool and click on the artboard and drag until the grid is the desired size. A box icon to the right of the Width field is. This is the reference point from where the grid will start drawing from. The corner highlighted is the point of origin.

- **Default Size**: This sets the width and height of the overall grid size.
- **Concentric Dividers**: This is where you can control the number of circular concentric dividers in the grid. The Skew value determines how the concentric dividers are weighted toward the inside or outside of the grid.
- **Radial Dividers**: This controls the number of radial dividers between the center and the circumference of the grid. The Skew value determines how the radial dividers are weighted toward the inside or the outside of the shape.
- **Create Compound Path From Ellipses**: This converts the concentric circles into separate compound paths and fills every other circle.
- **Fill Grid:** This selection fills the grid with the selected fill color. If unchecked, the fill is set to none.

SMART SHORTCUT

While drawing a Polar Grid, hold the Shift key to create a prefect circular grid.

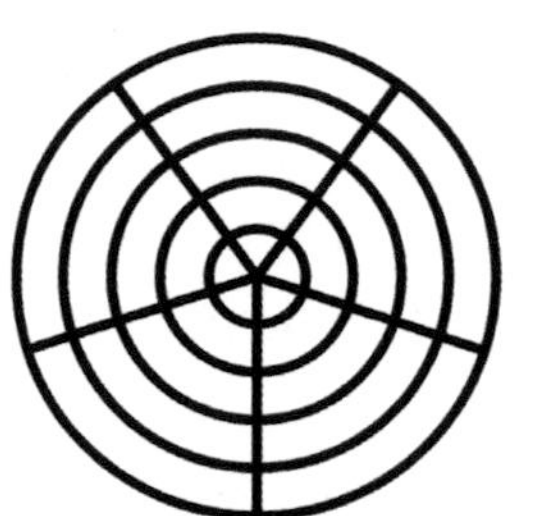

To change the number of concentric dividers while drawing the grid, use the up arrow or down arrow to increase or decrease the number.

To change the number of radial dividers while drawing the grid, use the left arrow or right arrow to increase or decrease the number.

Editing Shapes and Lines

Rotate Shapes

Select a shape or line with the Selection tool. Move the Selection tool beyond an outside corner and a rotate symbol will appear. Click and move the cursor to rotate the object. Hold the Shift key to constrain the object to 45-degree increments when rotating.

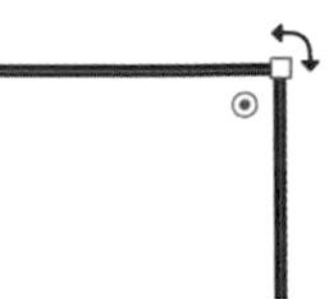

Resize Shapes

Select a shape or shapes with the Selection tool. Move the Selection tool on an outside corner or a midsection on any side of the shape and the resize arrow will appear. Pull the cursor in to the center or out from the center to resize it. Hold the Shift key to constrain the object to maintain the proportions while scaling.

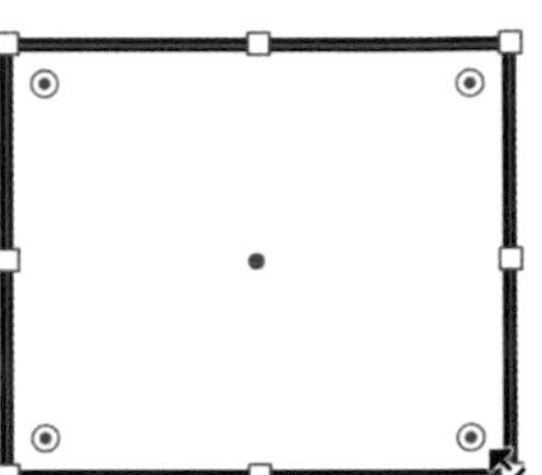

Duplicate Shapes

You can copy a shape by selecting the shape and choosing Edit Menu > Copy/Paste.

A much faster way is to select the shape or shapes to with the Selection tool and hold the Option key. Click on the selected shape(s) to duplicate them. Note that the cursor will double up when you hold the Option key and click and drag.

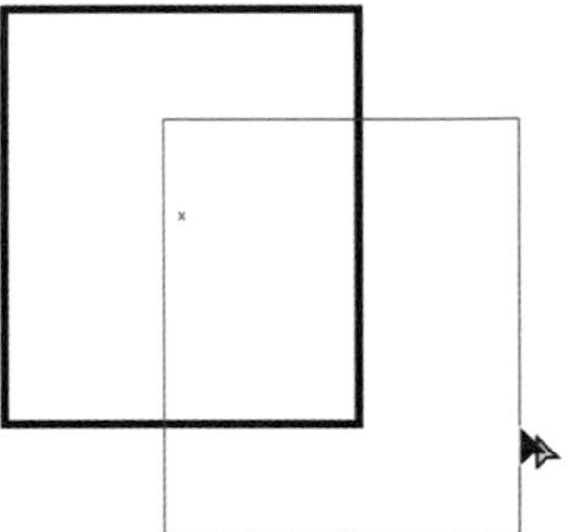

Flip Shapes

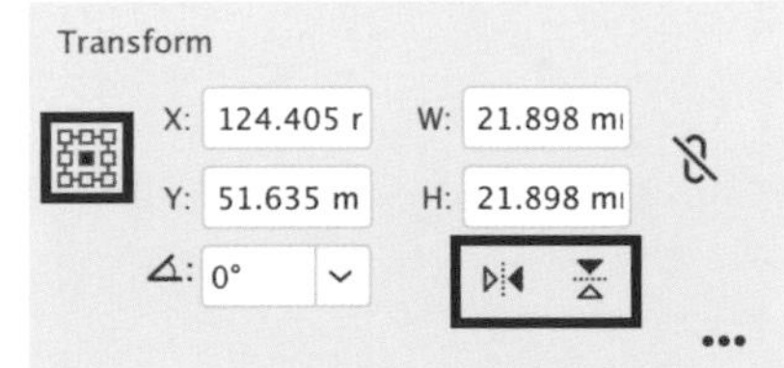

You can flip a shape vertically or horizontally by selecting the shape and clicking on the flip shape icons in the Transform section of the Properties panel—but not in the Transform panel itself.

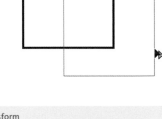

When using the flip action, set the reference point in the Transform section to determine where the flip will originate from. Currently in this example, the reference point is in the center as highlighted in red. If you are new to Illustrator this may not seem like a big deal, but for the seasoned user, this is a nice improvement. The previous ways to flip a shape were not so easy to do nor comprehend.

Properties Panel

The Properties panel is a recent addition to Adobe Illustrator. This is where you can view and control settings based on your current selected object or task. The Properties panel is under Window > Properties.

Most, but not all of the settings and controls for each item is shown in the Properties panel area. Additional settings and controls are accessed by clicking the ellipses in the lower-right corner, which can be an extra step when working with objects. Some of these controls are also available in the Transform panel, which shows all of the settings and controls without the extra step of clicking to reveal the additional features.

I use the Transform panel for most of my shape editing, scaling, and sizing because it saves time and it gives me specific editing for each shape and all its attributes in one location.

The Properties panel has lots of information in it based on the object or type you have selected. This panel has taken over for the control bar that was the main access for Information for previous versions of Illustrator. With today's users, the information in the program has grown and laptops cannot show the entire length of the control bar, leaving information out of reach. The Properties panel solves this problem and also adds many actions and shortcuts that would normally be accessed through the menus.

The Properties panel will change based on what and how things are selected. If I was working on shapes, I would see the information in the panel at the right. Working on type, I would see all the information relevant to type editing.

If you use other Adobe products, you will find this Properties panel in the newer versions as well.

Transform a Rectangle

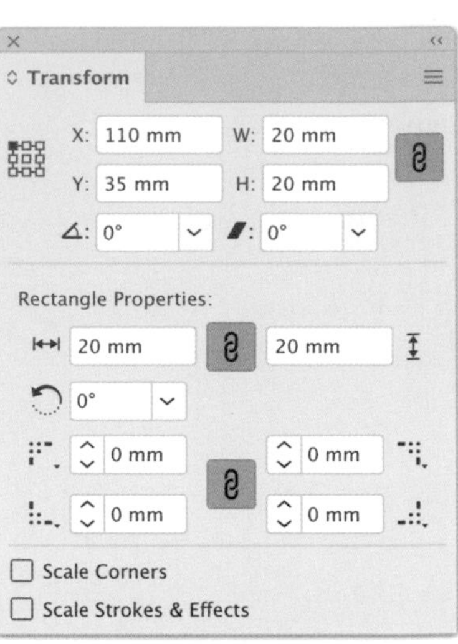

The Properties panel and the Transform panel are two separate panels. The Properties panel contains a Transform section that is nearly identical to the Transform panel but does not have the Skew field. The Transform panel also lacks the flip features that the Properties panel has.

Click on the dropdown menu in the upper right of the Transform panel (shown) to choose the option of having the Transform panel open automatically when you draw a shape.

- **Reference Point** is the point where a shape is measured from, scaled from, and flipped from.
- **X and Y** references the location, based on the reference point, of the object on the artboard. This is measured from the upper-left corner of the artboard to the selected reference point on the shape.
- **Width and Height** refer to the size of the object. These fields can be input separately if the link to the right of the fields is unlinked. Values can be entered into these fields or you can click on the field and use the up arrow or down arrow to increase or decrease the value. Hold the Shift key while using the up or down arrows to change the values, and the numbers will increase or decrease in increments of ten.
- **Rotation** of the shape can be entered into the field either from the dropdown menu to the right of the field or manually.
- **The Skew** of the shape (not available in the Properties panel) can be entered into the field either from the dropdown menu to the right of the field or manually. Skew is based on the selected reference point as well.
- **Corner Options** (corner widgets) can be adjusted in these fields. Click to unlink the link icon in the center of the fields to set each corner independently. The corner choice can be accessed through each corner dropdown menu and the size of that corner can be adjusted in its corresponding fields. Setting the corner radius to zero removes the corner options and the corner becomes squared.

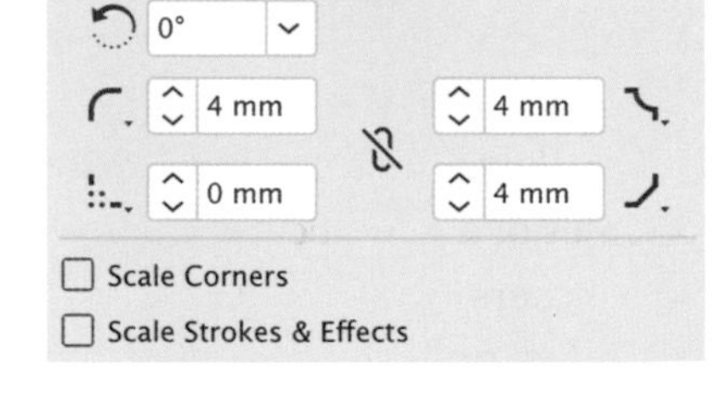

- **Scale Corners** allows the rectangle to be scaled larger and smaller, keeping the corner sizes in proportion to the shape. As the shape is scaled larger, the corners will also scale larger.
- **Scale Strokes & Effects** makes the stroke weight scale with the shape. The larger the shape, the heavier the stroke weight will become. For the times when you do not want the Stroke Weight to change, uncheck the Scale Strokes & Effects box before scaling the shape up or down.

Transform an Ellipse/Circle

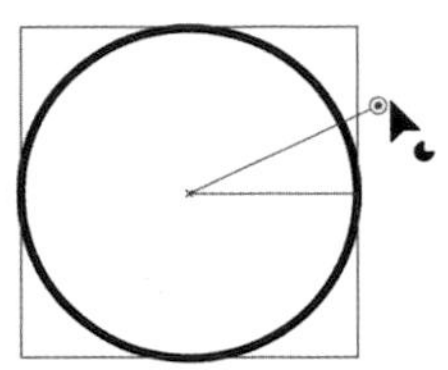

In the Ellipse Properties section, you can create a pie shape from a circle. This can be done manually on the ellipse/circle by clicking the handle that sticks out from the right side of the shape and moving it around the circle. This is the Pie Angle.

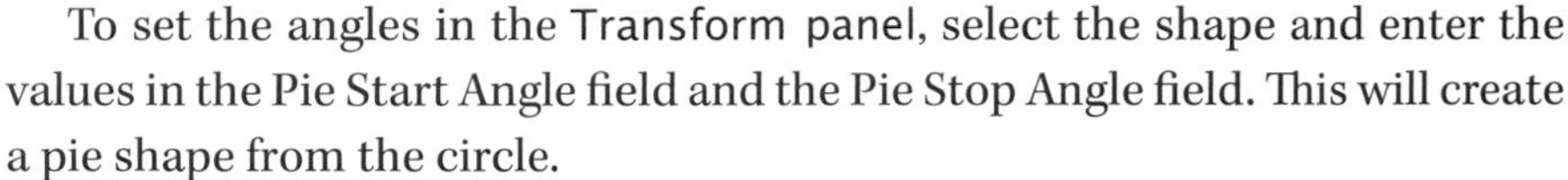

To set the angles in the Transform panel, select the shape and enter the values in the Pie Start Angle field and the Pie Stop Angle field. This will create a pie shape from the circle.

Creating a Simple Pie Chart

Ellipse Properties:
15 mm | 15 mm
0°
35° | 300°

Setting the angles to get the right pie shape can be done with math. And that math can be calculated directly in the value fields without a calculator. Here is how it works: For instance, if you want a pie chart that represents 65%, enter the equation **65*3.6** in the Start Pie angle field. The percentage of the pie you want to show is 65, and 3.6 is the number you multiply that percentage by to get the calculation of the degrees of a circle.

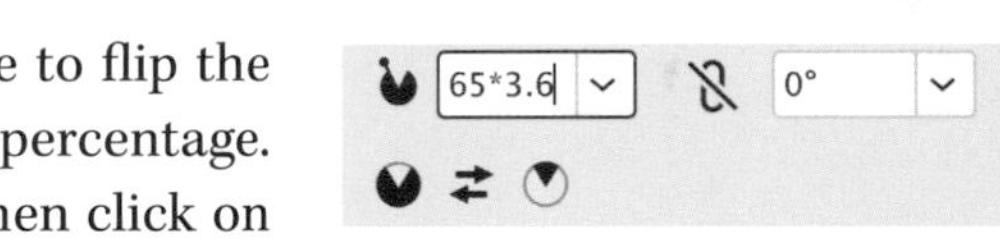

Use the invert arrows below the Pie Start Angle to flip the shape of the chart to get the inverse of the pie's percentage. Here is a great trick: select the pie, hold Option, then click on the invert arrows to duplicate that shape and flip it, making two perfect-fitting pieces of the pie.

I have this on my blog in case you want to reference it: https://www.jasonhoppe.com/blog/infographic-series-easiest-pie-chart-ever

Transform a Polygon

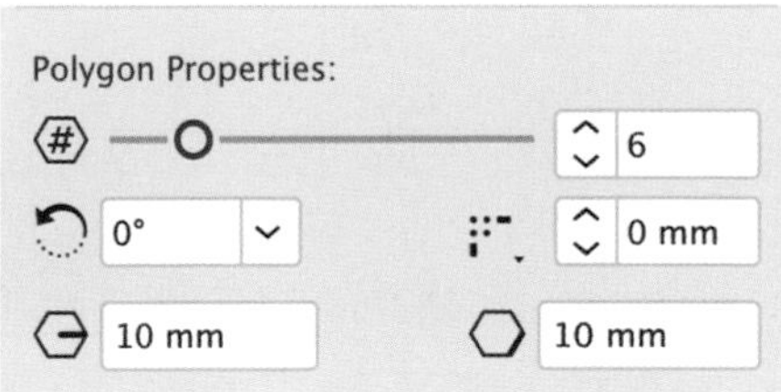

In the Polygon Properties section, you can edit the number of sides of a selected polygon.

Corners can be edited to create rounded, inverse rounded, or beveled corners.

The radius and the side length can be edited as well. The maximum numbers of sides that can be selected is 20.

Shape Creation
Advanced Construction
Editing & Transformation
Effects & Graphic Styles
Type & Text
Working with Color
Output

Transform a Star

The star is the one shape that Illustrator does not support editing for in the Transform panel other than basic size, location, and rotation. If you draw a star, the Properties section of the Transform panel will display no shape properties.

Transform a Line

With a line selected, the Line Properties Section in the Transform panel allows editing of the line length, rotation, size, and location.

Precision Editing

The Transform panel allows for precision editing of shapes and lines. Each field in the panel can be a calculation field. Using the standard add, subtract, multiply, and divide equations, you can change the size, location, rotation, and skew of any shape easily.

Select the shape and enter in the calculation in the field and hit return. The calculations will be performed from the selected reference point based on the reference point icon to the left of the fields.

For example, you want the shape to be 115% larger, select the field and enter +115% after the value and the calculation will be done. If you want it to be half the size, divide it by 2. Yes, it really is that easy. And once you discover this, try this in other Adobe applications (it works there too).

Grouping Items

Grouping items in Illustrator can be quite useful, although you may not need to group items all the time. You may want to move or scale objects without grouping them. If you select several objects at once, you can scale, transform, or rate the objects without grouping them together.

Grouping several objects is handy when you have a logo or artwork with many pieces. By grouping the objects together, you can simply select any part of the object and all the objects will move together. To group objects, select the object with the Selection tool and choose either Object > Group. Groups can also be nested or grouped within other objects or groups to form larger groups. The Layers panel is how we dig deeper into nested groups.

To ungroup objects, select the object to be ungrouped and choose Object > Ungroup.

SMART SHORTCUT

To group items together, select the objects and choose Command + G.

To ungroup items, select the objects and choose Shift + Command + G.

To select an item in a Group, use the Group Selection tool in the toolbar. 

Selecting Objects

Selecting objects with the Selection tool seems quite simple, yet sometimes it does not appear to work like you would expect.

Choose the Selection tool and click on or in an object to select it. However, there are instances when you click in an object and it won't select so you have to select the edge or the stroke of the object. The reason for this is the fill of the object. If an object has no fill color (white is a fill color) then the object is considered empty and you cannot click inside it because there is nothing there. If the object is filled with a color (even white), you can click anywhere on the object to select it.

The way I avoid clicking on an object is to use the Selection tool and drag a selection marquee around part or all of one or more objects. As long as the selection marquee touches any part of the object, it is selected.

Hold down Shift and click the objects if you want to add them to objects you have already selected or press Shift and click on the selected objects to remove them from what you have selected.

Selecting objects using the Layers panel is a great way to select items when your Illustrator files get more complicated. In the Layers panel, click on the object you want to select. You may need to expand a layer or group by clicking on the arrow to the left of the top layer (I call it a twilry) to expand the list of sublayers.

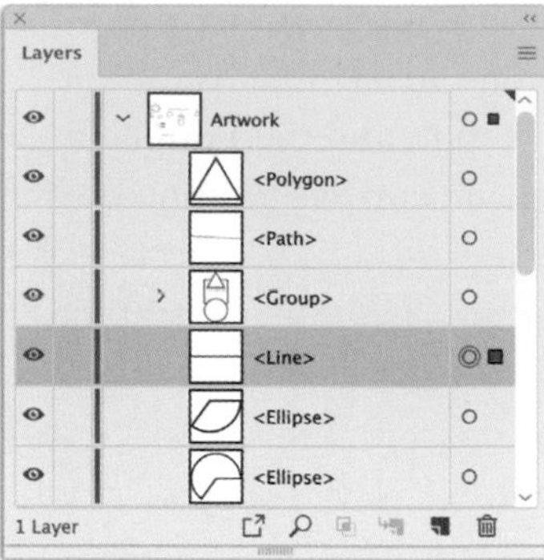

To select an object, click on the circle to the right its name. This will select the object in the document. You can use Shift + click to add to the selected objects or remove objects from the selection.

Isolation Mode

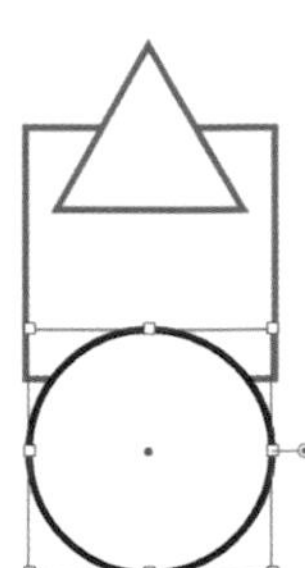

There are times when you do not want to ungroup artwork to get access to a in that group. Artwork can be complicated and ungrouping it may create other headaches down the road.

Isolation mode is great for getting into a group without ungrouping everything. There are three ways to get into Isolation mode. Double-click on the item in the group to isolate it from the rest of the group. The shape will be selected and all the other shapes in the group and in the document will be grayed out. You can now edit this shape, move it, or delete it from the group.

You can enter Isolation mode by selecting the group, right-clicking on the group, and choosing Isolate Selected Group.

To exit out of Isolation mode, hit the Escape key or click on the back arrow at the upper left of the document window.

You can enter and exit Isolation mode via the group in the Layers panel here. You can select the group from the layers, then click on the dropdown menu on the upper-right side of the panel. Choose Enter Isolation mode to enter into the group, then choose Exit Isolation mode from the same dropdown menu when finished.

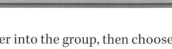

Locking/Unlocking

Locking objects prevents you from selecting, editing, and moving them. Single objects, groups, or entire layers can be locked.

To lock an object, select the object and choose Object > Lock > Selection. This will lock that one object. To unlock it, go to the Layers panel and click on the lock icon to the left of the layer name. In the Layers panel you can click on the space to the right of the eye icon to lock that layer.

Layers
Layer 1
<Path>
<Group>

To unlock all objects in the document, choose Object > Unlock All. Or you can click on the lock icons that appear in the Layers panel individually.

Symbols Versus Groups

When you have artwork that you want to use numerous times in an Illustrator file, you can create the artwork, group it, and then duplicate it as many times as you need. Or you can use symbols to make global edits to the shape if you choose to edit the object later.

Let's use this illustration of a box of popcorn to describe symbols. A symbol is an object that you can reuse in a document, but it has advantages over copying and pasting the artwork each time you need it.

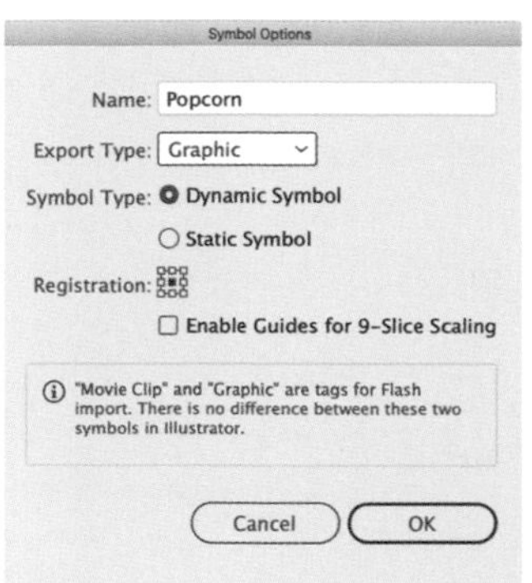

Open the Symbols panel from the Window menu. Select the popcorn you want to turn into a symbol. From the drop-down menu on the upper-right side of the Symbols panel choose New Symbol.... Name the Symbol, choose Graphic from Export type, and click on Dynamic Symbol. Click OK to add the symbol to the Symbols panel.

To use the symbol in your artwork, drag the popcorn symbol from the Symbols panel into your artwork as many times as needed; these are called instances.

Symbols

Each popcorn symbol instance is linked to the popcorn symbol in the Symbols panel. You can scale, rotate, and edit each instance in the document without affecting the original symbol. If you want to change every symbol, you can edit the symbol in the Symbols panel by double-clicking on it, making the edits, and when done, hit the Escape key.

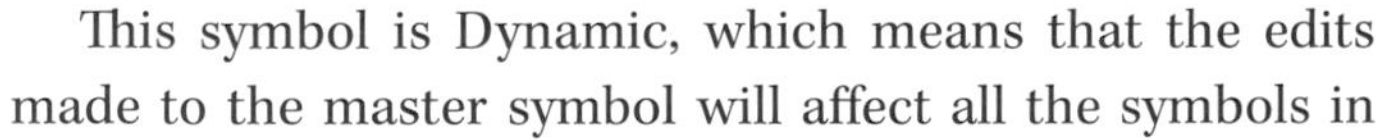

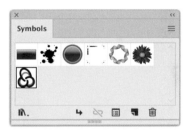

This symbol is Dynamic, which means that the edits made to the master symbol will affect all the symbols in the document; yet it will retain all the individual edits done to each instance. Using symbols can save time and reduce file sizes when used in place of copying and pasting multiple sets of grouped items.

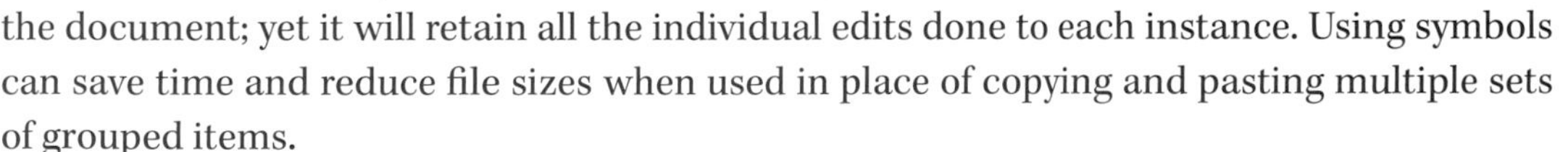

3 Advanced Construction

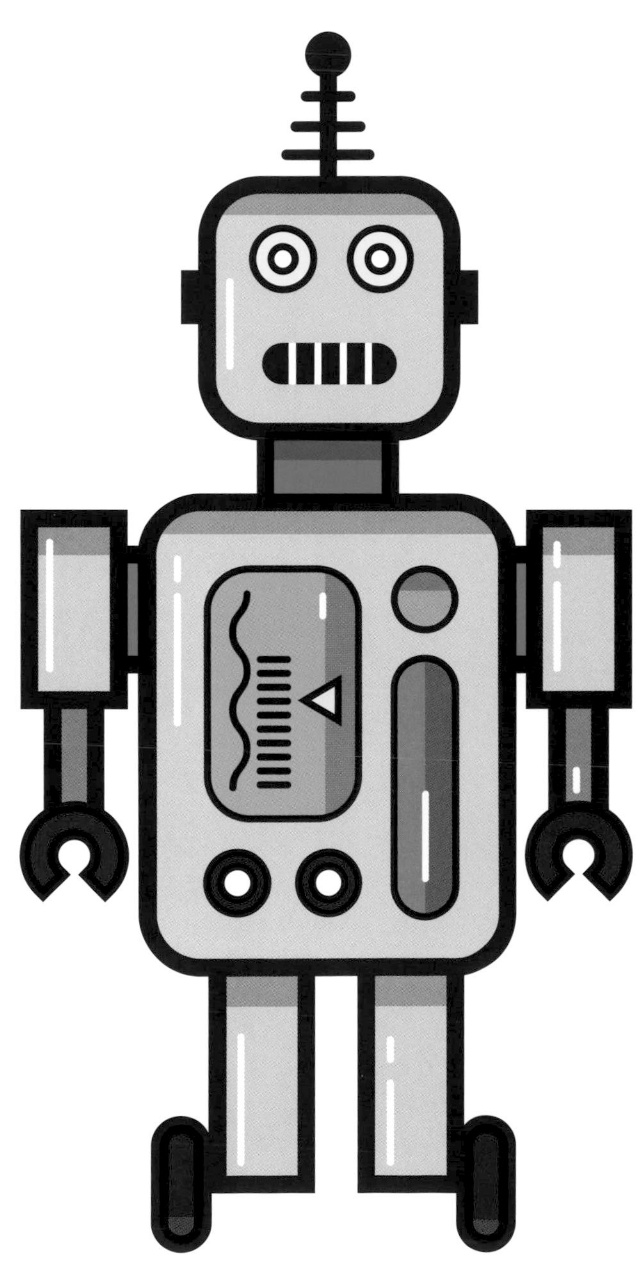

Pathfinder Panel

In teaching people how to use Illustrator for the past 20 years, I have realized that many people feel they cannot draw. While people do have differing levels of natural drawing ability, everyone can learn how to create anything they can imagine by using basic shapes and the Pathfinder panel. The Pathfinder panel is the gateway through the mental block of being afraid to draw and opening the mind to creativity. My weekly blog (www.jasonhoppe.com/blog) shows everyone how incredible the creation process can be even if you don't feel like a natural artist.

The Pathfinder panel is used to combine shapes together or subtract shapes from each other to create new shapes. This panel is the basis for creating most shapes and objects. The Pathfinder panel can be found under Window > Pathfinder.

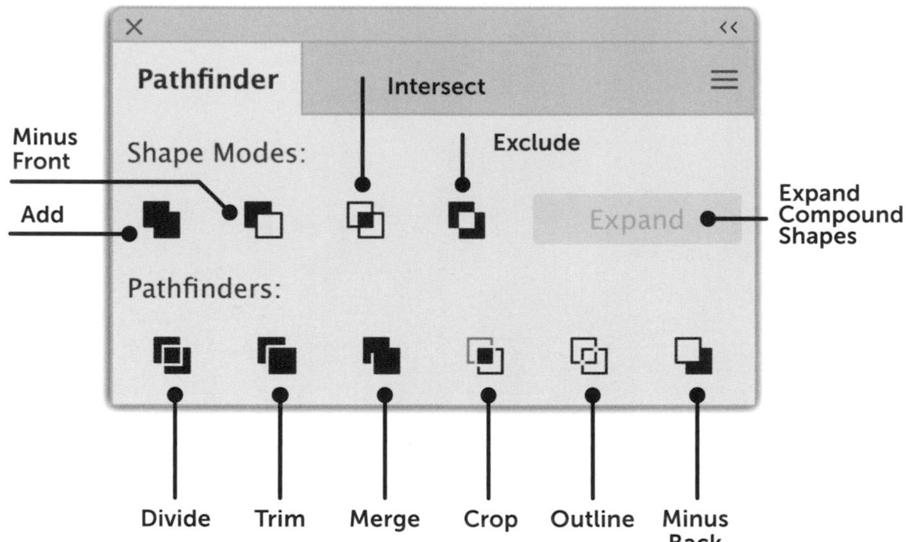

Shape Modes Section

The four shape modes in the Shape Modes Section are Add, Minus Front, Intersect, and Exclude. These shape modes work only with closed shapes; open lines or paths will either produce no results or undesirable results.

- **Add**: Select two or more shapes and click the Add icon. All the shapes will combine together into a single, merged object. The resulting shape takes on the fill and stroke attributes of the topmost object.

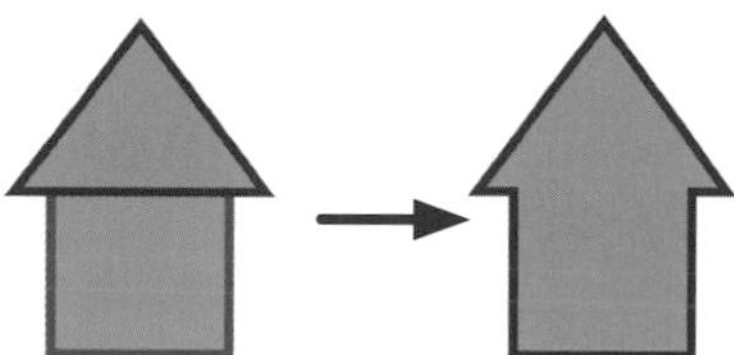

- **Minus Front**: Select two or more shapes and click the Minus Front icon and all the shapes in front of the backmost shape will be subtracted, leaving a single object.

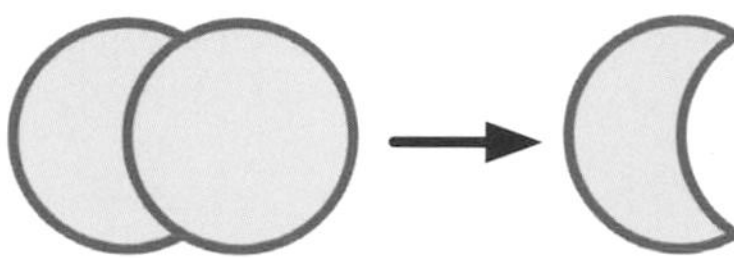

- **Intersect**: Select two or more shapes and click the Intersect icon to combine any overlapping areas and merge all the shapes into a single object. Where there is an even number of objects overlapping, the overlapping area is removed. Where there is an even number of objects overlapping, the areas are filled.

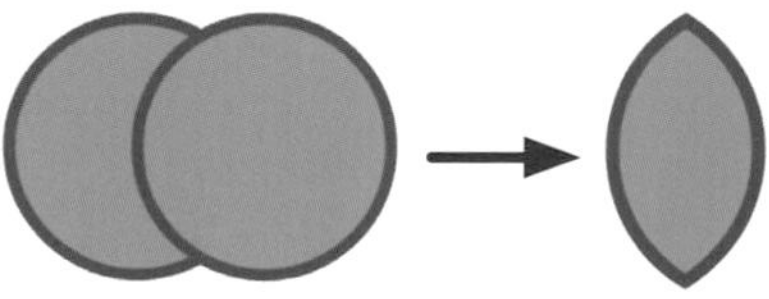

- **Exclude**: Select two or more shapes and click the Exclude icon to remove any overlapping areas and merge all the shapes into a single object.

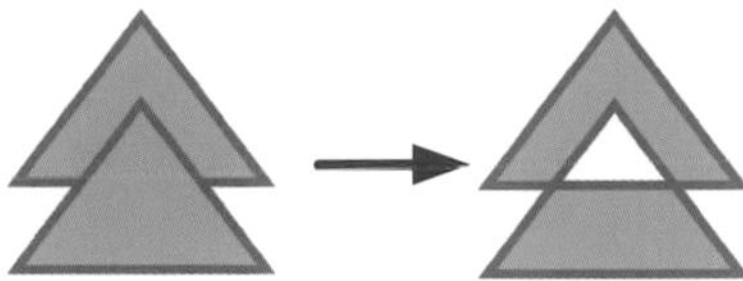

Pathfinders Section

The six shape modes in the Pathfinder Section are Divide, Trim, Merge, Crop, Outline, and Minus Back. These shape modes work with closed shapes, open lines, and paths.

- **Divide**: This allows open or closed shapes to be divided into separate sections where they overlap. The end results are grouped together.

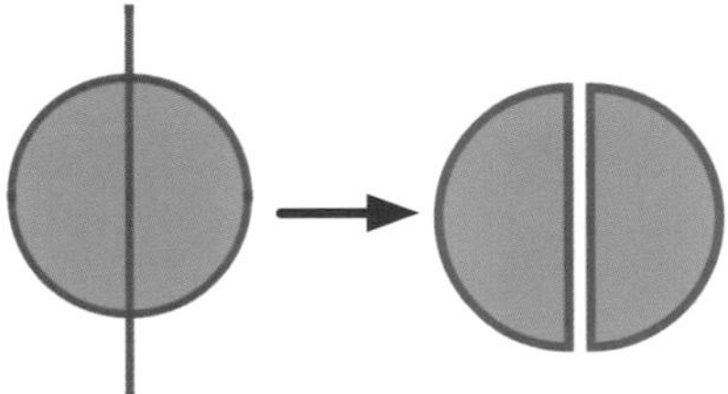

- **Trim:** This removes the part of a filled object that is hidden and eliminates any strokes on all the objects. Objects that are the same color do not merge together as they do in the Merge function.

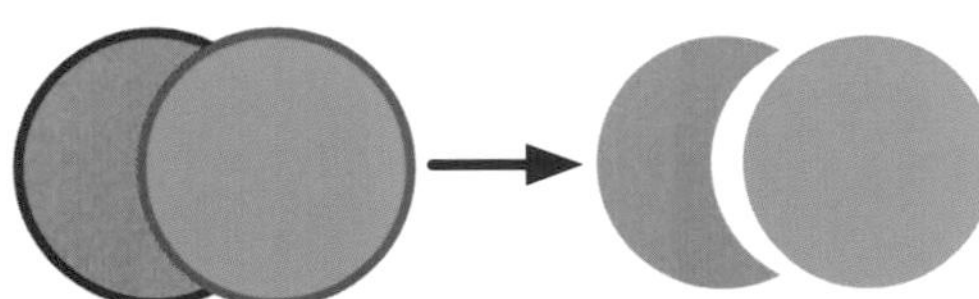

- **Merge**: This works the same way as the Trim function by removing the part of a filled object that is hidden, eliminates any strokes, and merges any overlapping objects filled with the same color.

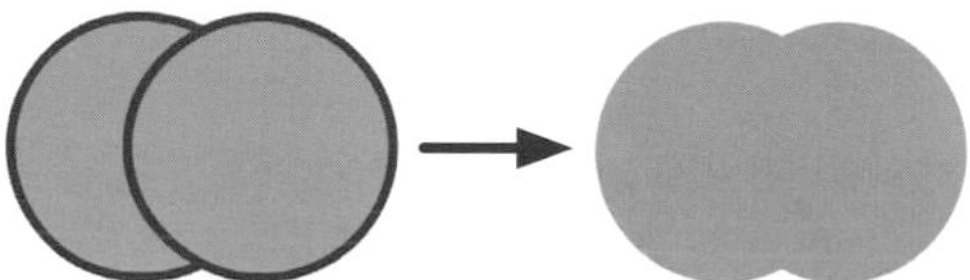

- **Crop**: This divides shapes where they overlap, then deletes all the parts of the artwork that fall outside the boundary of the topmost object and removes all strokes. The topmost object acts like a window in which the other shapes appear.

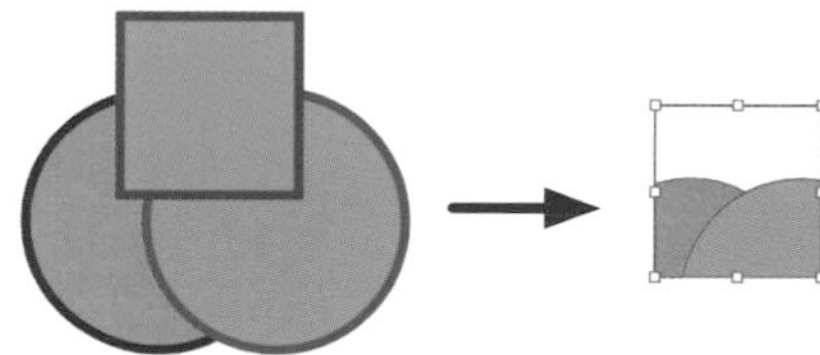

- **Outline**: This Pathfinder function may seem like it does nothing. Its results look like those from the Divide mode, only they are individual line segments instead of shapes.

- **Minus Back**: This subtracts the objects in back of the topmost object.

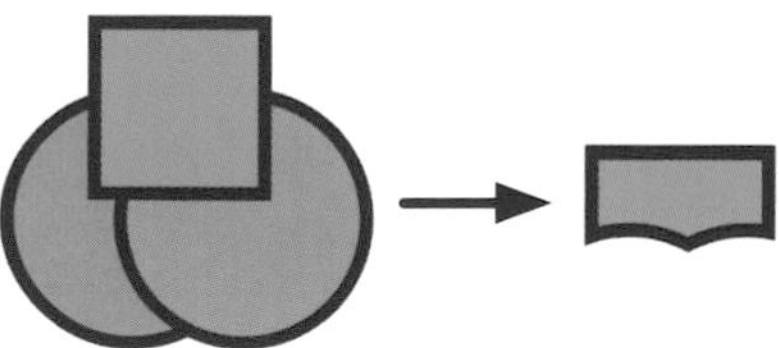

Expand Compound Shapes

The Pathfinder Expand button is grayed out under normal Pathfinder operations, unless the shape is turned into a Compound Path. When any Pathfinder operation is completed, the resulting item is one shape. A compound shape is made up of multiple shapes with the result being one visible shape that retains the ability to be edited.

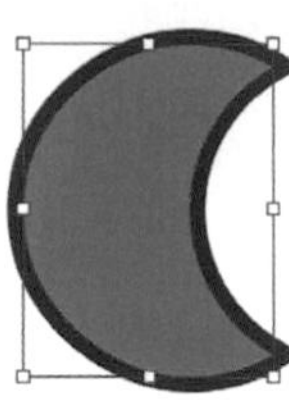

Minus Front

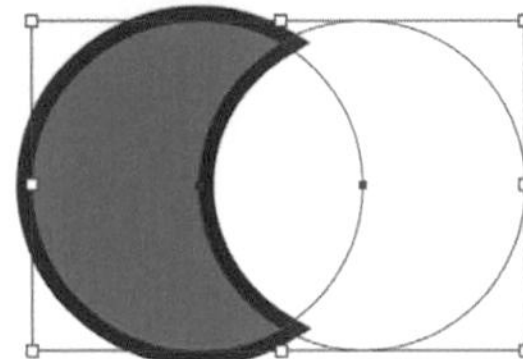

Minus Front Compound Path

To create a compound shape using the Pathfinder modes, select two or more shapes and hold Option then click on the desired Pathfinder mode.

The Expand button will be active once a compound shape is created. Clicking the Expand button will turn the compound shape into a single shape without the capability to be edited.

Pathfinders Options

Pathfinder options are under the dropdown menu in the upper right of the Pathfinder panel.

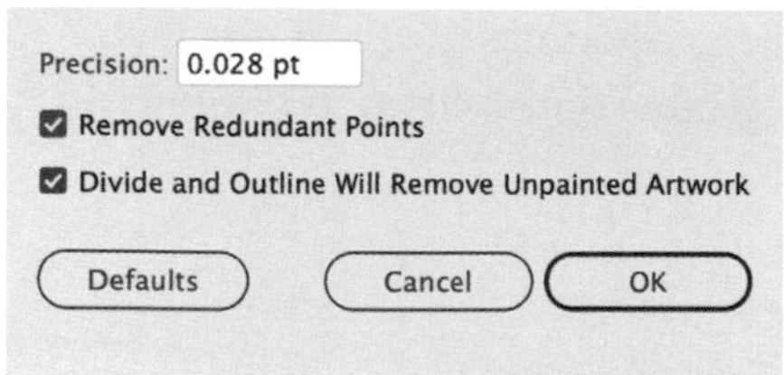

- **Precision**: This controls how precisely Illustrator draws the paths. A smaller number will more closely represent the original shapes. A larger number, will create a much less precise version of the original shape.
- **Remove Redundant Points**: This features removes any points that are on top of one another from the result of the Pathfinder functions.
- **Divide and Outline Will Remove Unpainted Artwork:** When you use either the Divide or Outline mode, the results may leave unwanted shapes and bits that have no fill or stroke. Click this box to automatically remove them.

Shape Builder Tool

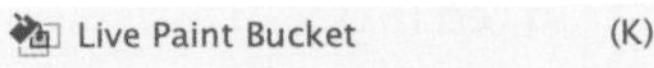

The Shape Builder tool is great for building shapes in Illustrator. It can create many of the same results that the Pathfinder panel can but there are functions and features that only the Shape Builder tool can perform. It can add and subtract shapes and lines quickly and directly on the artwork. One great feature is the ability to remove lines from a section where shapes overlap.

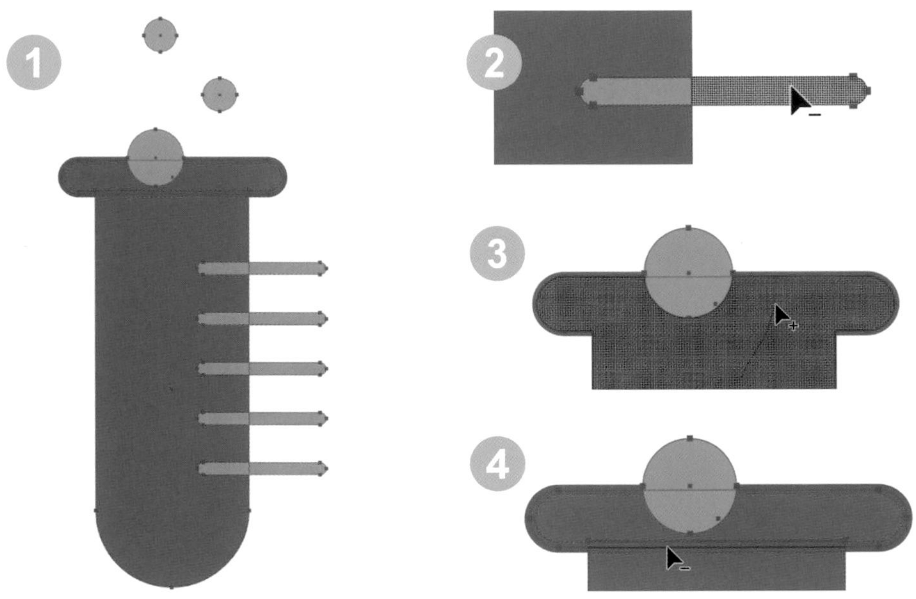

Select the artwork with the Selection tool (**Step 1**). Select the Shape Builder tool.

Delete Shapes

Move the Shape Builder tool over a shape or area you want to delete. Hold the Option key and the cursor will show a minus sign next to it (**Step 2**). Click on the shape that is highlighted with the mesh to delete it.

Add Shapes

Move the Shape Builder tool over a shape or area you want to add (**Step 3**). The default for the Shape Builder cursor is to add, so the cursor will show a plus sign next to it. Click on the shape and drag it over areas you want to add together.

Delete Lines

To delete a line rather than a shape or area, hover over a line and the line segment will highlight in red (**Step 4**). Hold the Option key and click on the highlighted line to remove it.

This is the result of creating with the Shape Builder tool. This could be done with the Pathfinder panel, the proper selection of items would be time consuming and the Pathfinder cannot remove line segments from overlapping artwork like the Shape Builder can.

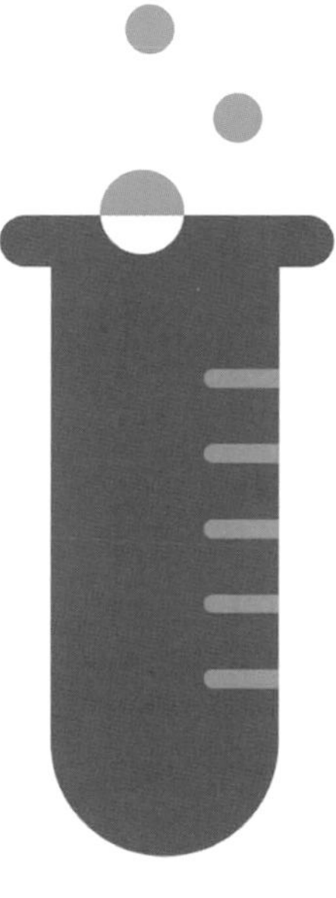

Appearance Panel

The Appearance panel lets you make changes to properties like opacity, blending mode, stroke, and fill. It also gives you control over the effects that are applied to objects. The fill and stroke colors, as well as other attributes, are also in the Properties panel and the control bar. The Appearance panel gives the most comprehensive control over object attributes.

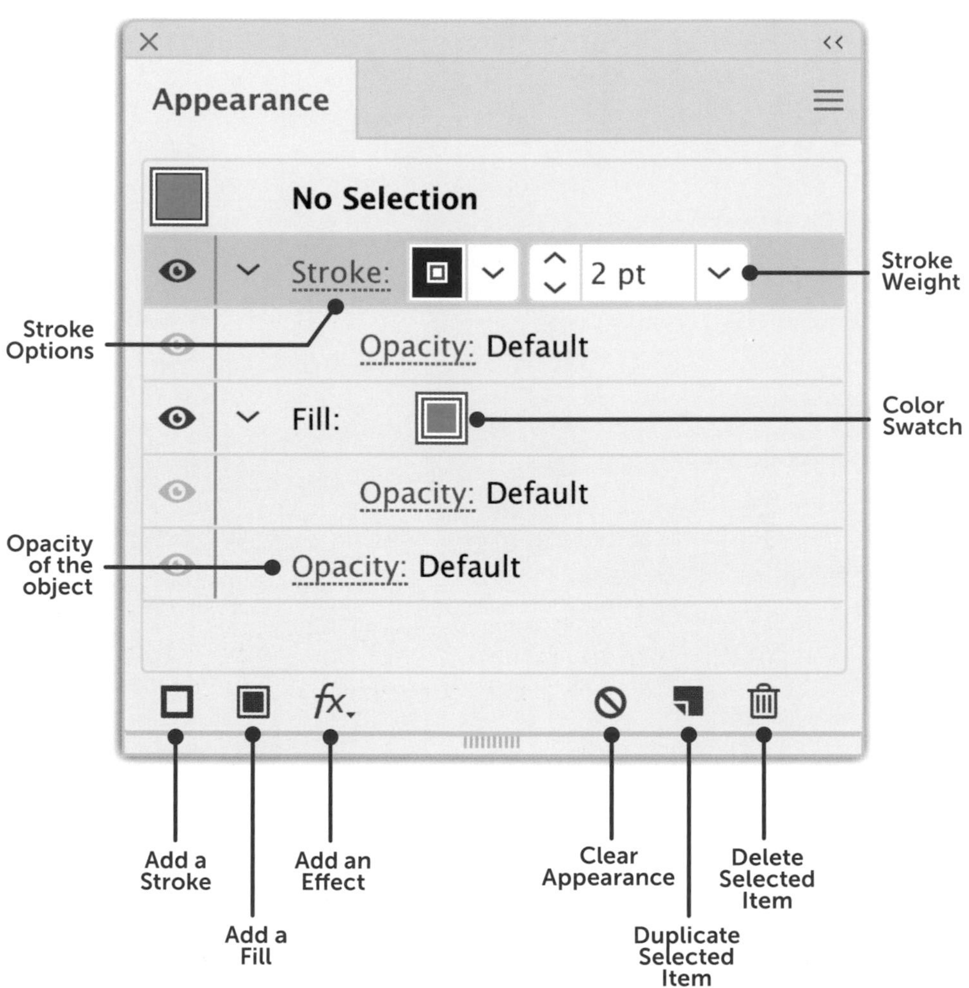

Stroke Panel

Stroke attributes appear when you click on the underlined Stroke link in the Appearance panel. The Stroke panel can also be opened as a free-floating panel under Window > Stroke.

Stroke weight is controlled by entering the value in the stroke entry field or by clicking on the dropdown menu next to the stroke value. Choose from the list of preset stroke weights or by clicking on the up or down arrows to the left of the stroke field.

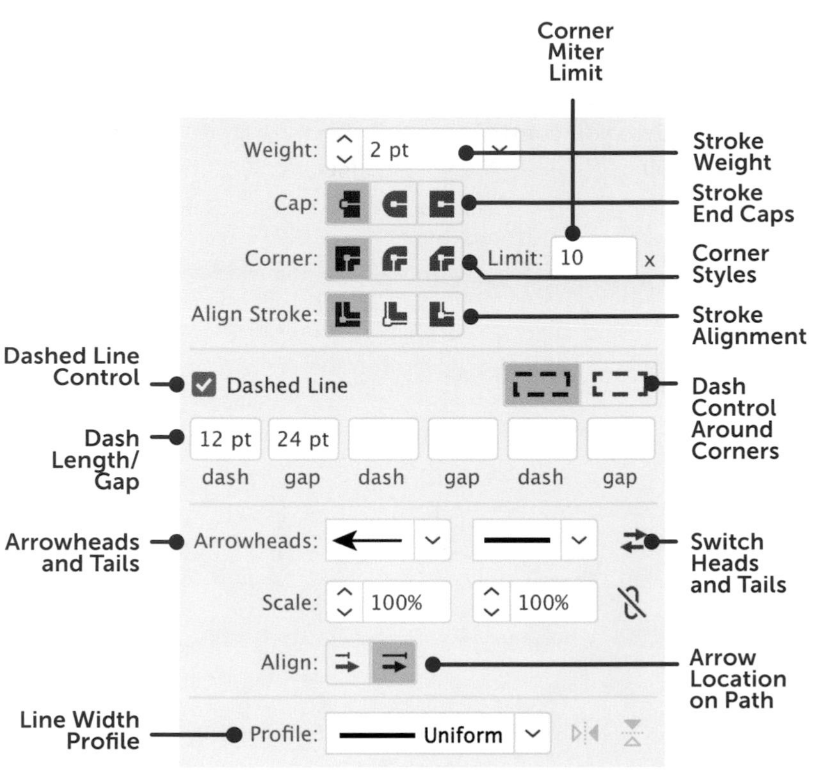

Stroke End Caps are how the ends of an open line appear. The default is to have a Butt Cap where the line ends at a perpendicular point. The second option is the Round Cap (which I call the "hot dog," based on how it looks) at the end of a line. The third option is the Projecting Cap, which adds an equal amount of line stroke to the end of the line.

Butt Cap Round Cap Projecting Cap

Stroke corner styles are for closed shapes. There are three styles: Miter Join, Round Join, and Bevel Join. These styles apply to the stroke only and not to the shape. If these are applied to a shape with no stroke, nothing will appear. Use corner widgets to apply these styles to the shape's corners.

Miter Join Round Join Bevel Join

Miter limit settings are for shapes set with Miter Join corners. Objects that form tight angles with strokes applied may not display as expected. The tighter the angles in the shape, the higher the miter limit needs to be to have the stroke corners display without the ends appearing flat or cut off.

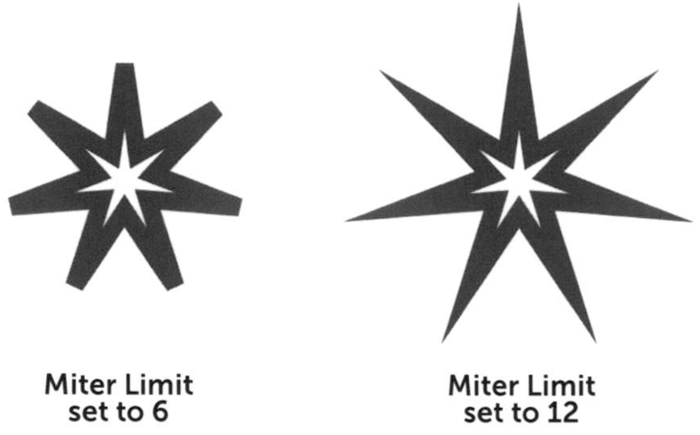

Miter Limit set to 6 Miter Limit set to 12

Stroke alignment options are for aligning the stroke of a closed shape. The default alignment is centered, which applies the stroke equally to the inside and outside the object shape. Align Outside adds the stroke to the outside of the shape. Align Inside adds the stroke to the inside of the shape. When a stroke is added to a shape, the size of the object doesn't change, but the shapes appear to be different sizes. Illustrator does add the stroke weight into the measurement and alignment of shapes. Measurement and alignment are based on the shape, or bounding box of the shape, which can make them appear misaligned. The Align Stroke is not available on open shapes or lines; only on closed shapes.

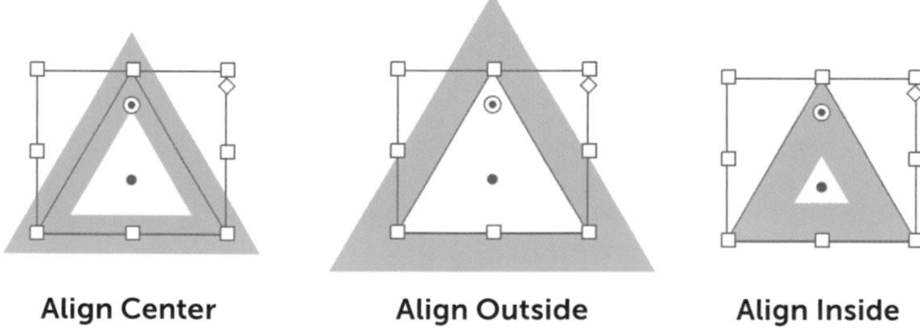

Align Center Align Outside Align Inside

Dashed lines can be created by checking the Dashed Line checkbox and entering in the length of the dashes and the gaps between. If you want the line to be consistent, there's no need to fill in the rest of the fields. If you want a more random dashed line, fill in the fields with uneven values.

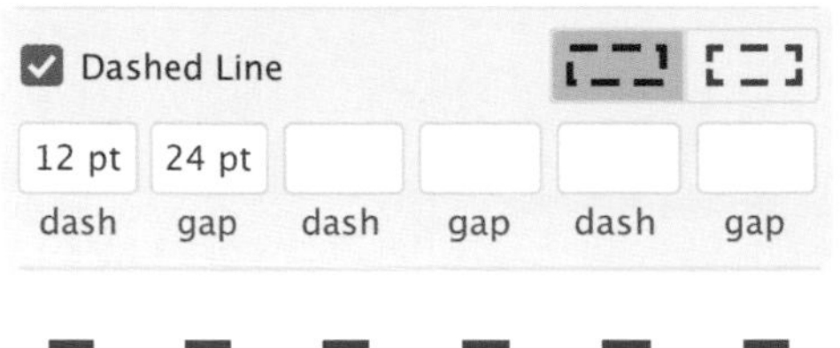

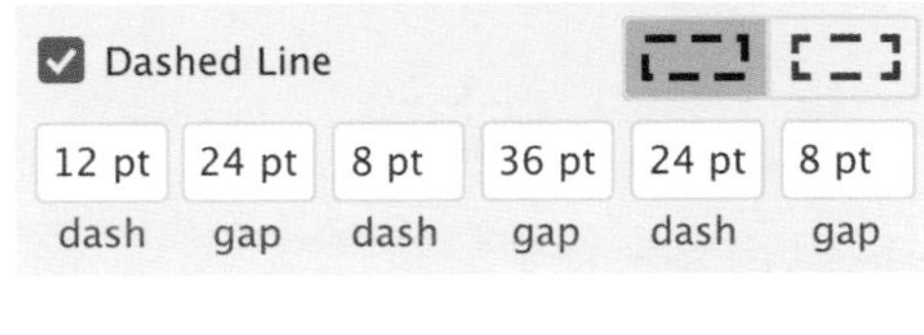

Illustrator has a very unique way of creating dots. A dotted line is a line with rounded caps placed end to end. The dash length needs to be set to 0 pt so the left and right cap appear to form a perfect dot. Make sure you set the cap to rounded or you will see very thin lines and no dots. Setting the dash to 1 pt or more will make a longer dash instead of dots.

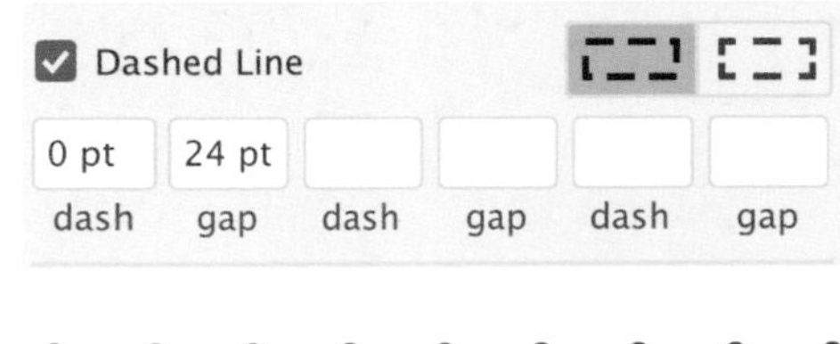

When applying a dashed line to a stroke around an object, the end and corner alignment may not be ideal. Click on the Align Dash icon to better line up the dashed line to the corners. This adjusts the gap and dash spacing and length to fit better visually.

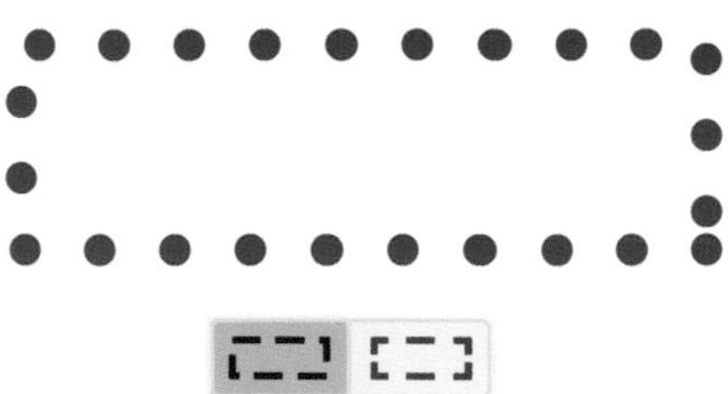

Preserve exact dash and dash/gap lengths

Align dash to ends and corners and adjust dash/gap lengths

Arrowheads and tails can be added to lines or strokes. Select the line and choose Arrowhead or Tail from the dropdown menu. The size of the heads and tails may appear quite large compared to the stroke weight. Choose the Scale to resize the head or tail larger or smaller. To swap the head with the tail, click on the double arrow to the left or the arrowheads drop-down menu.

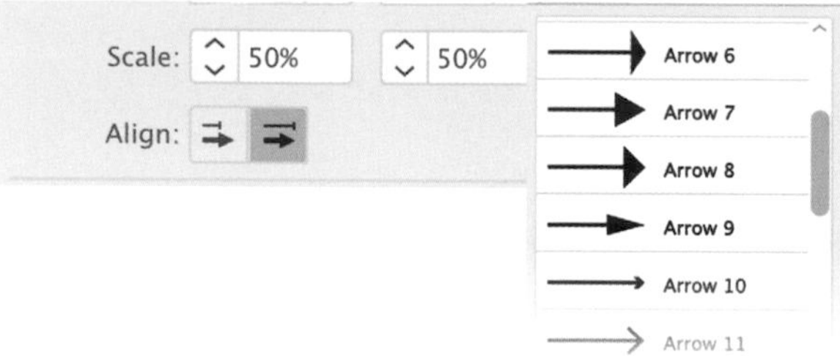

The arrowhead and tails can be set to end at the end or extend beyond the end of the path. If the line length is measured, the arrowhead and tail will not be included in the measurement if they extend beyond the end of the line.

Opacity

Opacity can be applied to objects to allow underlying objects to be visible. Any artwork in Illustrator can have an opacity applied to it to create visual effects or interaction of colors and shapes. In the Appearance panel, the opacity and fill of a stroke can be controlled independently on an object.

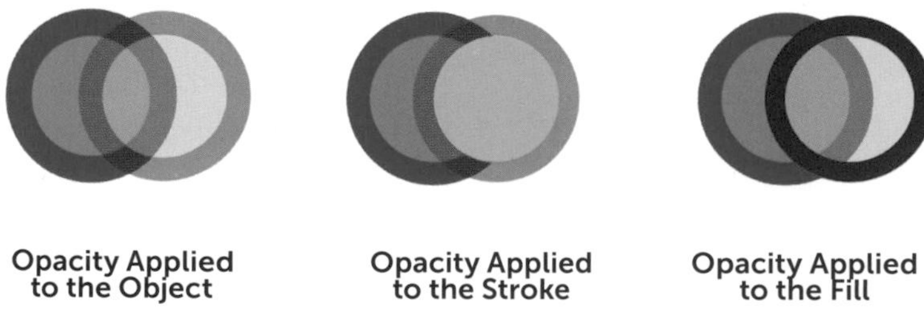

Blend Modes

Blend modes allow the interaction and blending of colors between shapes. Normal mode is the default. It allows no blending of colors. Multiply mode multiplies the colors together to create a darker color. Multiply makes white transparent and all other colors translucent. Screen mode multiplies the inverse of the colors together to create a lighter color. Screen makes black transparent and all other colors translucent.

More on the Blend modes can be found here at the Adobe website: https://helpx.adobe.com/illustrator/using/transparency-blending-modes.html

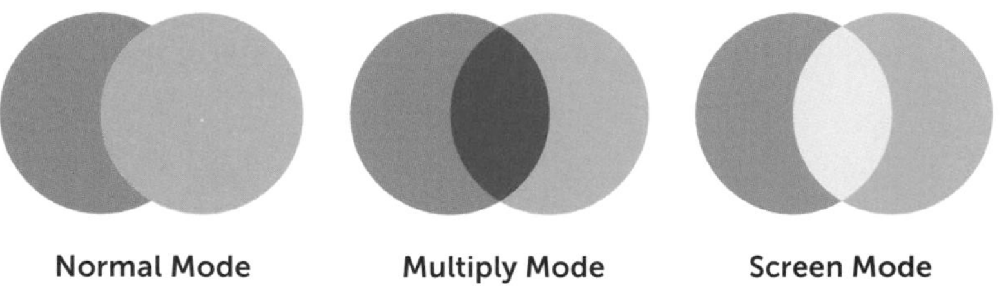

Normal Mode Multiply Mode Screen Mode

Fill and Stroke Color

Clicking on the Stroke or Fill color swatch will open the Swatch panel. From the Swatch panel you can choose from a list of colors or create your own.

Create a new color by clicking on the dropdown menu in the upper-right corner and choose New Swatch.... Choose the color mode and use the sliders to create a color. Click OK.

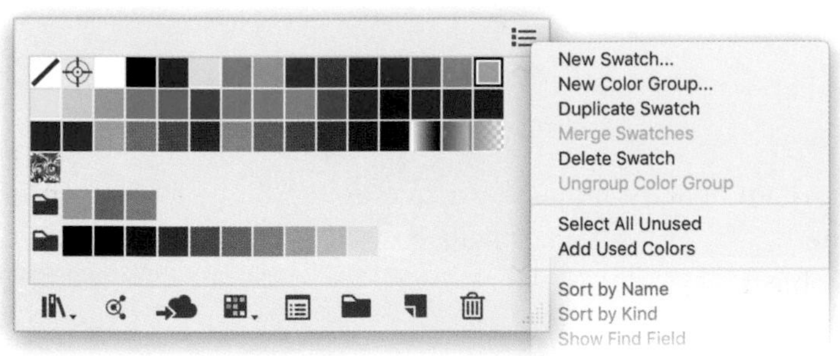

Live Paint

Live Paint allows you fill in areas with color even if those areas are not closed shapes. By creating shapes and converting them to Live Paint Objects, you can use the Live Paint Bucket tool or the Live Paintbrush tool to add colors to areas or strokes. Much like coloring in areas of a coloring book or a sketch with pencils, Live Paint detects any fillable area and applies a color.

Paint with Live Paint

The fillable areas of a Live Paint Object are called edges and faces. An edge is the section of a path where it intersects with other paths. A face is the area enclosed by one or more edges.

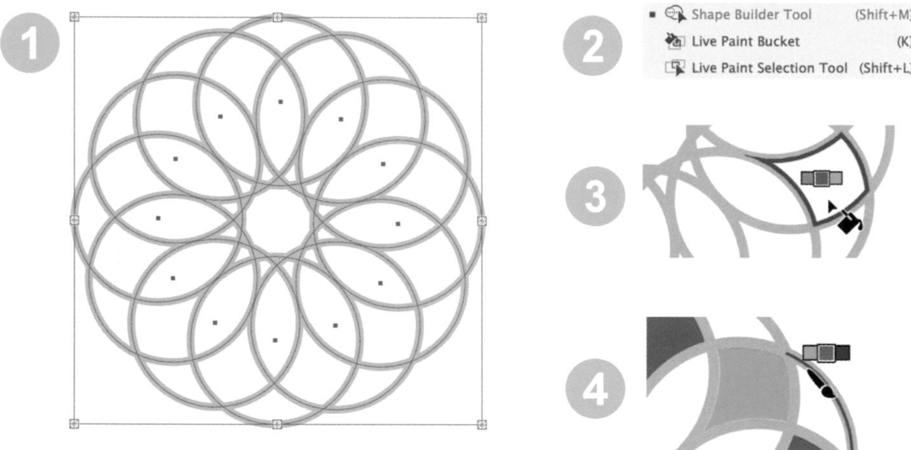

1. To turn your creations into a Live Paint Object, select all the objects you want to include and choose Object > Live Paint > Make.
2. Select the Live Paint Bucket tool from the toolbar.
3. Move the Live Paint Bucket tool over an area you want to add a fill color to. Above the Live Paint Bucket tool are three squares that show a preview of the swatch colors. Use the left or right arrow to cycle through the colors to choose one to add to the face. Click on that face to fill it with the chosen color.
4. Move the Live Paint Bucket tool over a line segment you want to add a stoke color to and hold the Shift key to activate the Live Paint Brush. Use the left or right arrow to cycle through the colors and choose one to add to the edges. Click on that line segment to apply the color to the edge. The Live Paint Selection tool allows you to change the color or stroke weight (of an edge).

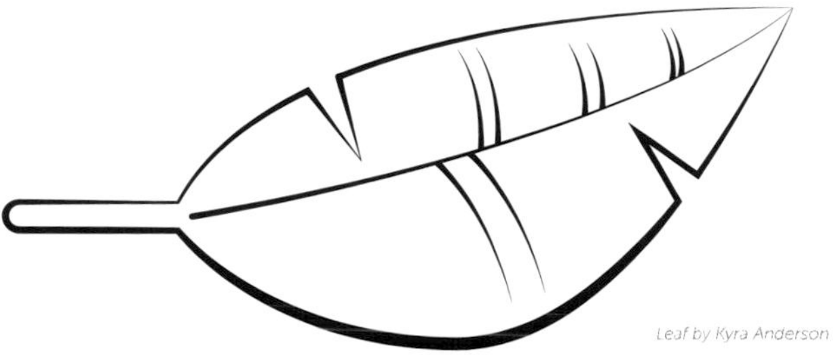

Leaf by Kyra Anderson

Live Paint Objects with Open Edges

In some instances the edges of a Live Paint Object will not be closed. The openings are called gaps. Gaps can be closed with an invisible edge or can be drawn in automatically using the Gap Options.

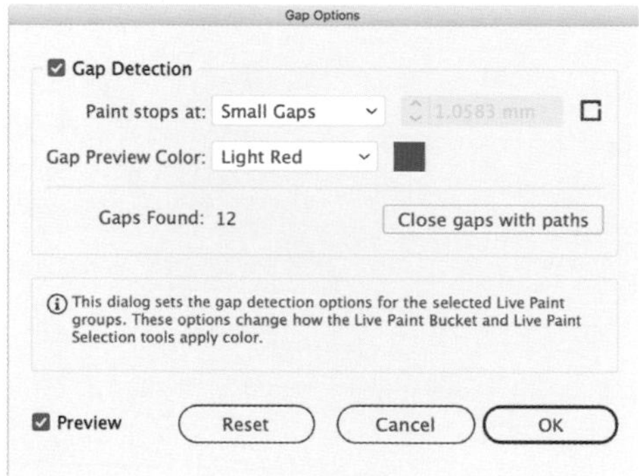

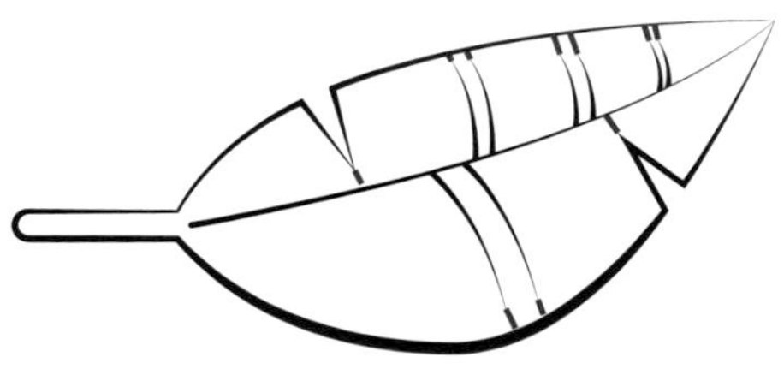

To highlight the gaps in a Live Paint group, choose Object > Live Paint > Gap Options.

Paint stops at specifies the size of the gap that paint can't flow through. You can choose Small, Medium, Large, or Custom defined gaps.

Close gaps with paths inserts actual paths into your Live Paint group to close gaps. When the gaps are closed with paths, it may appear gaps are still exist. The reason for this is that the newly generated paths have no stroke attributes applied to them so they appear to be nonexistent. Use the Live Paint Selection tool to select the path and change the attributes in the Appearance panel.

Expand a Live Shape

Live Paint Objects can be left as they are, but they are not editable using many of the Illustrator tools. To make the object editable, it needs to be Expanded. Select the Live Paint Object and choose Live Paint > Expand. This will break out the faces into filled shapes and the edges into stroked lines. All the shapes will be grouped together. Chose Object > Ungroup to ungroup the filled shapes and lines.

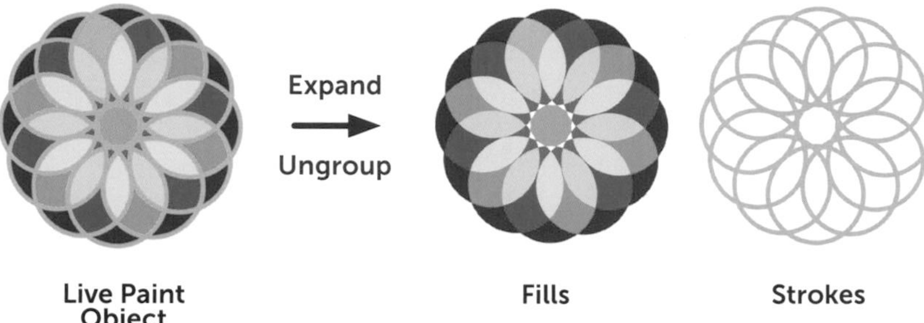

Image Trace

Creating vector textures with Image Trace is very common. Image Trace lets you convert pixel- (raster-) based images into vector artwork by placing an image in Illustrator and applying Image Trace options for desired effects. JPG, PNG, and PSD are acceptable formats.

Original Image

Black and White

Grayscale

Color

Image Trace has three basic conversion modes: Black and White, Grayscale, and Color. There are many options for each conversion mode, allowing you to achieve a wide range of artistic conversions.

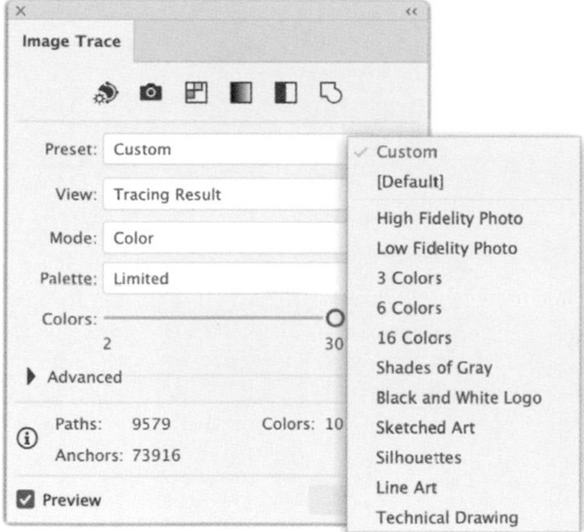

Tracing Presets are in the dropdown menu. These offer basic presets that can be further edited. This is a good place to start.

- **View** shows the view of the traced object. Your choices are to view the tracing result, source image, or outlines. You can click the eye icon to overlay the selected view over the source image. The recommended view is to show Tracing Results to actively see the results of the options and choices applied.
- **Mode** choices are Black and White, Grayscale, and Color Rendering.
- **Palette** displays the number of grays in Grayscale mode or the number of colors in Color mode. In Color mode the choices are Automatic, Limited, and Full Tone, or you can choose colors from the Document Library.

Black and White Mode

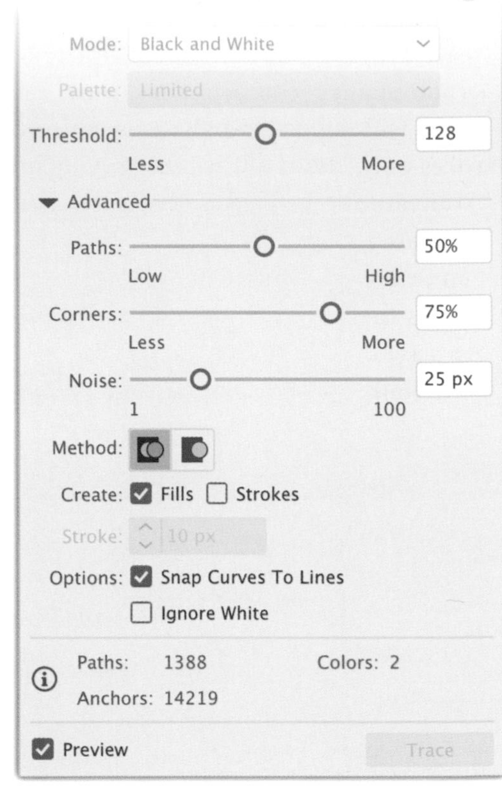

- **Threshold** acts like an exposure setting when used in Black and White mode. Less Threshold will create art that has more light areas, but it may remove portions of the artwork. More Threshold will darken the artwork and may cause areas to fill in.
- **Paths** controls how loose or tight the paths will be drawn around the artwork. Low creates less paths and more angles and is less exact. More creates more points on a path and tries to more accurately represent the artwork's curves and lines.
- **Corners** control the corners' points at a sharp bend. More will turn sharp points into a corner point. Less will reduce the number of corner points. This function does not smooth out the corners, it only controls which sharp areas will be actual corner points.
- **Noise** controls how much small detail will be included or removed from the image. The original artwork scan or image may have picked up texture or details from the paper it was drawn on—that texture is noise. A low number keeps the small details; a high number removes the small details.

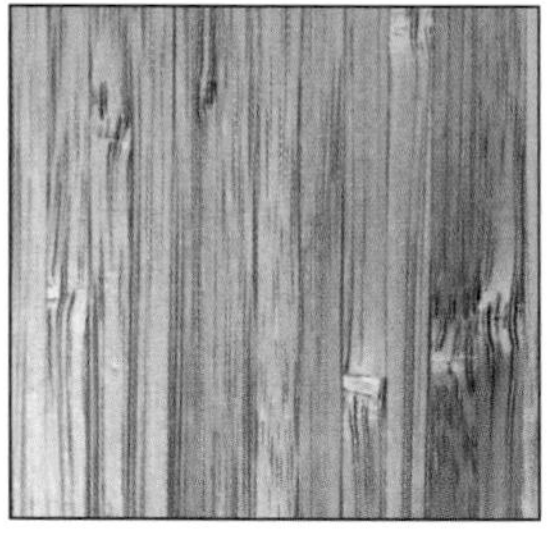
Original image

More Threshold

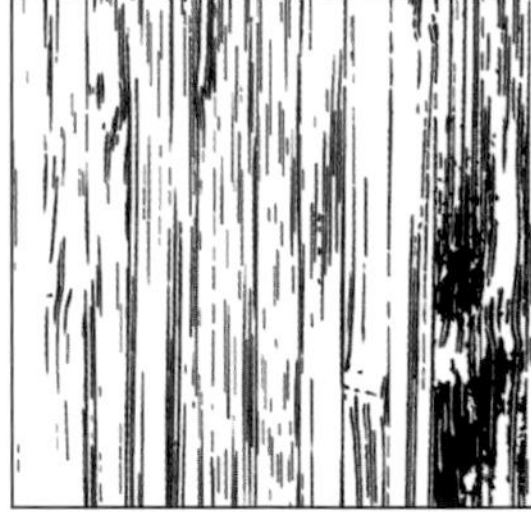
Less Threshold

Paths—Low
(less acurate)

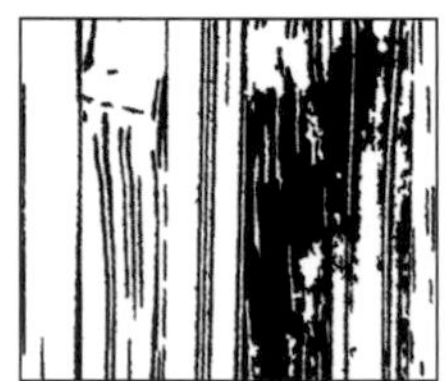
Paths—High
(more acurate)

Noise—Low
(more small details)

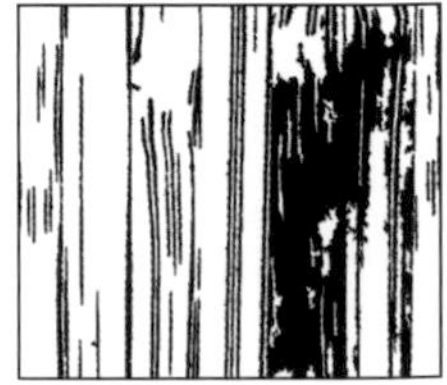
Noise—High
(less small details)

- **Method** removes any overlap rendering of paths that abut each other.
- **Create Fills/Strokes** sets how the artwork will be rendered in vector form. The default is fills and that is recommend because it will represent the artwork's shapes more accurately. Strokes will replace the fill areas with stroked lines and render the artwork as lines with strokes only. The result will be very different from the original artwork. The weight of the stroke can be specified in the Stroke section below the Create Fills/Stroke option.
- **Snap Curves To Lines** takes slightly curved lines and replaces them with straight lines. This works well if your artwork is slightly rotated—this will snap lines to horizontal or vertical in the tracing results.

Create: Fills

Create: Strokes

Ignore White
not checked

Ignore White
checked

- **Ignore White** keeps the white areas from rendering as white filled shapes. This is a common issue when tracing art as the white areas will render as white filled shapes. Check the Ignore White box and it will not fill the white areas with a white fill color.

> **VERY IMPORTANT TIP: EXPAND!**
> There is no place on the Live Trace panel that will allow you to render your artwork as the vector. To convert the Live Trace artwork to a vector, click the Expand button at the bottom of the Properties panel or choose Object > Live Trace > Expand. Once the artwork is expanded, the original image is removed from Illustrator. The results will be the vector tracing of that image.

Applying Color to a Black and White Vector

Once your Image Trace artwork has been expanded into vector artwork, you can fill the vector shapes with a color. Once you choose a Fill color for the artwork, the color is applied as a gray value. The solution for this in the Color panel (Window > Color). Click on the color dropdown menu and choose a color mode. This will convert the artwork mode to color and the color you chose will appear correctly.

Grayscale Mode

Grayscale converts an image into shades of gray. Settings range from 0 to 100. A low setting minimizes the number of grayscale steps for a high-contrast look, whereas a high setting makes the image look more realistic.

Low setting

High setting

Color Mode

Color mode converts images using preset color profiles, including, Automatic, Limited, Full Tone; or you can choose colors from the Document Library.

- **Automatic** renders the image using the colors that are automatically picked to match the color range in the image. The number of colors can be controlled by using the Less or More Sliders in the Image Trace panel.
- **Limited** sets the range of color from 2 to a maximum of 30 colors based on colors in the image.
- **Full Tone** converts an image into its full range of colors. This also creates a much more complicated rendering because the number of paths created may be much higher than an image converted using the Automatic mode.
- **Document Library** converts an image into a range of colors that is specified in a Color Library. The colors in the Library are mapped to the tonal ranges of the image.

Automatic

Limited
(10 colors)

Document Library

4 Editing and Transformation

Editing and Transformation

Creating simple shapes and objects in Illustrator is straightforward. Trying to create more complex shapes can be a bit more work. Understanding how to get the desired results is what this chapter clarifies. I take the approach that everything can be built with basic shapes. Instead of tracing an object to create it, you can add, subtract, unite, knockout, and divide objects to achieve your desired creations.

Illustrator has Live Objects that can be edited using the Transform panel. The Live Objects attributes can be edited on many shapes to round edges, change the rotation angle, change the number of sides, and easily edit the size.

Corner widgets can be edited on any shape that has corners, such as rectangles, squares, polygons, and stars. Corner widgets can be edited directly on the shapes using the Direct Selection tool or in the Transform panel.

Reshaping of objects can be done with the Direct Selection tool. By directly selecting a point or path, shapes can be edited beyond what can be done in the Transform panel. Turning a rectangle into a parallelogram, creating longer points on a star, or making a triangle taller can all be done with the Direct Selection tool.

Creating curved lines can be done using the Curvature tool or the Pen tool. The Curvature tool allows for either creation of a shape or editing of an existing shape. A straight line can be easily edited or curved with the Curvature tool. Corners on a shape can be converted to smooth curves and smooth curves can be converted to corner points as well. The Pen tool offers more features but requires a bit more practice to understand. It, too, can convert straight lines into curved lines and convert corner points (and back).

Further editing of shapes can be done with the Free Transform tool. This tool allow for editing shapes in perspective as well as free-form editing without the need for the Direct Selection tool.

The Pencil tool is used for quick rendering of ideas or sketching. It allows for the creation of precise or smooth lines.

For more precise rotation, scaling, and transforming, there are the Scale tool, Rotate tool, Shear tool, and Reflect tool. These tools allow for calculated actions, spacing, and copying of the resulting outcomes.

For shape creation beyond the editing a shape, the Pathfinder panel allows for the addition, subtraction, knocking out, and dividing of a shape. The Shape Builder tool has many features of the Pathfinder panel with the added ability to remove lines from overlapping shapes, as well as the quick addition or deletion of overlapping shapes.

All these tools and features make creating, editing, and manipulating shapes possible, so you can achieve anything you can imagine.

Corner Widgets

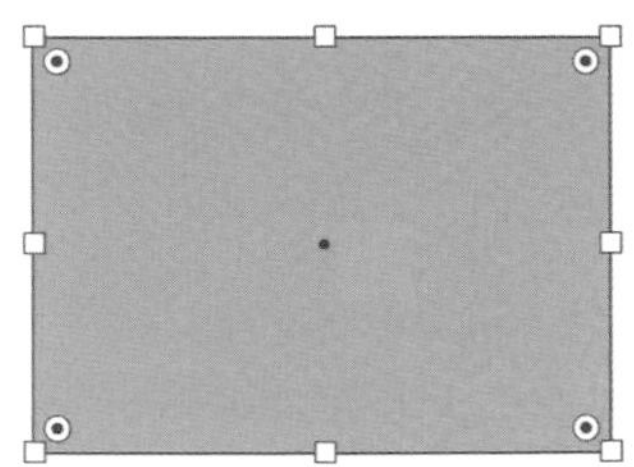

Shapes with corners have widgets that you can use to alter the shape of the corners, including Round, Inverted Round, and Chamfer. Corner widgets appear on all corners of a rectangle, and at the top point of a polygon. Star shapes have corner widgets that are activated by selecting the shape with the Direct Selection tool.

Editing Corners

Draw a rectangle and select the shape with the Selection tool. The corner widgets will appear as targets in each inside corner. If the targets appear small, there is a setting in Illustrator's preferences for setting the size of anchor points and corner widgets. Choose Preferences > Selection & Anchor Display... to bring up the options. Set the slider to the desired size and click OK. If the corner widgets are not showing when you select a rectangle, choose View > Show Corner Widget.

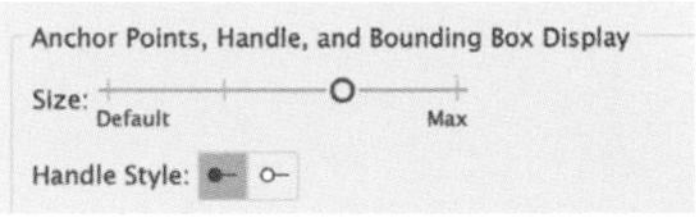

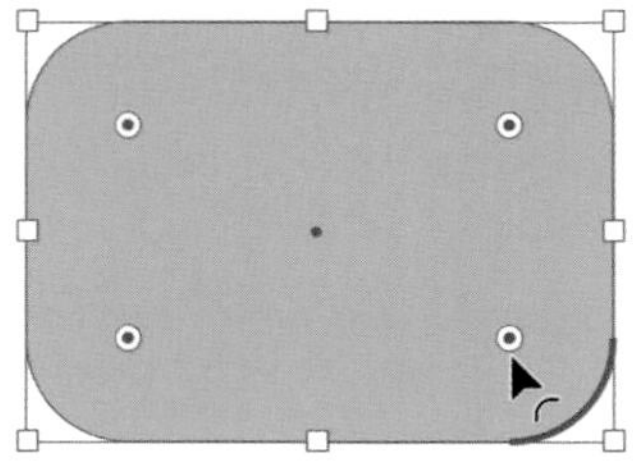

Select the corner widget

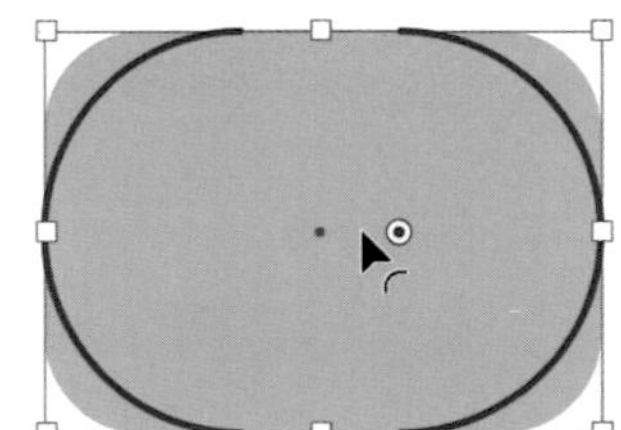

Corner widgets are pulled in to maximum corner "widgetness"

Click on a target and pull it into the center of the shape. This will result in a rounded corner, which is the default setting. Once you pull the corner into the center the maximum amount, a red line will appear alerting that you have achieved maximum "widgetness"—the point where the corner cannot be pulled in anymore.

When you've reached this maximum capacity, you will not be able to scale the shape any smaller. To keep the corners in proportion when you alter the size of your shape, click on the Transform Corners box at the bottom of the Transform panel.

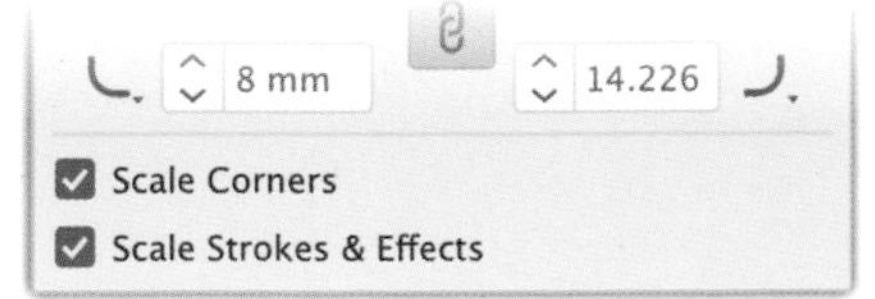

Scaling a shape larger with Scale Corners checked will keep the corners scaled in proportion to the object. Scaling a shape larger with Scale Corners unchecked will keep the corners a set size while the shape gets larger. This may change the look of the shape because the corners will remain a fixed size.

To change from the Rounded corner style to the Inverted Round or the Chamfer, Option+click on the target. Each Option+click will cycle through to the next style. Option+-click will also open the Transform panel where the corner options are available. To edit a single corner, double-click on the target to change it into a donut. The corner widget will be isolated from the rest of the corners and can now be pulled independently. Once you release the donut, it will turn back into a target and no longer be independent unless you double-click on it again. Corners can be selected with the Direct Selection tool to one or more corners.

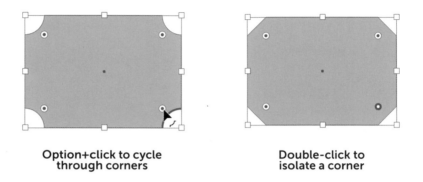

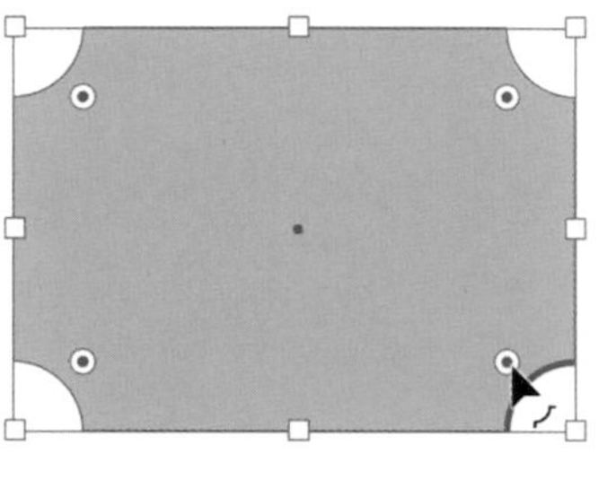

Option+click to cycle through corners

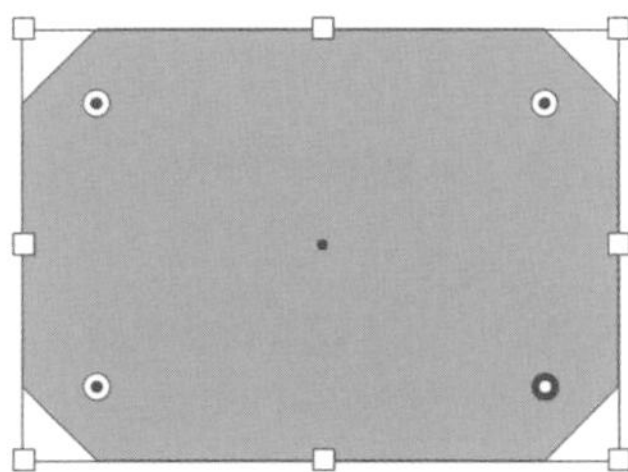

Double-click to isolate a corner

Corner options are in the Rectangle Properties section of the Transform panel. The corner options are available in the dropdown menu next to each corner. The radius of the corners can be edited independently if the link icon is unlinked. If the link is selected, the corner radius will be in sync but the corner styles will still be independent of each other. To return the corners to their default (square) corner, set the value of the corner to zero.

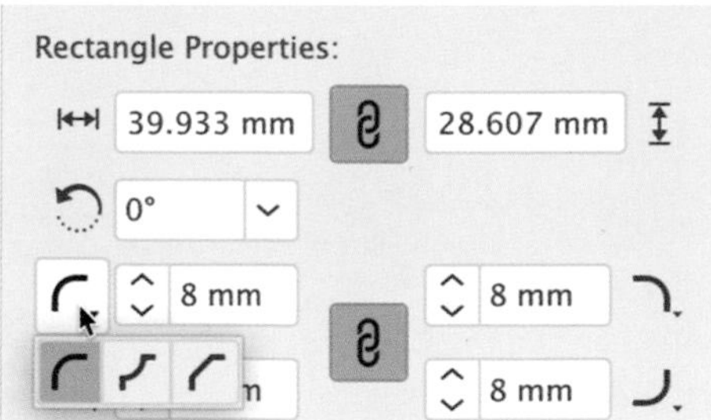

The Direct Selection tool can be used to select and edit individual corners. You can click and drag the Direct Selection tool over the corners to activate them or click on one corner, then Shift + click on more corners to select them.

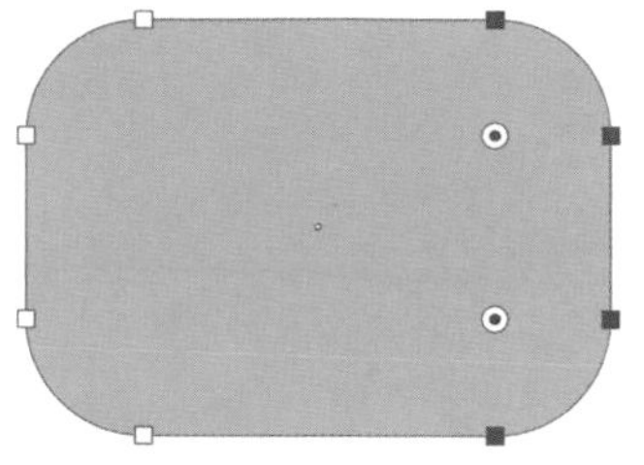

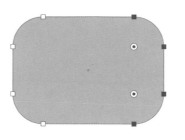

Direct Select corners

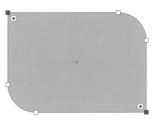

Direct select one corner, Shift + select the other corner

When selecting a Polygon with the Selection tool, only one corner widget will appear.

By clicking on the target and pulling it in toward the center of the shape, all the corners will round at the same radius. If you select the polygon with the Direct Selection tool, all the corner widgets will appear. You can click and drag the Direct Selection tool over the corners to activate them or click on one corner, then Shift + click on more corners to select them.

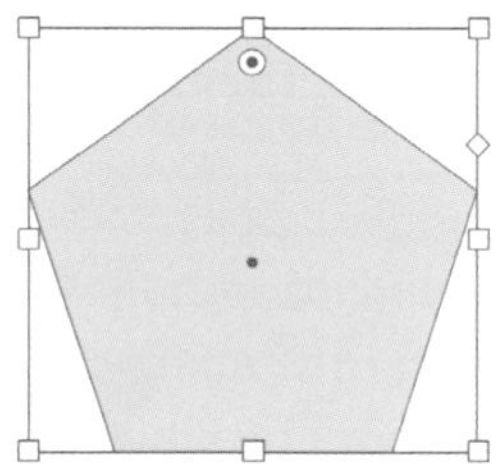

Selection tool

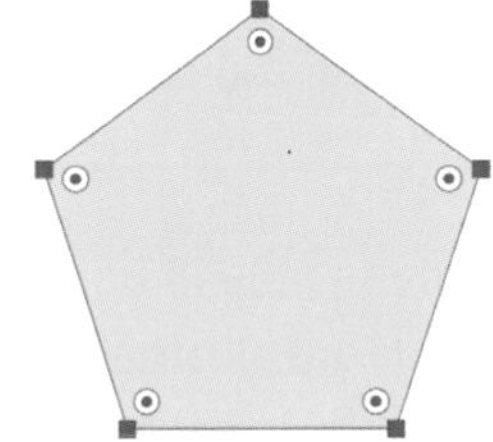

Direct Selection tool

Polygon Live Shape Properties include editing the radius, side length, angle of rotation, and corner styles and radius of the shape. If the polygon is stretched and the sides are not equal, the Properties will create a button to make all sides equal.

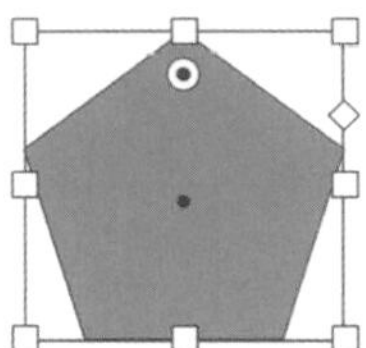

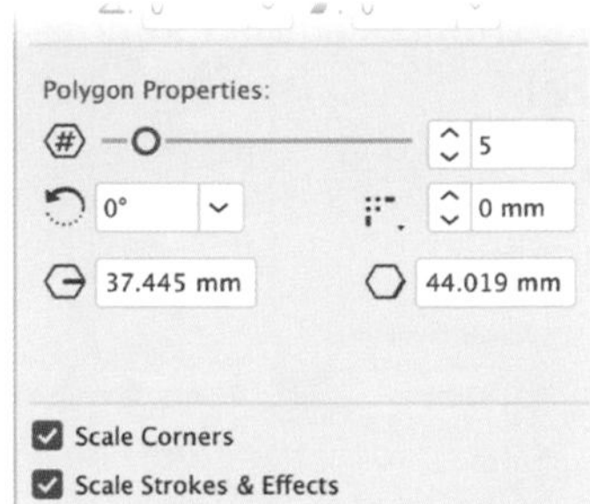

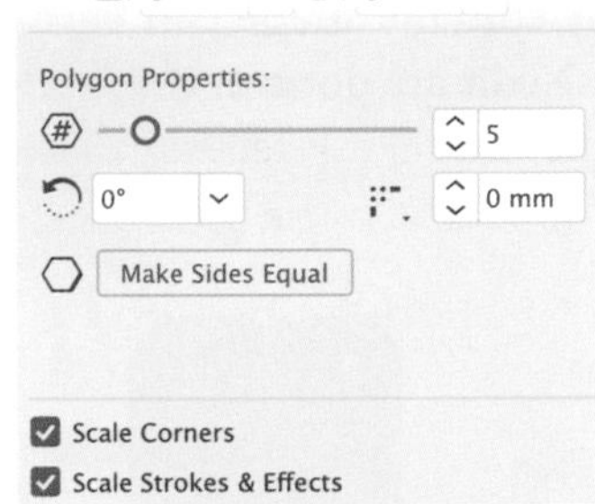

Shape Editing

Direct Selection Tool

The Direct Selection tool lets you directly select individual anchor points or paths by clicking on them. When selecting an object or shape with the Direct Selection tool no bounding box will appear, just the points and line segments.

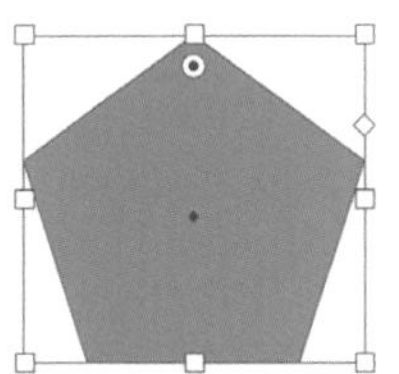

Selected with the Selection tool (showing bounding box)

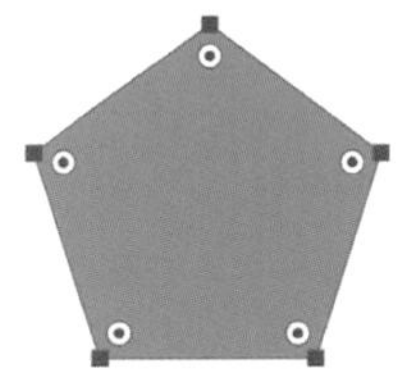

Selected with the Direct Selection tool (all points selected)

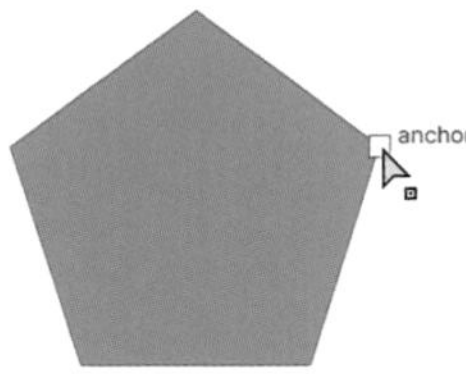

Directly selected an anchor point

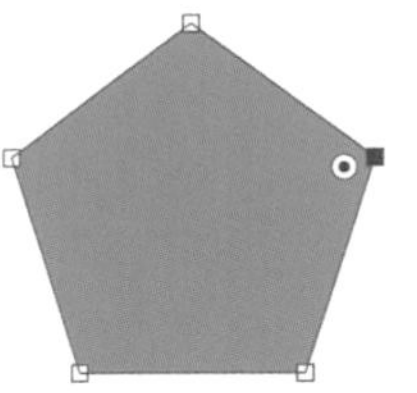

One anchor point selected

Selected and Non-Selected Points

To fully understand the Direct Selection tool, we need to look at the anchor points on a shape. When the anchor points on a shape have a solid blue outline, they are selected. When the anchor points on the shape have a blue outline, they are active but not selected. If a bounding box is active around the shape, you are using the Selection tool, which does not allow direct editing of a line or anchor point. A very common issue that arises while using the Direct Selection tool is selecting a shape that is already selected with the Selection tool. All the points will be active and the shape will act like you are using the Selection tool. Click off the shape and then click back on it with the Direct Selection tool and the points can be activated individually.

Smart Guides help in the direct selection of an anchor point or line segment (path). When you hover over a corner point or line segment, the Smart Guides show what you are selecting. You can turn your Smart Guides on under View > Smart Guides. Smart guides also help with moving anchor points or lines along a horizontal or vertical path because they help keep the movement constrained along an axis.

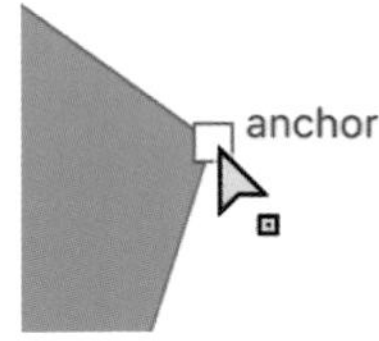

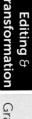

Moving Selected Points or Paths

Use the Direct Selection tool to select and move a path. The same can be done with an anchor point. If you direct select an anchor point or path, you can move the point with the arrow keys as well, you do not need to move the point with the mouse.

This helps when you just want to move the anchor point a small amount. Holding Shift while using the arrow keys to move the path or anchor point will move it 10x the standard amount.

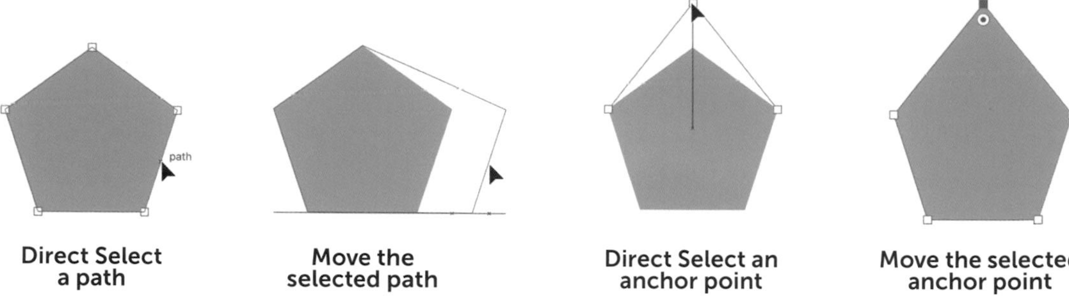

Direct Select a path | Move the selected path | Direct Select an anchor point | Move the selected anchor point

Corner Points and Smooth Points

There are two different points types in a shape; a Corner Point and a Smooth Point. Corner points make a square corner. Smooth points appear as a square anchor points with pull handles. The pull handles of a smooth point determine the direction of the curve by the direction and amount of curve in the line, depending on the length of the handle.

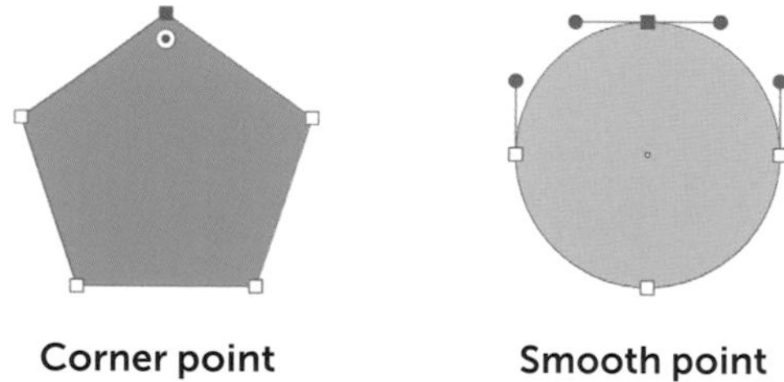

Corner point | Smooth point

A corner point can be converted into a smooth point and a smooth point can be converted into a corner point. In the Properties panel, there is a convert anchor point section to change from one point type to another.

Convert:

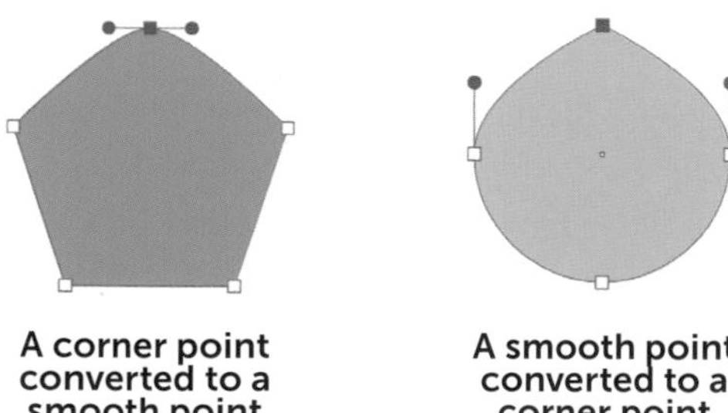

A corner point converted to a smooth point | A smooth point converted to a corner point

Transform Tool

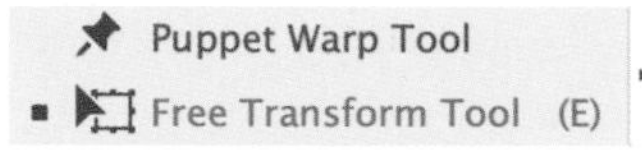

The Free Transform tool allows different types of transform-ations of shapes. The four modes are Constrain, Free Transform, Perspective Distort, and Free Distort. This tool has been in Illustrator for many version and there are other methods of performing the tasks that the Free Transform tool performs, with the exception of the Perspective Distort. The Free Transform tool is nested with the Puppet Warp tool. You have to select the shape first, then choose the Free Transform tool to activate it. A floating panel will open showing the four modes of transformation, it may pop up under the toolbar so you may have to hunt for it.

- **Constrain** keeps the shape in proportion when scaling: the same action can be performed using the Selection tool and holding Shift while resizing the shape.
- **Free Transform** allows for free scaling of the shape: the same action can be performed using the Selection tool to resize the shape without holding Shift.

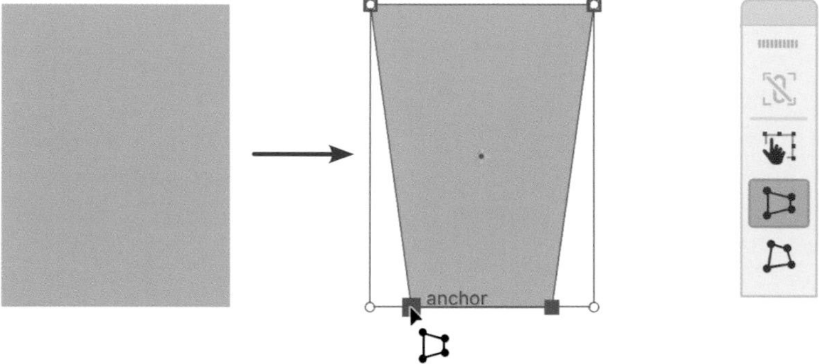

- **Perspective Distort** allows transforming of the shape in perspective. Click on a corner and pull in or out to control the perspective of the shape symmetrically. This function is unique to the Free Transform tool.
- **Free Distort** allows transformation of the shape by selecting the corners and moving them. This action can be done using the Direct Selection tool to select a corner point or path and move it around.

Closed Shapes and Open Shapes

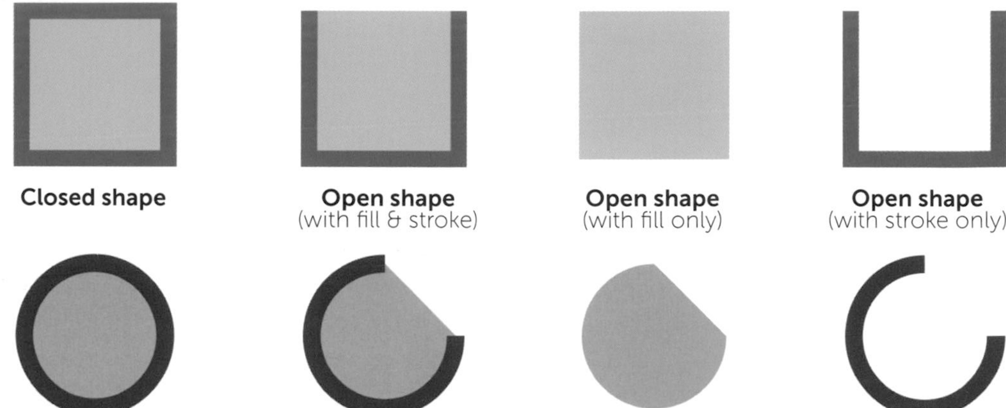

Closed shapes are fully enclosed by the point and line segments that connect to form one continuous path. This path defines the shape. The shape drawing tools create a closed path and the Line Segment tool and Arc tool create an open shape or path. An open shape has end points that are not connected to each other. Shapes that are closed or open may look closed if they are filled with a color and stroke attributes are not applied. An open shape can be created by drawing a closed shape and using the Direct Selection tool to select a line segment on the shape and delete the segment.

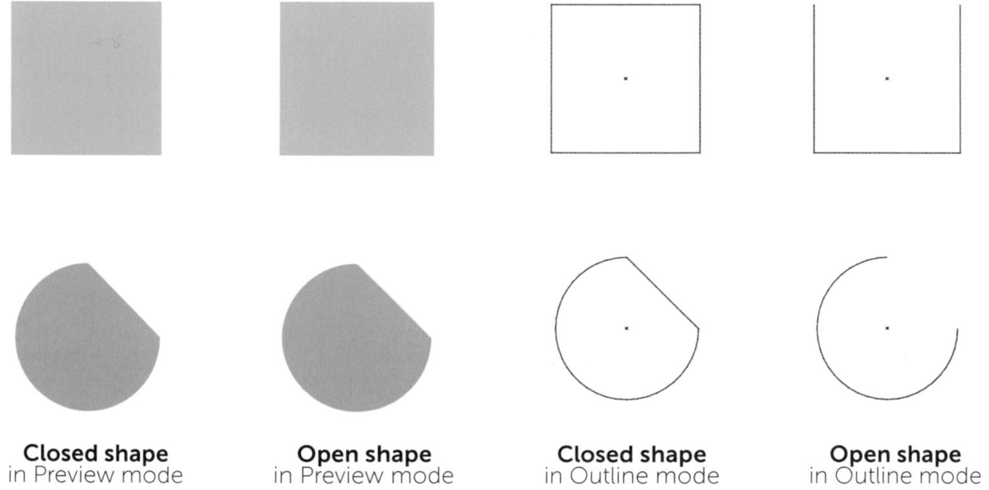

When a stroke color or weight is applied to an open shape, the stroke will appear only along the edges of the path. When a fill is applied to an open shape with a stroke attribute, the fill will stop at the point where the two open points of the path stop. This may appear as an abrupt end to the fill in some shapes. Depending on the end points of the open path, it may not be clear if a shape is closed or open.

Outline Mode

To see the outlines mode, choose View > Outline. This view is helpful in determining if a shape is a closed or open. It is easier to edit a shape when the fill and/or stroke is not visible. To return the view to the normal or Preview mode, choose View > Preview. When in outline mode, the shapes can only be selected by clicking on the path, the center point, or clicking and dragging over the shape to select it. Clicking inside the shape will not select it.

Joining Lines

When open shapes or lines require closure, you can join the ends to close the path, turning the line into a closed shape. You can also join more than one open shape or line together. Select an open path and choose Object > Path > Join (the shortcut is Command + J).

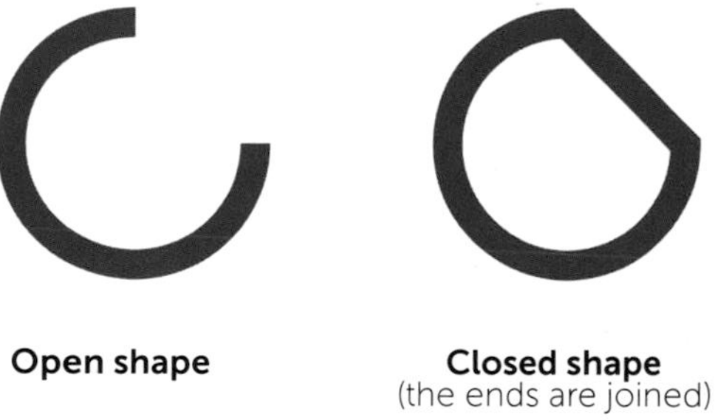

Open shape **Closed shape**
(the ends are joined)

To join more than one open shape to another open shape, select the shapes and choose Join. This will join one set of open ends. Apply Join again to join all the remaining open ends. This second Join command joins the rest of the open paths to their nearest neighbor points.

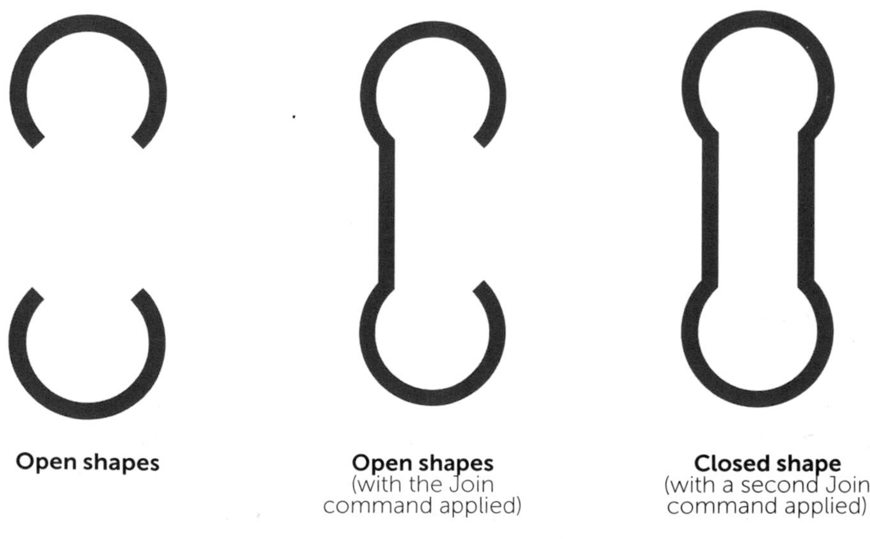

Open shapes **Open shapes**
(with the Join command applied) **Closed shape**
(with a second Join command applied)

To join open ends in a specific order, use the Direct Selection tool to select the two end points you want to join, then choose Join. To close the shape, choose the remaining two end points with the Direct Selection tool and choose Join.

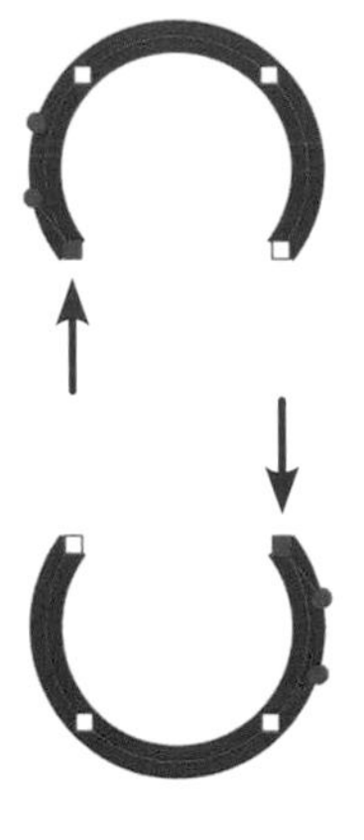

Direct Select two end points

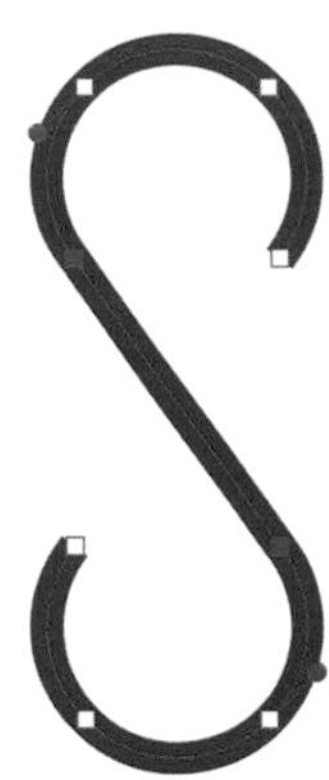

Join the end points

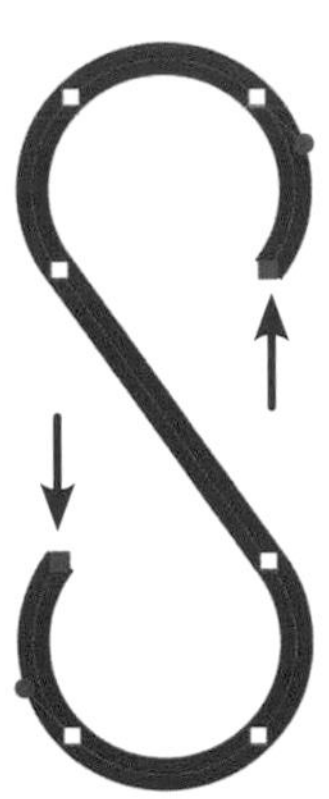

Direct Select two end points

Join the end points

Curvature Tool

The Curvature tool is an improved, simplified Pen tool that makes drawing easier and more intuitive. The Curvature tool has many of the same features as the Pen tool, such as edit, add, or remove smooth or corner points. You don't have to switch between different tools to get to these modes, which speeds up the drawing process.

By default the Curvature tool draws curves. It can also be used to curve a line or line segment on a shape or open path.

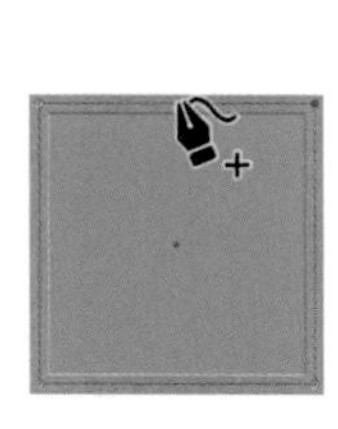

Click on a path

Pull the path

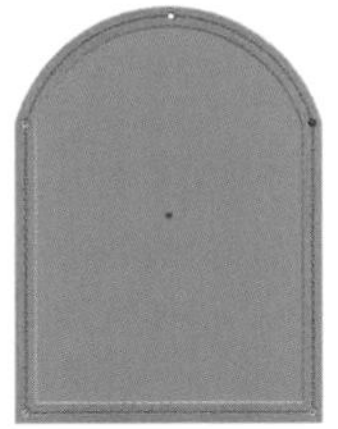

The Curvature tool creates a curved line

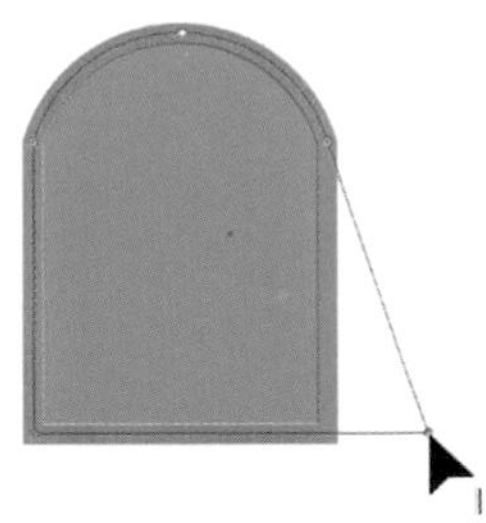

Click to move an anchor point

Click on a path with the Curvature tool, and it will create a point that can be pulled any direction to create a curve. Click on an anchor point to move it just like you would using the Direct Selection tool.

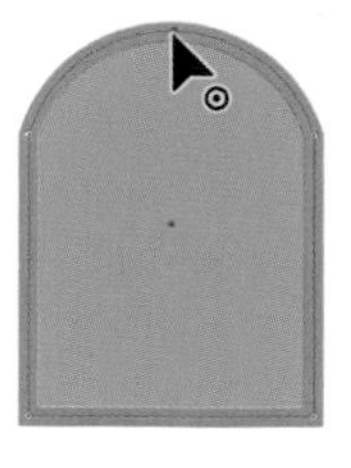

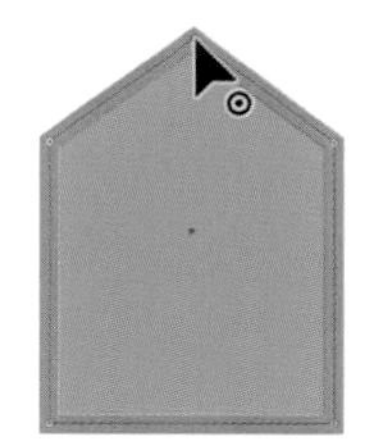

This converts a smooth point into a corner point

Click to move the anchor point

Click on a point and delete

Convert Corner to Smooth Points

Any point on a path or shape can be converted using the Curvature tool. Click on a point to select it, then Option+click on it to convert it. Option+clicking again will convert it back to its previous point. Delete a point by clicking on it then pressing the Delete key. The point will go away and the curve or line connected to the point will be deleted, and the shape will close itself.

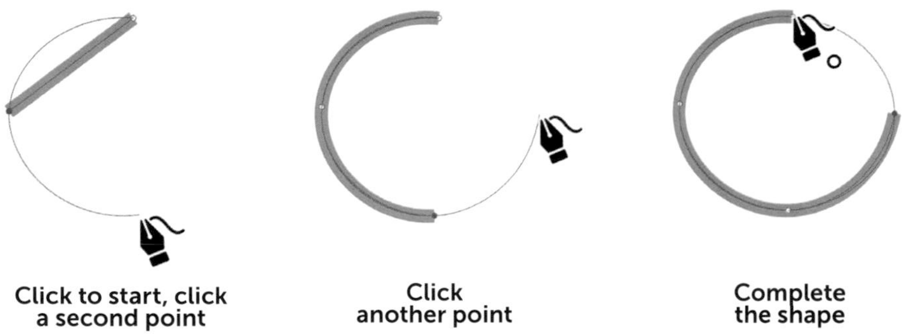

Click to start, click a second point

Click another point

Complete the shape

Draw Curved Lines

The default mode for the Curvature tool is a curve. This may not seem apparent when you first click and start drawing. Click to create a starting point, then click again to create another point. This creates a straight line but do not worry, the third point you create will curve the line around the second point. All the subsequent points created will curve the line based on the last two points created. Close the path by clicking on the original point. The cursor will have a small circle in its lower right indicating that clicking on the point will close the shape.

Anchor Points, Handle, and Bounding Box Display
Size: Default Max
Handle Style:
☑ Highlight anchors on mouse over
☐ Show handles when multiple anchors are selected
☐ Hide Corner Widget for angles greater than:
Enable Rubber Band for: ☑ Pen Tool ☑ Curvature Tool

Rubber Band preview allows you to see the next potential curve before you click. This is a setting that can be set in preferences under Selection & Anchor Display.

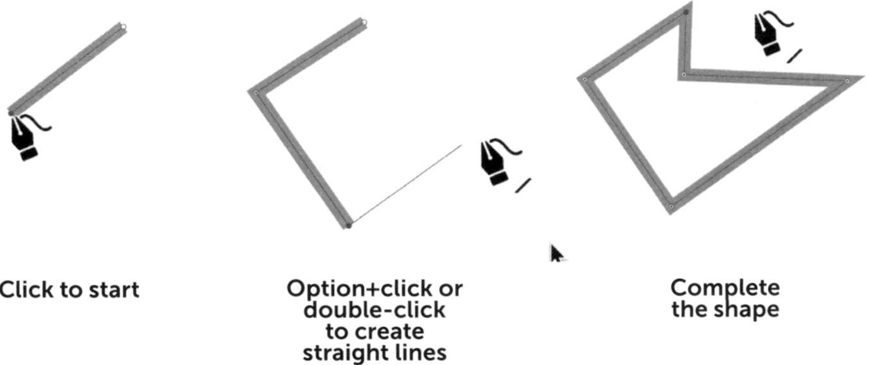
Click to start
Option+click or double-click to create straight lines
Complete the shape

Draw Straight Lines

When you create the first two points with the Curvature tool, it creates a straight line. To keep drawing straight lines, hold Option and click to create points. You can also double-click on a point and draw a straight line from it. All the lines between the points will be straight lines. Close the shape by clicking on the starting point.

Pen Tool

Love It or Hate It...

I love the Pen tool; its one of my most favorite tools of all time. The Pen tool has its followers and its opponents, but I believe you should know how it works even if you never use it. Once you know how the Pen tool works, you will have a better understanding of how it works in other Adobe applications. It also has certain features that the Curvature tool does not have, such as pull-handle editing and the option to change the handle's direction.

Pen Tool Shortcuts

The Pen tool has four tools in one. Pen tool, Add Anchor Point, Delete Anchor Point, and Convert Point. Using the Command or Option key allows you to select between the tools instead of choosing them from the toolbar each time you want to use it.

Select/Move a Path

To select a path, choose the Selection tool. You can copy, move, or transform a path. All the points on the path will be solid and will all move at the same time.

Move Points on a Path

To move points on a path, choose the Direct Selection tool and select the path. The point on the path will be hollow. Only the points selected by the Direct Selection tool will move.

Start Drawing with the Pen Tool

When starting with the Pen tool, it will have an asterisk to show it is ready to create lines.

Drawing Straight Lines

Using the Pen tool, click a point then hold Shift and click another point in the direction you want the lines to be drawn. It will be constrained to a straight line between the two points.

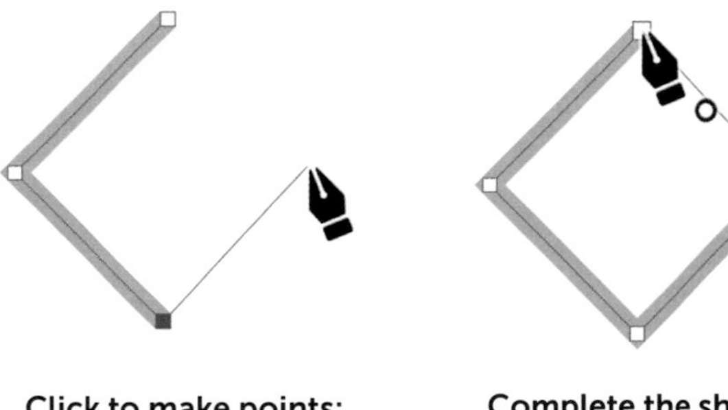

Click to make points; the lines will be created between the points

Complete the shape at the starting point of your creation

To draw a line at a 45-degree angle, hold the Shift key down when drawing. Click to create points, and a straight connecting line will appear between them. To close the path, click on the original point and the Pen tool will show a circle, indicating the closing of the path.

Drawing a Curved Line

Click and pull to create points with (bezier) handles that will allow you to make curved lines. The more you pull, the longer the handles will be, and the more curved the resulting line will be. The direction you pull the point determines the direction the curve will follow.

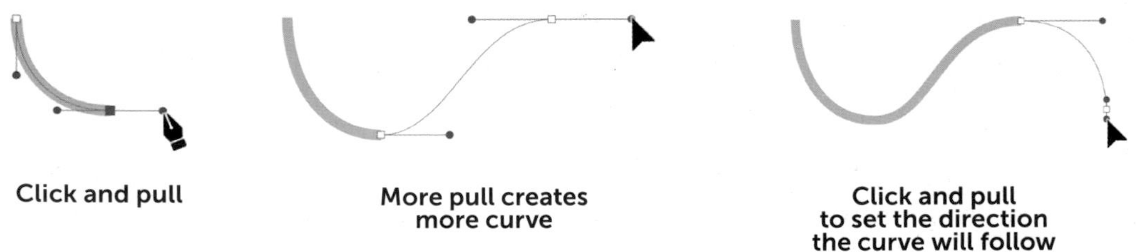

Click and pull

More pull creates more curve

Click and pull to set the direction the curve will follow

Adding and Deleting points

With the Pen tool, hover over an existing line segment and the cursor will automatically change to the Add Anchor Point tool. To delete a point, hover over an existing active point and the cursor will automatically change to the Delete Anchor Point tool. Click to remove the point and the shape will close.

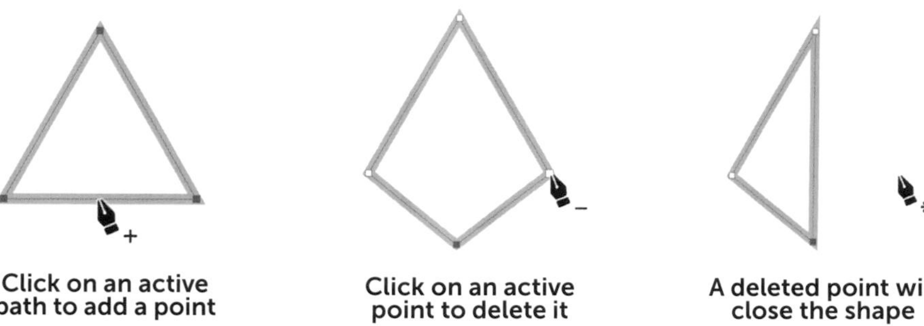

Click on an active path to add a point

Click on an active point to delete it

A deleted point will close the shape

Move a Point on a Line

To select and move a point, hold down the Command key and the Pen tool will change to the Direct Selection tool.

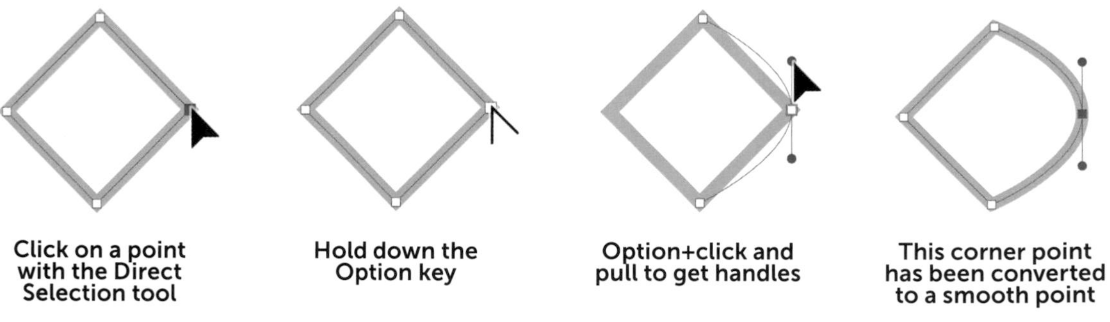

Click on a point with the Direct Selection tool

Hold down the Option key

Option+click and pull to get handles

This corner point has been converted to a smooth point

Create a Handle From a Point

To make handles from a point that has none, hold down Option; the Pen tool will change into the Convert Anchor Point tool. Click on the point and pull. This converts a corner point to a smooth point, revealing the handles to edit the curve.

To remove handles from a point, hold down the Option key; the Pen tool will change into the Convert Anchor Point tool. Option+click on the point and the handles will go away, creating a corner point.

Move Handles on a Point/Change Direction

To select and move a handle independent of the other handle at the point, use the Direct Selection tool to select the point to reveal its handles. Press Option to turn the Pen tool into the Convert Anchor Point tool. Click on the handle (not the anchor point) and pull or rotate it to change its direction, or pull the handle back into the point to remove it.

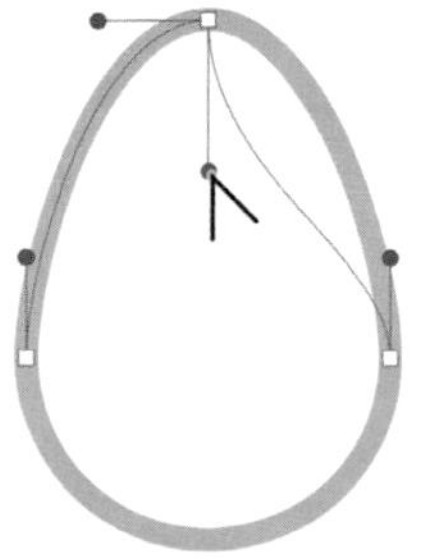

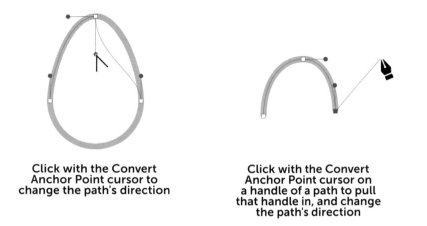

Click with the Convert Anchor Point cursor to change the path's direction

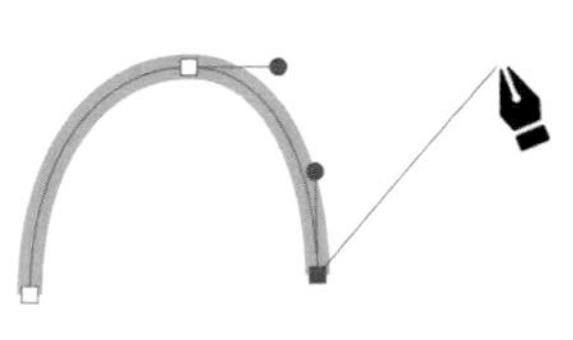

Click with the Convert Anchor Point cursor on a handle of a path to pull that handle in, and change the path's direction

Straight Line Followed by a Curve

While drawing with the Pen tool, to follow a straight line with a curved line, click the last anchor point again and drag the Pen tool the direction you want to create the curve.

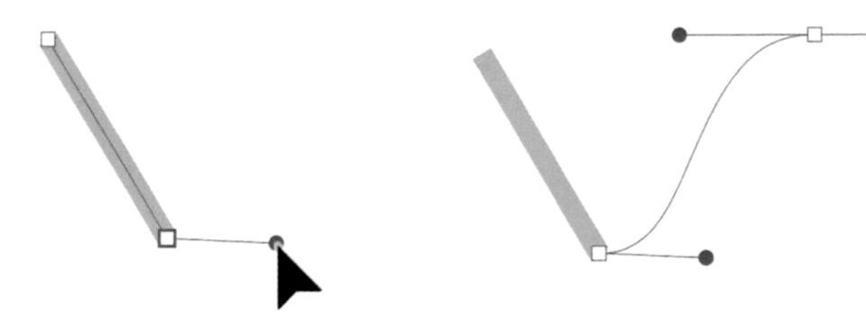

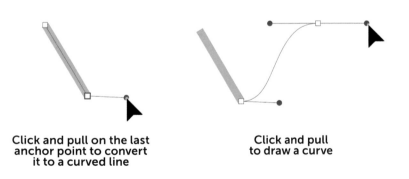

Click and pull on the last anchor point to convert it to a curved line

Click and pull to draw a curve

Curve Followed by a Straight Line

After drawing a curve, position the Pen tool over the last anchor point drawn, and the Pen tool with the convert-point icon will apear. Click the anchor point to convert the smooth point to a corner point. Reposition the Pen tool where you want the straight segment to end, and click to complete it.

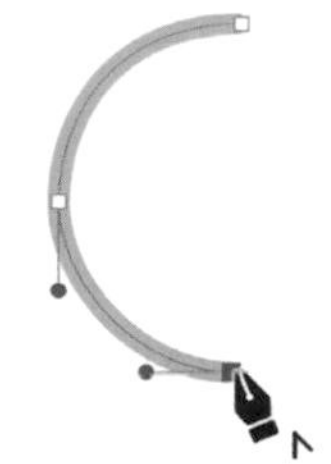

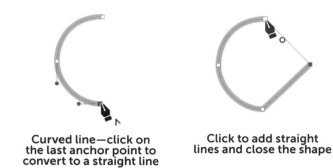

Curved line—click on the last anchor point to convert to a straight line

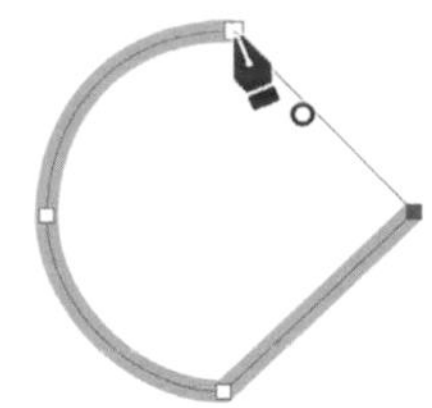

Click to add straight lines and close the shape

Path Direction

When creating or drawing a shape, the path has a direction. When drawing with the Pen tool, pay attention to the direction you draw—clockwise or counterclockwise—because this will matter when you covert a corner point with the Pen tool or edit handles on a smooth point. When using the Convert Anchor Point tool, click and pull in the direction the shape was drawn. If you pull in the opposite direction, you will get handles that cross over each other. To correct this, simply pull the handles in the opposite direction so it follows the direction of the path. If you find you frequently pull points in the opposite direction, you can change the direction of the path with Object > Path > Reverse Path Direction.

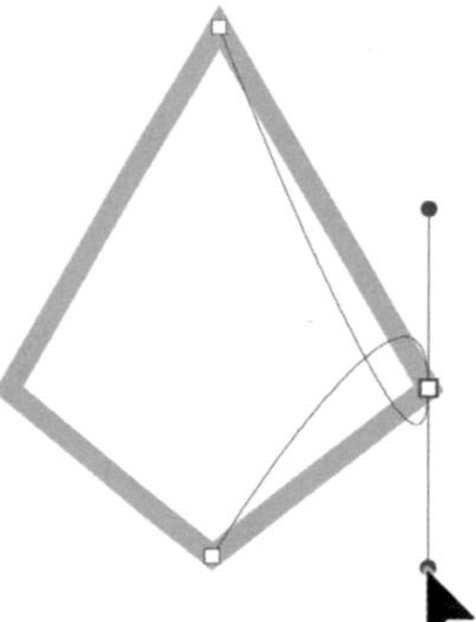

These handles were pulled against the direction of the path

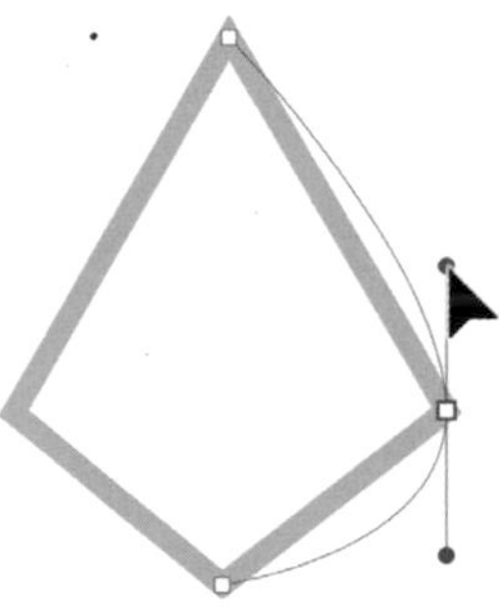

These handles were pulled with the direction of the path

Join Two End Points or Two Open Paths

Select the paths to make them active. Click on one end of the path with the Pen tool. The Pen tool has the Convert Anchor Point (triangle) symbol next to it. Click on another end point and the Pen tool will show the merge path icon (box with ends), and it will connect your path.

Pencil Tool

The Pencil tool lets you draw freeform open and closed paths as if you were drawing or sketching. This simplicity makes the Pencil tool easy to use and edit because there is no pulling of handles to create curves. If you like to sketch with a pencil on paper, this tool is similar to that quick-sketch look and action.

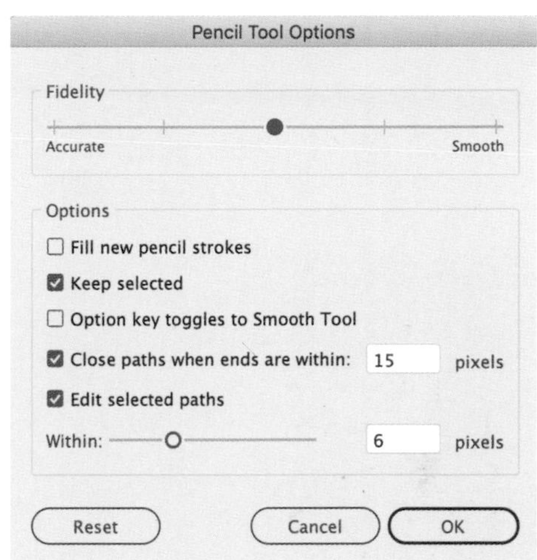

The Pencil tool has options that can be set to fit the needs of your drawing style. To access the options, double-click on the Pencil tool in the toolbar.

- **Fidelity** sets how accurate or smooth the path made with the pencil will be. The accurate end of the slider will produce more points and be more true to the inputs from the mouse or stylus. The smooth end of the slider will remove any small bumps or movements and produce a more refined, but less detailed, version of the shape you are drawing.
- **Fill new pencil strokes** fills the path, either open or closed, with the selected fill color.
- **Keep selected** keeps the path selected once you are done drawing the line, allowing you to add to the end of an open line by clicking on a point and continuing to draw, adding to the existing line.
- **Option key toggles to Smooth Tool** switches to the Smooth tool while drawing when Option is held down.
- **Close paths when ends are within:** allows you to set the distance between the end point and the beginning point of the line so that it will automatically close when the points are within that distance value.
- **Edit selected paths** allows you to click on any active path and begin drawing from that point, removing all the later points.

Drawing Freeform Lines

Select the Pencil tool, click, and draw to create a freeform line. To close a shape, while drawing, return to the original start point. The Pencil tool will have a circle next to it, indicating the closure of the path.

Drawing Line Segments

The Pencil tool can create open line segments that are straight or freeform. Hold the Shift key while drawing with the Pencil tool to constrain the line into straight segments vertically, horizontally, or at a 45-degree angle. To draw a straight line that is not constrained, hold the Option key.

To draw multiple line segments that are straight and connected together, draw with the Pencil tool while holding Option. Draw the line, then Option+click on the end of the line segment to draw another line with the Option key held down.

Transform Shapes and Lines

Movement

To move an object a determined distance, select the object with the Selection tool and hit Return to open the Move dialog. The position of the selected object can be moved horizontally or vertically by entering in the values in the appropriate boxes. You can also move the object at an angle, and move the object a selected distance. To copy the object, click the Copy button. To repeat the last movement, use Command + D to duplicate the last move or copy.

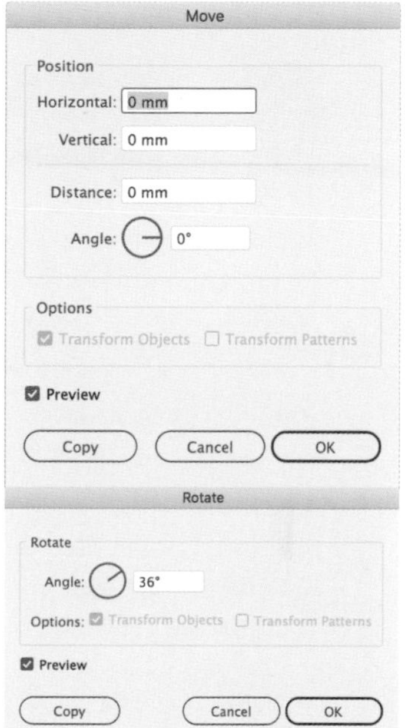

Rotation

To rotate an object using the Rotate tool, select the object and choose the Rotate tool. A set of crosshairs will appear at the center of the object. With the cursor, select and rotate the object around the center point. To open the Rotate dialog, double-click on the Rotate tool in the toolbar. Set the value of rotation and click OK or copy the object by clicking the copy button.

To set a point of rotation different from the default (the center of the object), select the object with the Selection tool and choose the Rotate tool. Click where you want the center point to be and the crosshairs will appear at that new rotate point. The point of rotation does not need to be on or in the object. The object will rotate around that newly set point.

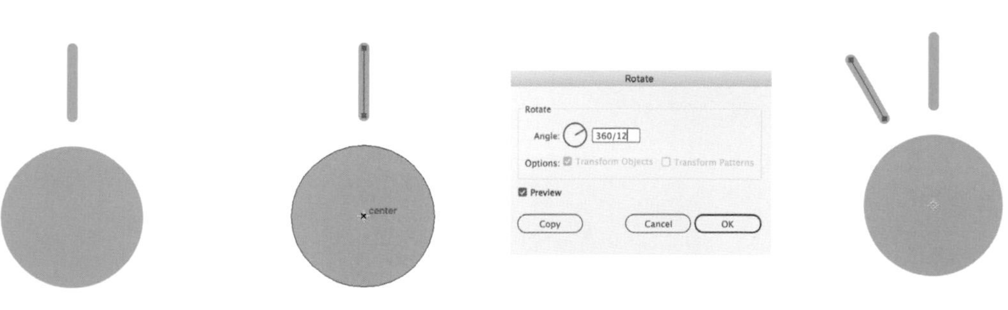

To create a sun, draw a circle and a line, and select the line

Select the Rotate tool and Option+click on the center of the circle

Enter in the value of rotation and click Copy

Use Command + D to duplicate the rotate command

To rotate an object and copy it, select the object and then choose the Rotate tool. Option+-click will set the point of rotation *and* open the dialog box to enter in the rotation value. To set the number of items to equally rotate around the center point, you can do math inside the angle field. In this example, I want to create 12 rays around the sun that are equally spaced. There are 360 degrees in a circle, so enter in 360/12 and click Copy. This will create a duplicate ray at 30 degrees (360°÷12 rays = 30°). Use Command + D to continue to duplicate the last step. Repeat Command + D until you have all the rays of sun desired.

Scaling

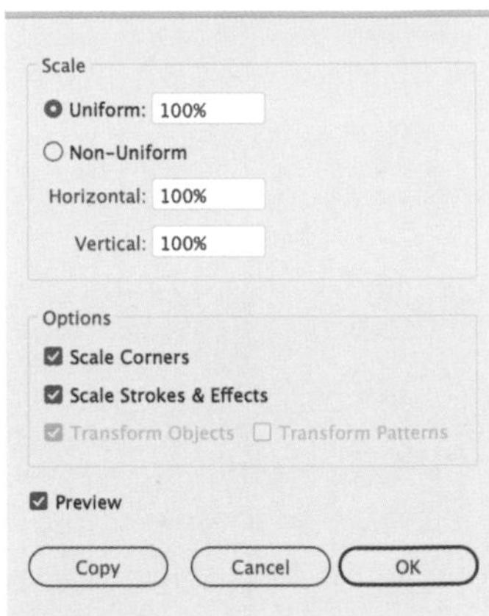

To scale an object using the Scale tool, select the object and choose the Scale tool. A set of crosshairs will appear at the center of the object. With the cursor, select and pull in or out from the center point to scale the object freely. Hold Shift while scaling to constrain the shape proportionally.

To open the Scale dialog box, double-click on the Scale tool in the toolbar. Set the value to scale the object uniformly or set the horizontal and vertical values to scale the object nonuniformly. If the object has stroke and/or effects applied to it, set the options to choose to Scale Corners and Stroke & Effects. Click OK or copy the object by clicking the Copy button.

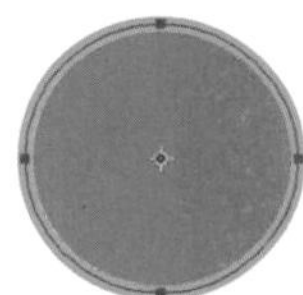

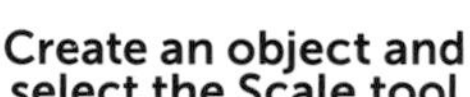
Create an object and select the Scale tool

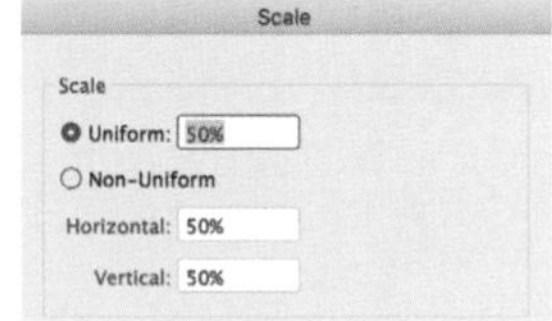

Double-click on the Scale tool, enter a value, and click Copy

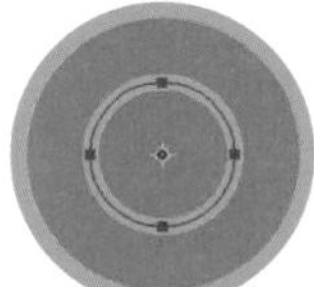

The result is a 50%-smaller shape on the center point

To set a different point of scale from the default (center of the object), select the object with the Selection tool and then choose the Scale tool. Click where you want the center point to be and the crosshairs will appear at that new scale point. The point of scale does not need to be on or in the object. The object will scale along that newly set point.

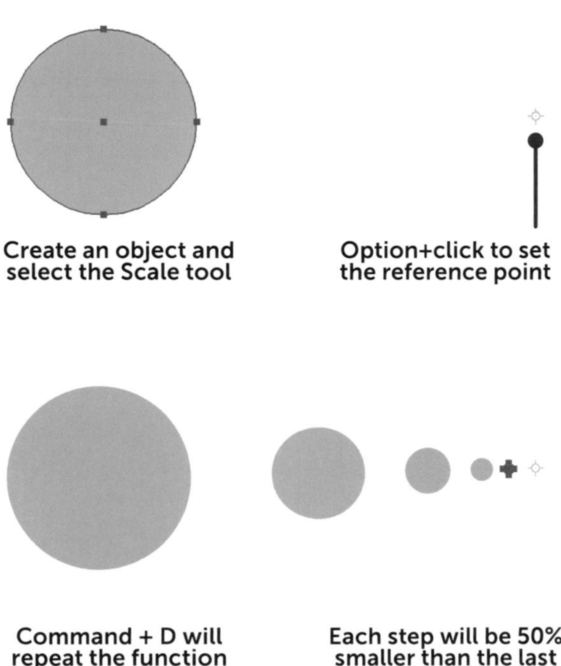

Create an object and select the Scale tool

Option+click to set the reference point

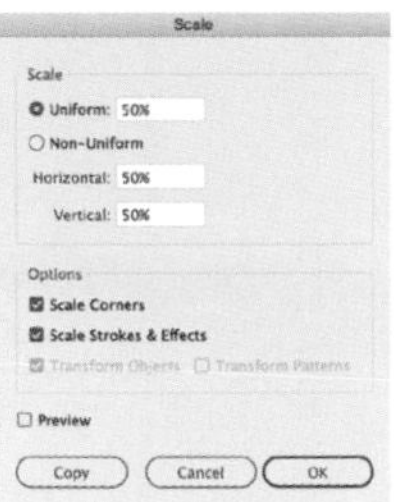

Enter a value and click Copy

Command + D will repeat the function

Each step will be 50% smaller than the last

Reflection/Flip

To flip an object, select it and choose the Reflect tool. A set cross-hairs will appear at the center of the object. To open the Reflect dialog, double-click on the Reflect tool in the toolbar. I like to think of reflect as a mirror, and setting the axis is like setting the location of that mirror. Set the point of reflection and click OK, or copy the object by clicking the Copy button. The Properties panel now includes the flip vertical and flip horizontal icons, which is a welcome change when you want to flip something quickly.

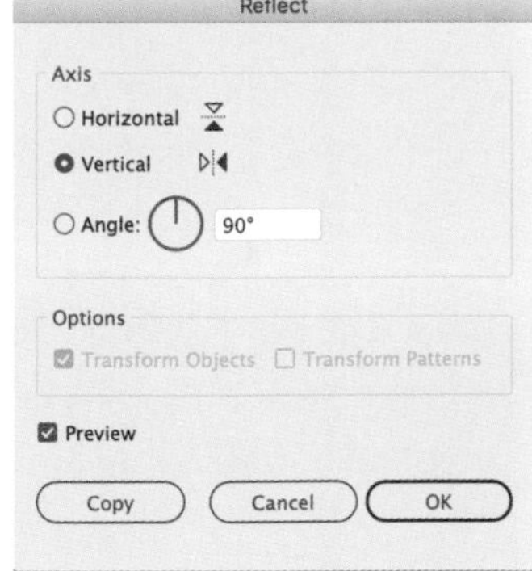

Shearing

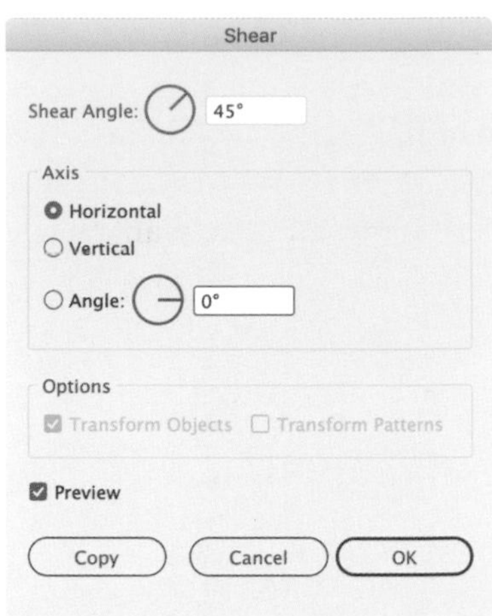

To shear an object, select the object and choose the Shear tool. A set of crosshairs will appear at the center of the object. That is the point that the object will shear from. With the cursor, select and pull the object to shear it vertically or horizontally from the center point. Trying to shear an object manually is a true test of patience because it will seem to shear in all directions. Open the Shear dialog box by double-clicking on the Shear tool in the toolbar. Set the Shear Angle value and choose which axis—horizontal or vertical—you want to shear along. Set the Angle value to shear both horizontally and vertically along the angle entered. Click OK or copy the object by clicking the Copy button.

When using the Shear tool, I select the object, then Option+click on a corner and shear from there. I think of it as putting a pin in that corner so all the movement happens from that stationary point. When I shear horizontally or vertically, the results are more predictable.

Direct Selection Tool

The Direct Selection tool (shortcut is A) is used for finer editing. Points and line segments can be selected using the Direct Selection tool, you can move, edit, and delete portions of a shape independent of the rest of the shape.

When selecting and object with the Direct Selection tool, its necessary to understand what is active on the shape and what is not active. A very common issue that people experience is selecting a shape with the Direct Selection tool and not being able to edit a point or a line segment independently. If you select the shape with the Selection tool first, then switch to the Direct Selection tool, all the points on the shape are active, so the entire shape moves as one rather than each point individually. If all the points are selected, you can click on a point with the Direct Selection tool to select a single point.

An active point will be solid; a nonactive point will be open or not filled. That was so you can tell what is and is not selected. I get in the habit of selecting the Direct Selection tool, clicking off the shape so it is not active, then clicking on the point or the path to select the specific items. With Smart Guides turned on (View > Smart Guides) you will be able to see what is being selected by hovering the Direct Selection tool over a point or path. The Smart Guide Tool Hint will

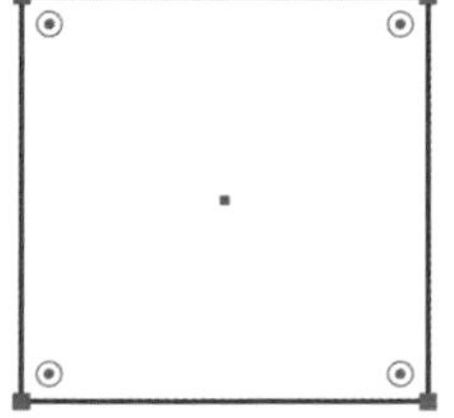

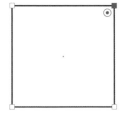

Object with all the points selected (solid squares)

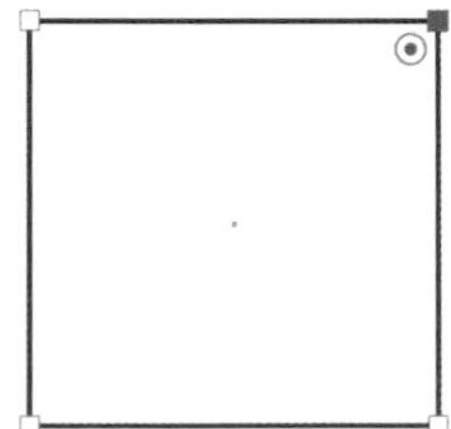

Upper-right point selected only; all other points are not solid

show what item will be selected.

Group Selection Tool

The Group Selection tool is nested with the Direct Selection tool. When there is a grouped set of items and you want to edit one without ungrouping the entire group, you can use the Group Selection tool to select one of the objects.

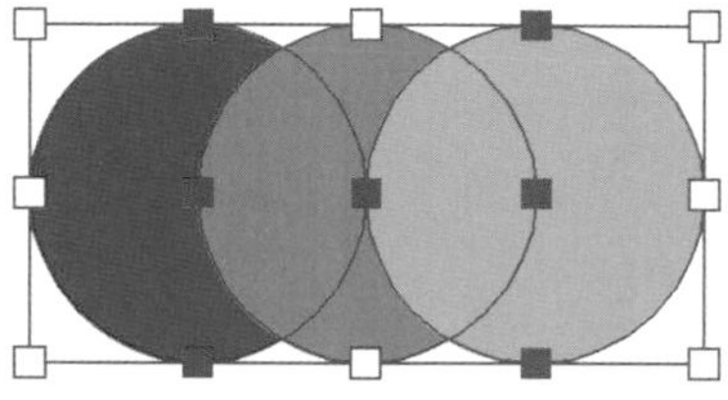

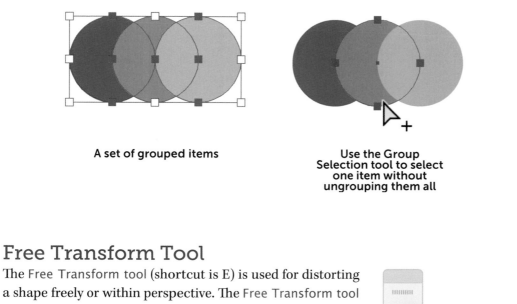

A set of grouped items

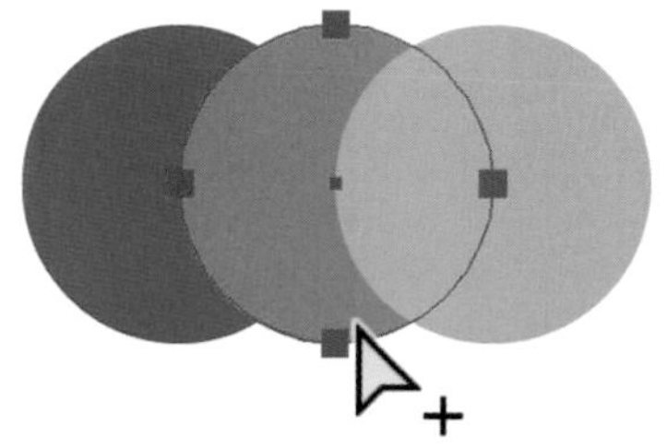

Use the Group Selection tool to select one item without ungrouping them all

Free Transform Tool

The Free Transform tool (shortcut is E) is used for distorting a shape freely or within perspective. The Free Transform tool has its own floating toolbar so if you don't see it when you select the Free Transform tool, it may be hiding behind another panel. There a four options in the Free Transform tool.

- **Constrain** locks the scaling of the object to keep it proportional.
- **Free Transform** allows the scaling of the object much like the Selection tool does.
- **Perspective Distort** allows scaling of objects within perspective.
- **Free Distort** allows the editing of objects much like the Direct Selection tool does.

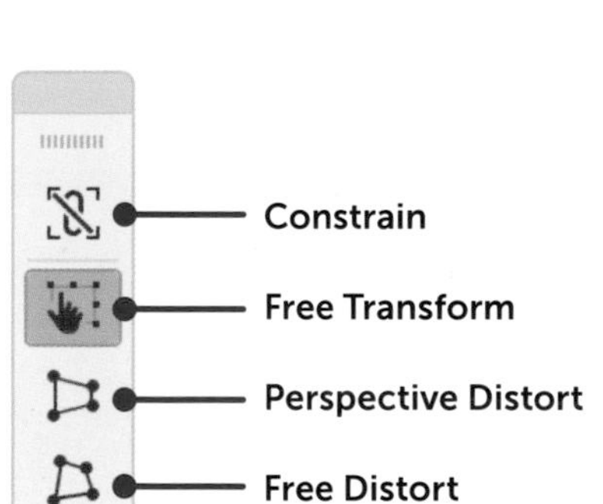

Scaling Corners

The Scale Corners checkbox allows the corners of a shape to be scaled in proportion to the size of the object. There are several places to control this

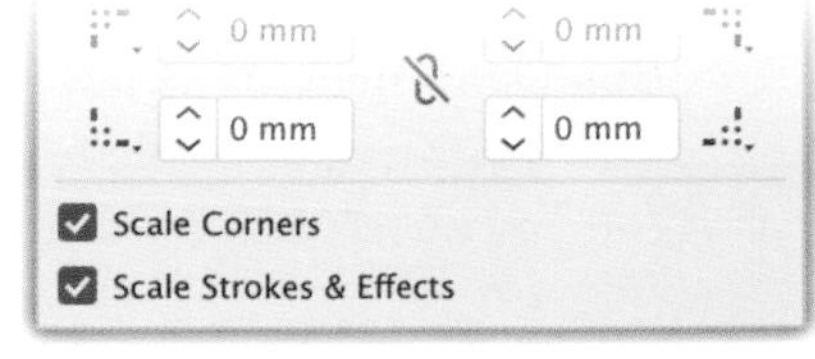

option: in Preferences > General, in the Transform panel; in the Properties panel, and in the Scale tool dialog box. When using corner widgets to round the corners of a shape, there is a point of maximum widgetness (indicated by red lines highlighting the corners). This is the point where the corners cannot be made any larger because they will run into the corner adjacent to itself. When this happens and you try to scale the shape smaller, the corners' size will prevent the shape from scaling smaller. To keep the corners in proportion with the object when scaling, click on the Scale Corners checkbox. If this option is unchecked and you have a shape with a 5mm corner radius, the corners will remain 5mm until the object is scaled down to the corners and will not scale any smaller.

The inside object scaled to 50% with the Scale Corners box checked. The corner radius is 50% smaller.

The inside object scaled to 50% without the Scale Corners box checked. The corner radius remains the same as the original shape.

Scaling Strokes & Effects

Scaling Strokes & Effects is an important feature when changing the size of objects with strokes and/or effects applied to them. Make sure the Scale Strokes & Effects checkbox is marked when scaling these kinds of objects so the stroke will scale up and down in proportion to the object. If the object is scaled up 250%, then the stroke weight will also increase by 250%. This may or may not be what you want. If you have a 10pt stroke on your objects and you scale it up and down, the result will be several objects with different stroke weights. However, when scaling objects down without this option checked, the stroke may overpower the fill. The object may look very different when the stroke overtakes the object.

An object scaled to 50% without Scale Stroke & Effects checked

An object scaled to 50% with Scale Stroke & Effects checked

Scaling effects is important as well. The effects when scaled small can overpower the object or change it significantly. This example has the Scale Corners box checked to keep the corners scaled in proportion to the object. If these were scaled smaller still, the one without the Scale Stroke & Effects box checked would get filled at the center of the shape.

Paths

Illustrator's basic building blocks are points and paths. Everything that we create in Illustrator starts from these and expands from there. Paths go by many names: paths, strokes, lines, line segments, or borders; yet they all refer to the same thing. These paths can be straight or curved and they can be open or closed (a shape).

Open Paths Versus Closed Paths

An open path is a line or line segment where both ends of the line are not touching. A closed path is what we know as a shape: a circle, square, rectangle, and polygon are all closed paths.

Outline Stroke/Expand

An open or closed path can be outlined or expanded to create a closed shape from a path. A stroke generally refers to a path that has a color and a weight applied to it; when a shape is described as an object with a fill and no stroke, that means there is a colored, gradient, or pattern fill, and no color or weight applied to the path around the object. Paths can have a color and weight applied to them, have end caps applied for rounded ends, and have arrow heads and tails applied; at this point we refer to those as strokes.

Paths and strokes are used interchangeably, yet that may not be the terminology that all people use in the same way. I try to make it more simple in referring to these paths (having no visible attributes) and strokes (having color and weight), but even I use the terminology interchangeably.

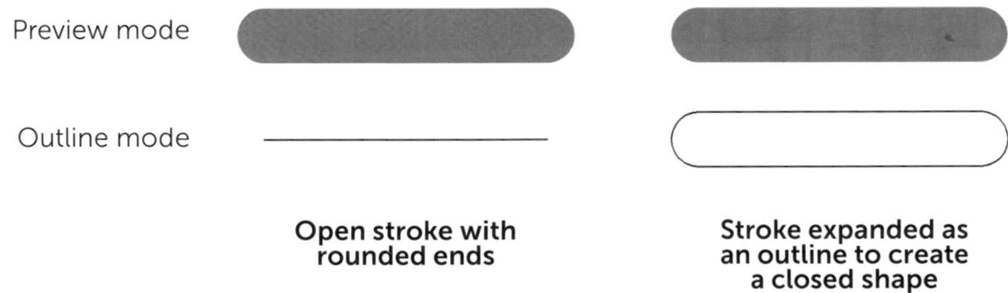

To convert a stroke to a shape, select the stroke and choose Object > Expand or Object Path > Outline Stroke. The example shows a stroke (left) and the stroke outlined or expanded into a shape (right) in Preview mode (top) and in Outline mode (bottom). The difference between a path and a shape are clear when viewed in Outline mode. The example shown has no stroke on the shape; when there is a stroke on the shape, the results will be an outline of the stroke of the object.

So what is the difference between Expand and Outline Stroke? In this next example, you will see the difference between Expand and Outline Stroke depending on the attributes of the strokes.

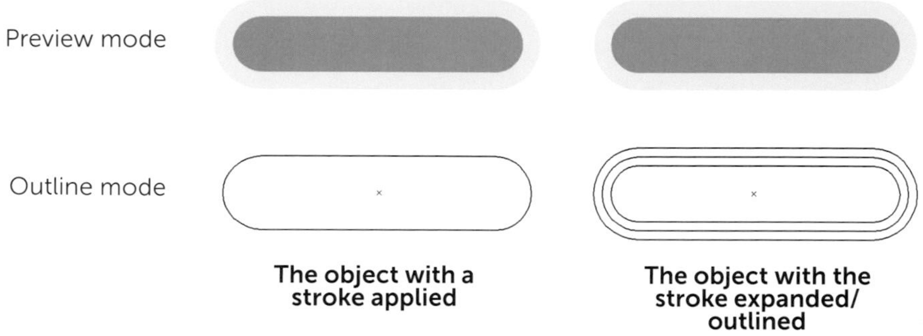

The object with a stroke applied

The object with the stroke expanded/outlined

When the object has a stroke around it, and either Object > Expand or Object > Path > Outline Stroke is chosen, the results will be a shape within a shape where the stroke has been expanded. If Expand is chosen, the dialog box will appear to expand the Fill and Stroke.

Expand
Expand
Object
Fill
Stroke
Expand Gradient To
Gradient Mesh
Specify: 255 Objects
Cancel OK

If Object > Path > Outline Stroke is chosen, the stroke will be outlined and no dialog box will appear.

When a Stroke is outlined or expanded, and you turn on the Outline mode to see the paths, the shape will appear to have three shapes around it. In this example there appears to be an outer shape, middle shape, and inner shape. This is due to the default setting of the stroke being applied to a path in the center of the stroke. This path has a 10pt stroke applied to it, which means 5 points of the stroke were applied outside the path, and 5 points was applied inside the path.

This example shows what the newly expanded stroke looks like (the outlined stroke has 50% opacity applied to it) when selecting the shapes. To have the stroke be outside the shape and not overlapping, set it to be outside before expanding or outlining it.

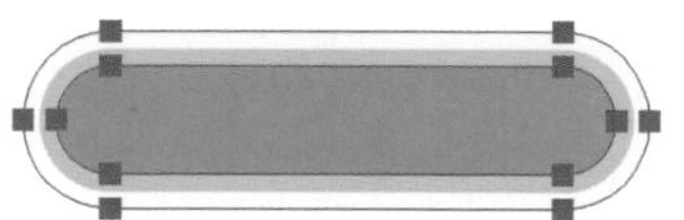

Offset Path

This is one of my favorite commands in Illustrator. Offset Path allows you to make a larger or smaller copy of your object matching the shape, curves, and angles. If you are duplicating and scaling a circle or a square you don't need this feature, but when you have a more complex shape, Offset Path comes in very handy. When an object is duplicated and scaled, the shape of the two objects won't match. With Offset Path, it will create a shape that mirrors the original shape, either larger or smaller. To create a larger shape than the original, set the offset to a positive number; to create a smaller shape, use a negative number.

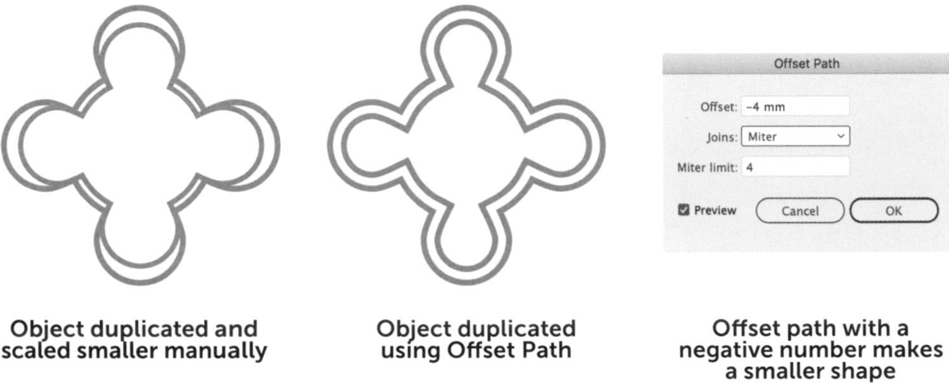

Object duplicated and scaled smaller manually

Object duplicated using Offset Path

Offset path with a negative number makes a smaller shape

- **Joins** modifies the type of the ends a shape has where a corner or tip appears when a path is Offset. There are three types: Miter is a pointed corner; Round is a rounded corner; and Bevel is a squared corner or a flat tip on the end of a line. The ends and corners can change when the shape is offset larger or smaller than the original shape.
- **Miter Limit** is how far the points can extrude from angles in the shape. Setting the miter limit is based on how tight of an angle the lines form. The tighter the angle, the higher the number in the miter limit needs to be in order for the Miter (pointed end of the line) to appear.

Join Paths

Joining paths is connecting two or more open paths together to create a longer path or create a closed shape.

Divide Paths

Dividing a path or an object can be done in several ways. A common way is using the Pathfinder Divide mode on the shapes and paths to create divided objects.

The Scissor tool will also divide or cut a path. To create a shape or a path, click on the Scissor tool (C), and click on the path or object to create a cut. If you click on an open path, it will cut the path into two sections. If you cut a closed path, click in two places to create a section which can be selected by the Direct Selection tool and moved or deleted.

The Divide Path function under Object > Path > Divide is easy to use as well, but it will divide or cut shapes, not open paths—even though it resides under the Object > Path menu. To divide a shape, use a path (either open or closed) as the “cutting” or dividing device.

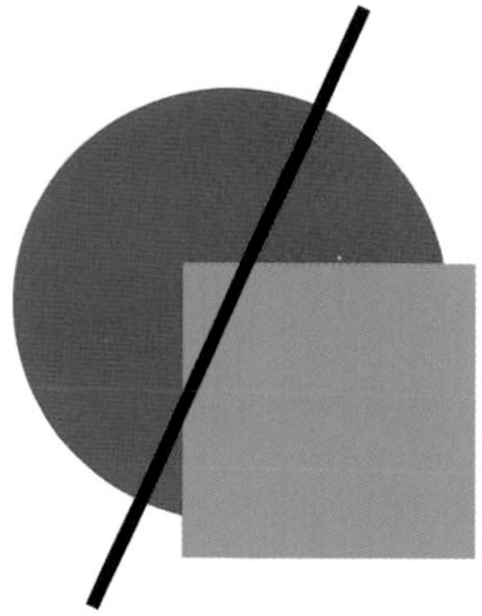

Create paths to be divided; draw a path that is the divider

Select the “divider” only and choose Divide Objects Below

Select the path you want to use as the cutting device (don’t select the objects you are dividing) and choose Object > Path > Divide Objects Below. This will result in the objects below being divided based on the inserted path. The path will disappear and the result will be the divided shapes. This is a nice feature instead of using the Divide function in the Pathfinder panel because you do not need to Ungroup the items after you have performed the Divide Objects Below command.

Compound Paths

Compound paths are closed paths that create a hole in another closed path. I call this the donut feature. We see compound paths everywhere in type, such as lower case “a.” The compound path of the “a” has a shape that creates a hole in the lower part of the letter. Compound paths can be a bit frustrating when you outline type to manipulate a type face for a logo or identity because the letterforms may not act the way you would like them to when editing the paths. Compound paths are also very helpful when adding effects to type using the Appearance panel.

Creating a Compound Path

Compound paths can be created in a few different ways. Select two or more closed paths and choose Object > Compound Path > Make. When two closed paths are turned into a Compound Path, the result will be a hole in the area where the two closed paths overlap. When there are more than two closed paths, the Even-Odd Rule applies. An even number of overlapping closed paths converted to a compound path will result in a hole where the paths overlap. An odd number of overlapping closed path will result in a hole where the first two objects overlap, and a filled area where the third closed path overlaps.

Using the Pathfinder Intersect mode will also create compound paths and the same

Even-Odd Rule applies.

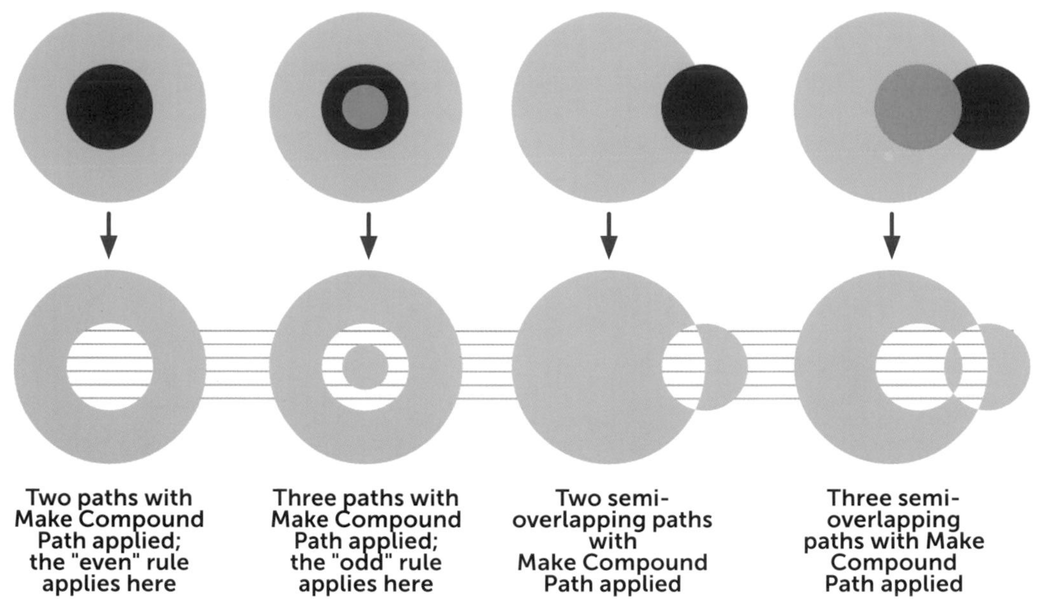
Two paths with Make Compound Path applied; the "even" rule applies here
Three paths with Make Compound Path applied; the "odd" rule applies here
Two semi-overlapping paths with Make Compound Path applied
Three semi-overlapping paths with Make Compound Path applied

A compound path will result in the knockout area being transparent, as shown by the example above (the lines are showing the areas that have been knocked out). Compound paths do not work when using a line, they only apply to shapes. If you want to include a line in a compound shape, it will need to be expanded or outlined and turned into a shape before it can be included in a compound path.

Edit a Compound Path

To edit a compound path, you can choose the Group Selection tool and directly select a shape in a compound path. You can double-click on the shape to enter into Isolation mode and isolate the shape from the other shapes to move or edit it. You can use the Direct Selection tool to select and edit a shape in the compound path. Editing a compound path using the Layers panel is not possible because only the compound path—not the individual paths that make up the shape—will show.

Release a Compound Path

To release a shape from its compound path, select the compound path and choose Object > Compound Path > Release. This will return the object back to its original shapes before the compound path was created. If the shapes were filled with different colors before they were turned into a compound shape, the shapes will all be filled with the original fill color after the compound path is released.

Type as a Compound Path

When type is outlined, the letters are listed as compound paths in the Layers panel, even if they are not made of two or more shapes. This is where the rules do not seem to apply like they would with other closed shapes. You can have a single closed path be a compound path but this isn't something you would do to a single shape since the result is a closed shape. However, when you outline type (Type > Create Outlines) to manipulate each character or apply an effect to all the characters in the Appearance panel, compound paths can be helpful; but they can also get in the way. Here is how the different scenarios work:

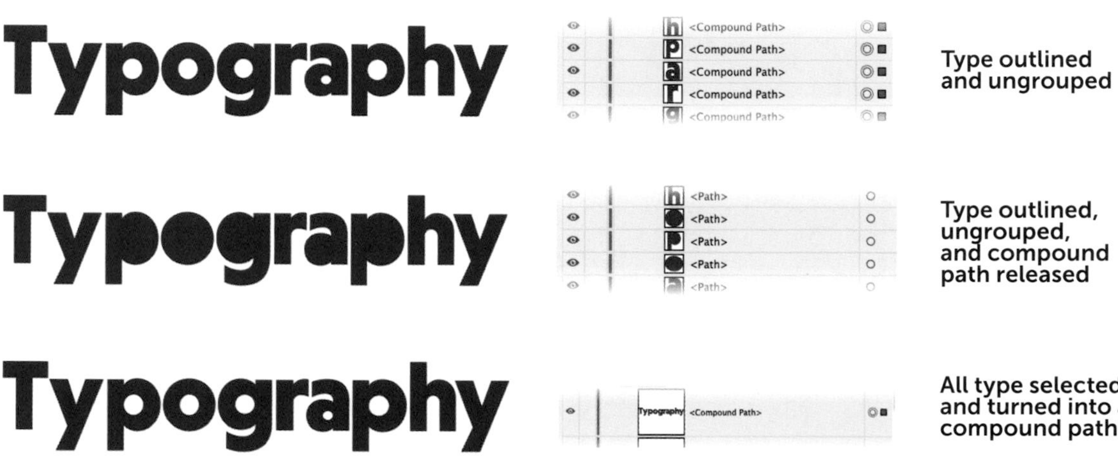

When the type is outlined and ungrouped, every letter is a compound path even if there is only one path making up the letter, as shown in the Layers panel. If all the letters are selected and a compound path is released, all the letters return to their normal paths and any letter that requires the compound path (such as the p, o, and g in this example) loses their compound path. All the other letters simply become closed paths.

If you select all the paths that make up the word and then choose Object > Compound Path > Make, the entire word will become a compound path and all the letters that require a compound path become a compound path. This step of making the entire word a compound path is important if you want to use the Appearance panel to add a fill, stroke, or effect to the entire word. If you do not make the outlined word a compound path, the Appearance panel will not show the fill and stroke attributes.

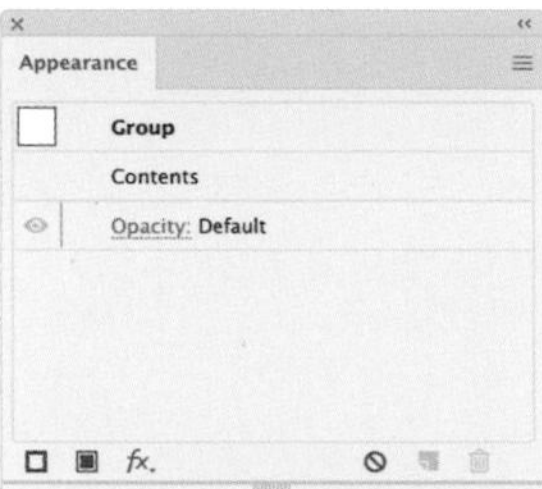

Type converted to outlines, not converted to a compound path

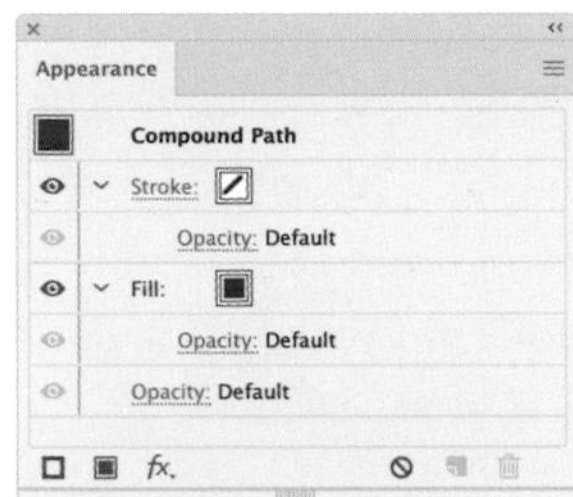

Type converted to outlines, converted to a compound path

Blend

The Blend tool allows you to blend paths or object together in three different ways. You can create impressive color blends with multiple steps between objects or paths.

Blend Options

The three blend options are Smooth Color, Specified Steps, and Specified Distance. To call up these options you can double-click on the Blend tool in the toolbar or choose Object > Blend > Blend Options.... You can also access the Blend Options at the bottom of the Properties panel. These options can be set before a blend is created or changed after a blend is created.

Smooth Color

Blending with Smooth Color calculates the number of steps required to make the blend. The steps are calculated to provide the optimum number of steps for a smooth color transition. If the objects contain the same colors, it will give you a stepped appearance much like Specified Steps.

Specified Steps

Blending with Specified Steps controls the number of steps between the start and stop of the blend. The number you enter does not include the first and last objects in the set; enter only the number of steps you want to appear between the first and last objects.

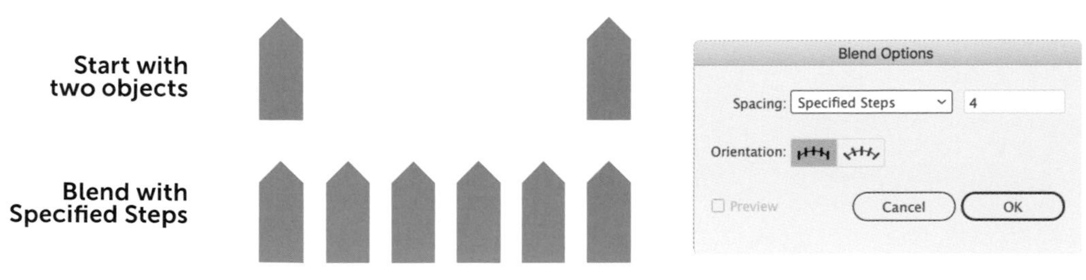

Preferences & Workspaces
Shape Creation
Advanced Construction
Editing & Transformation

Specified Distance

Blending with Specified Distance controls the distance between the number of steps you enter in the Blend Options dialog box. The Distance Specified is measured from the edge of one object to the corresponding edge on the next object.

Editing the Blend Spine

The spine of the Blend is the invisible line that connects the first and last object together. You can see this spine when you enter into Outline mode. The spine can be edited or moved with the Direct Selection tool, or you can use the Curvature tool to change the length, position, or curve of the spine.

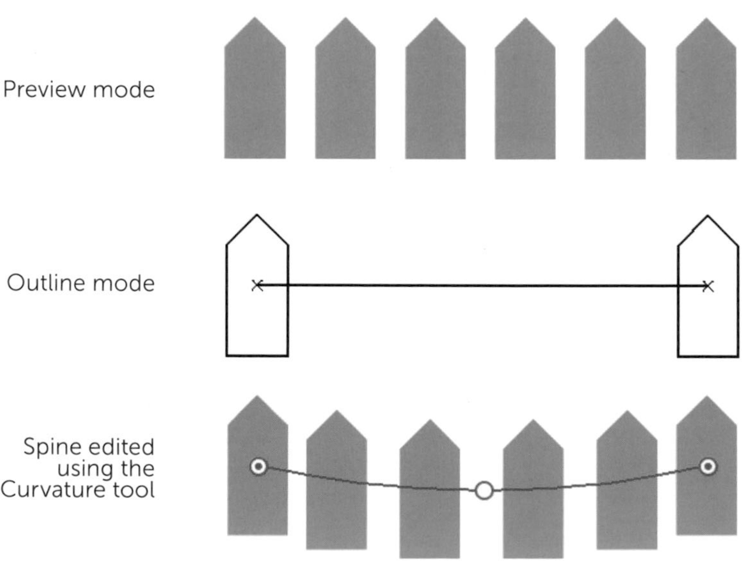

Orientation

Orientation refers to how the objects follow the spine. There are two options in the Blend Options panel. Align to Page (creates an arch) and Align to Path (creates an arc). Note how the spacing is affected by the Orientation, the level of curve, and the width of the spaces between the objects.

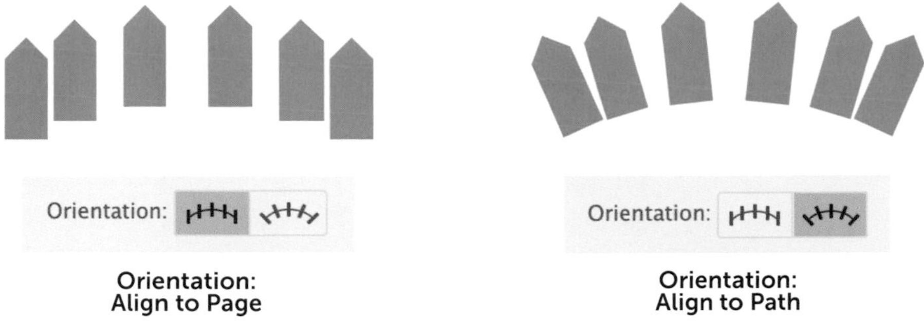

Orientation: Align to Page

Orientation: Align to Path

Replace Spine

The spine of the blend can be edited or it can be exchanged for another spine completely. Create a new path you would like the blend to conform to. Select your blend items on the new path, and choose Object > Blend > Replace Spine. The new path you have drawn will now be the spine of the blend, replacing the previous spine.

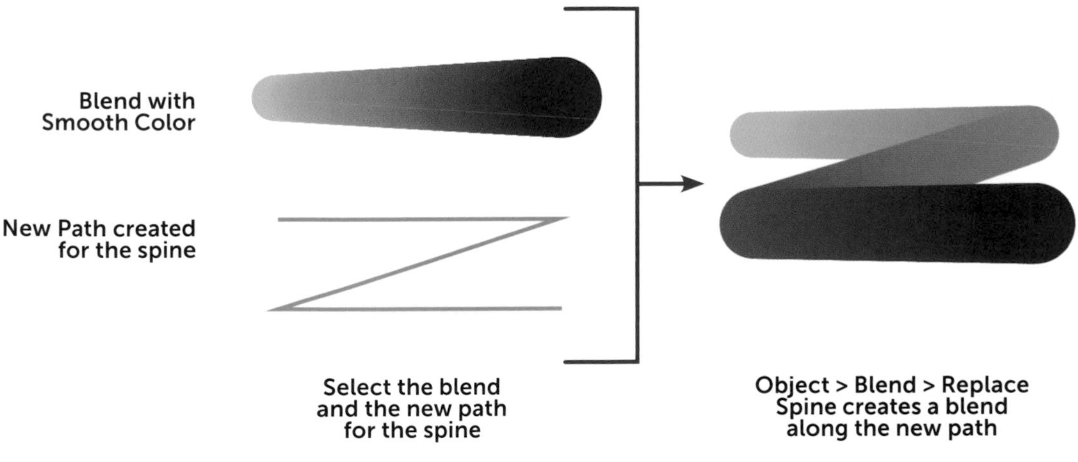

Select the blend and the new path for the spine

Object > Blend > Replace Spine creates a blend along the new path

Preferences & Workspaces | Shape Creation | Advanced Construction | Editing & Transformation

To reverse the order of a blend on its spine, select the blended object and choose Object > Blend > Reverse Spine. This will reverse the blend direction from start to stop. To change the stacking order of the blend so the path underlaps instead of overlaps, choose the Reverse Front to Back command.

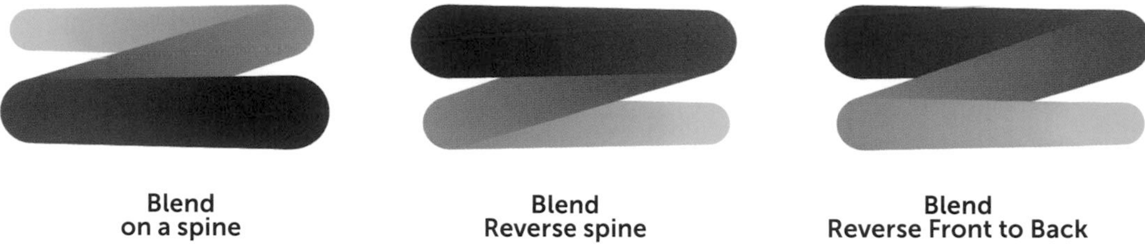

Blend on a spine | Blend Reverse spine | Blend Reverse Front to Back

Blend Three or More Objects

To blend three or more objects together, select the Blend tool in the toolbar. Click on one object with the Blend tool, then the next object, then the next object, until all the objects are blended together. The order you select the objects with the Blend tool affects the way they blend.

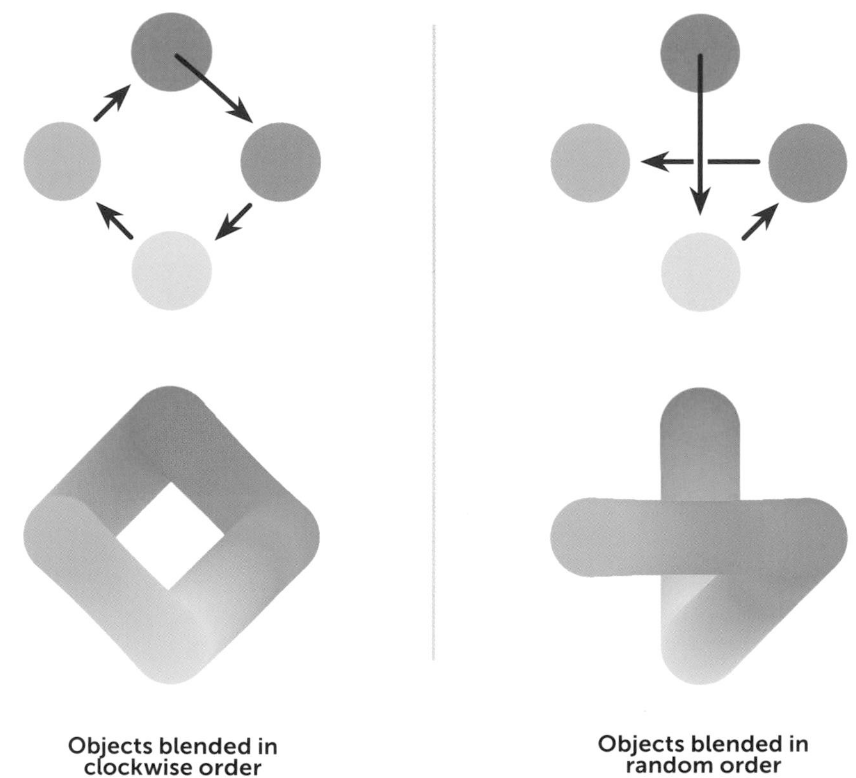

Objects blended in clockwise order | Objects blended in random order

Blending objects on top of each other can create a very realistic looking blend with highlights and shadows. It can produce dimensional-looking objects and effects such as this sphere with a specular highlight. The blended shapes can be different sizes and shapes, and have different fills and strokes.

Select the three shapes, then select the Blend tool
Click on the white shape, then the yellow shape, then the green shape

To create the sphere, draw the circles with different color fills. Select all the shapes and select the Blend tool. Click on the first shape, then second shape, and then the third shape. Set the Blend Options to Smooth Color to blend all the colors together seamlessly.

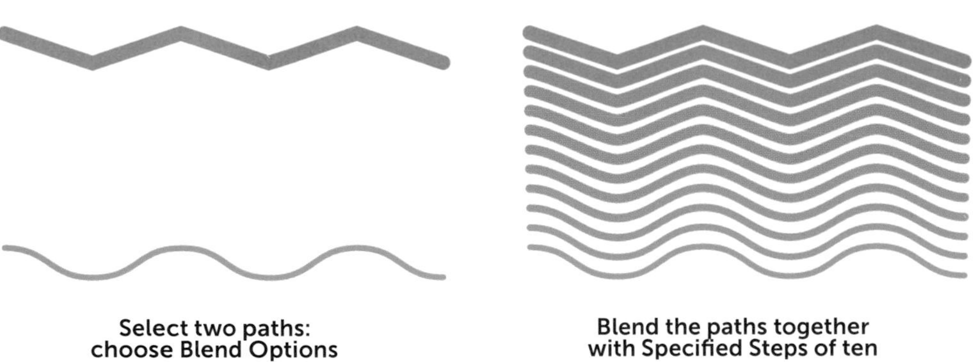

Select two paths: choose Blend Options
Blend the paths together with Specified Steps of ten

Release a Blended Object

To release a blended object back to the original objects, choose Object > Blend > Release. This releases the objects back to their original state before the blend was applied. In Outline mode, you will see that the spine that connects the objects together is visible. This path is no longer connected to the shapes and is independent.

Expand

Expanding a blended object will render each step of the blend as an individual shape and the blended shapes can no longer be edited using the Blend Options. To expand the blend, select the blended object and choose Object > Blend > Expand. Expanding the blend may create several shapes (possibly hundreds or more) depending on the extent of the blend. You can see the individual steps in Outline mode. If they appear as a black mass, zoom in to see each individual shape.

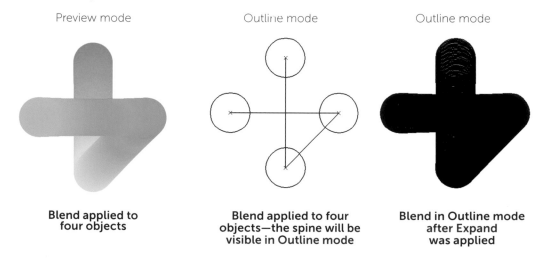

Blend applied to four objects

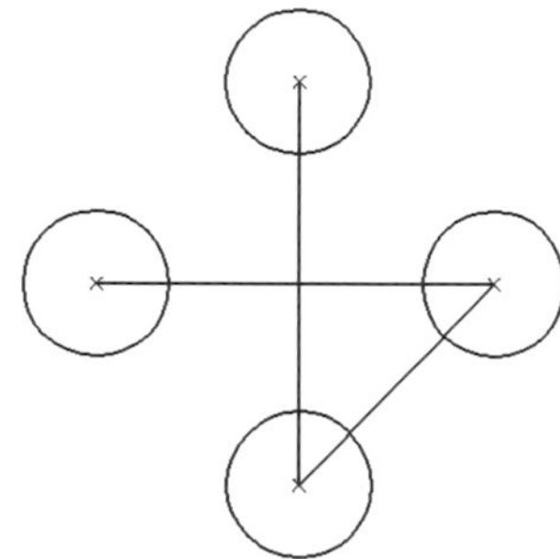

Blend applied to four objects—the spine will be visible in Outline mode

Blend in Outline mode after Expand was applied

Symbols

Symbols are different than an object that is assembled using shapes and paths, grouped together, then copied and pasted when needed in a document. A symbol is an object that can be reused. The advantages of using a symbol instead of copying and pasting a grouped object multiple times is that a symbol can be edited and all instances of the symbol will reflect the edits. Symbols can be scaled, rotated, and resized using the Symbol tools. Using symbols can save you time and reduce file size.

Creating Symbols

Create or use the artwork you want to make into a symbol. I usually group all the pieces of the artwork together so none are left behind when the symbol is created. Open the Symbols panel from the Window menu. Drag the artwork into the Symbols panel. A dialog box will open; name the symbol and choose Graphic from the dropdown menu.

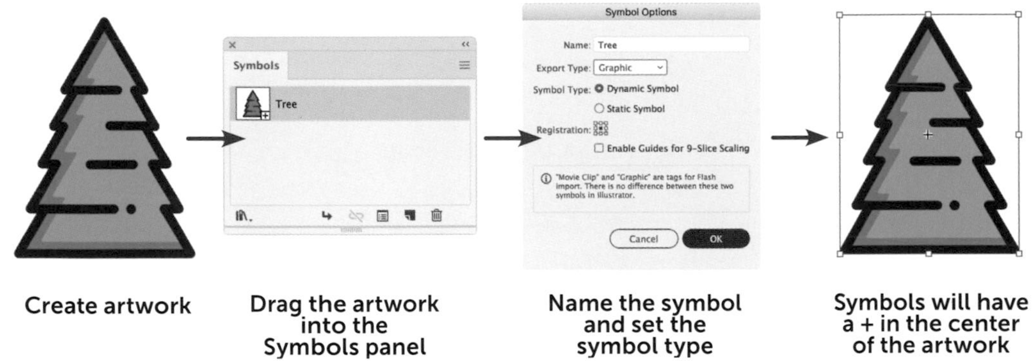

Create artwork — Drag the artwork into the Symbols panel — Name the symbol and set the symbol type — Symbols will have a + in the center of the artwork

Using Symbols and Symbol Tools

Each single occurrence of a symbol is called an instance. Single symbols can be added to the composition by dragging them from the Symbols panel onto the artboard. They can also be added by using the Symbols panel dropdown menu and choosing Place Symbol Instance. Each instance of a symbol can be scaled, rotated, flipped, and skewed and still retain the link to the original symbol. Again, this means when the original symbol is edited, all instances of that symbol will reflect the edits.

Editing Symbols

To edit a symbol, double-click on it in the Symbols panel, make any edits to the symbol and hit the Escape key. You can also click on the upper-left corner of the open window and click on the right arrow next to the name of the symbol to exit symbol editing.

At the bottom-left corner of the Symbols panel is the library icon. Click on the Library icon to open up a set of symbols that is part of Illustrator. There are numerous Symbol Libraries to choose from and all are easy to access.

Replace Symbol

Replacing a symbol with another symbol can be done by selecting the instance of the symbol and choosing a different symbol from the Replace Symbol dropdown menu in the control bar. There is no Replace Symbol menu in the Properties panel. Multiple instances of a symbol can be selected and replaced at the same time.

Break a Link to a Symbol

Breaking a Link to a symbol can be done from the Symbols panel dropdown menu. Select the instance of the symbol and chose Break Link to Symbol in the Properties panel to break a link. This will convert the instance of the selected symbol back to an editable object that is no longer linked to the symbol in the Symbols panel.

Symbol Tools

You can drag and drop the symbol from the Symbols panel and duplicate it on the artboard to create multiple instances just like you would with grouped artwork.

You can add multiple symbols quickly; select the symbol and choose the Symbol Sprayer tool from the toolbar. "Spray" the symbol like you would using a can of spray paint onto the artboard. Several other duplicate instances of the symbol will appear.

Nested with the Symbol Sprayer tool is the Shifter, Scruncher, Sizer, Spinner, Stainer, Screener, and Styler tools. All of these tools will edit the selected instances of the symbols on the artboard. To set the settings of each tool, double-click on the tool in the toolbar.

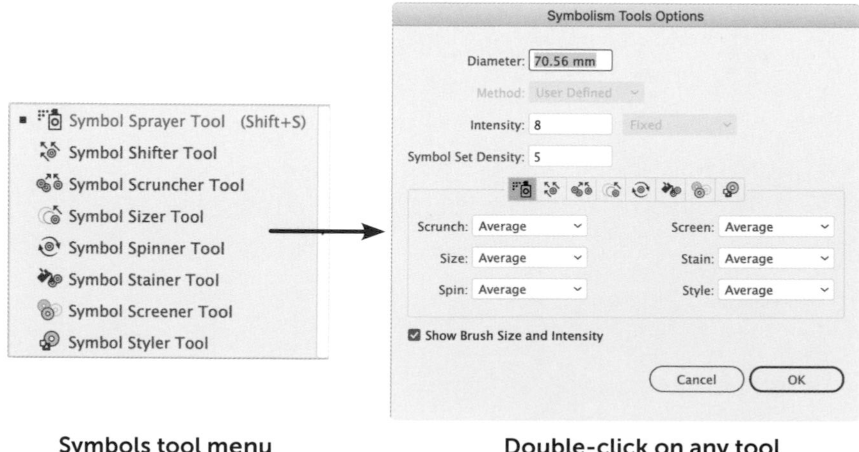

Symbols tool menu

Double-click on any tool to open the preferences

Once the Options for the Symbol tools are set, choose the symbol you want to spray on the artboard with the Symbol Sprayer tool. Spray away; the faster you move the mouse the more space will appear between each instance of the symbol. The slower you move the mouse the more the instances will build on each other.

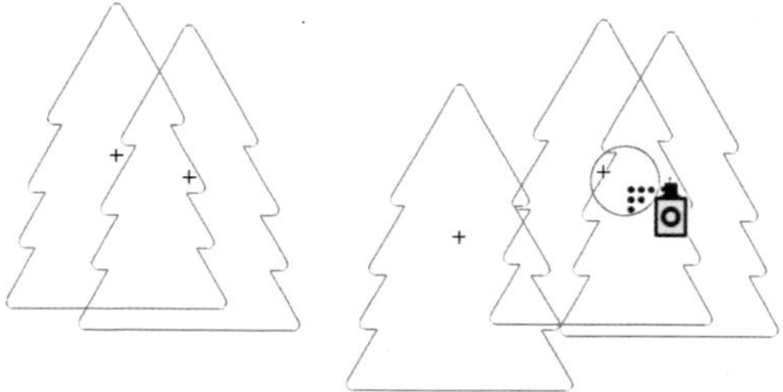

Spraying the tree symbol using the Symbol Sprayer

Symbols can then be edited using the Symbol tools

Dynamic Symbols

Dynamic Symbols are symbols that share a master shape, but individual instances of the symbol can have dynamically modified appearances, such as rotating, skewing, scaling, or flipping. If the master shape is modified, symbol instances reflect the edits, but maintain their own modifications as well.

Patterns

Creating Patterns

Creating patterns in Illustrator is a great way to add texture or visual interest to any object. Create and select the artwork that you want to turn into a pattern, then choose Object > Pattern > Make. The Pattern Options panel will open and the artwork will appear in the thumbnail window of the panel. If you open the Pattern Options panel before creating a pattern, the panel will be grayed out, which can be confusing. This is why I recommend using the Object > Pattern > Make. You can choose Make Pattern from the Pattern Options panel so the panel options will be active.

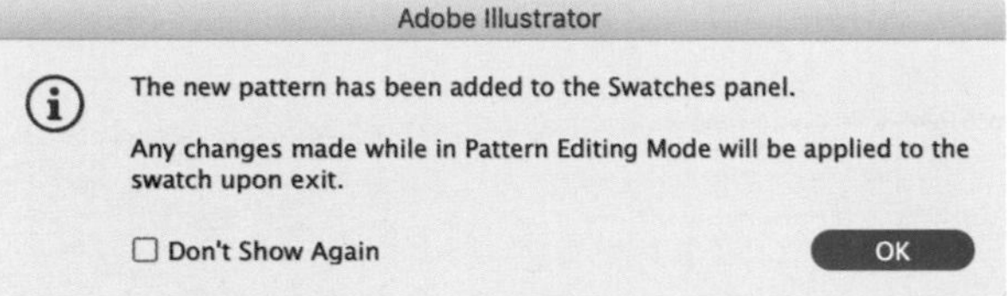

Once a pattern is created, a dialog box will appear stating that the new pattern has been added to the Swatches panel even though you have just begun the process. Any updates you make to the pattern will be updated in the Swatches panel when you are done editing it.

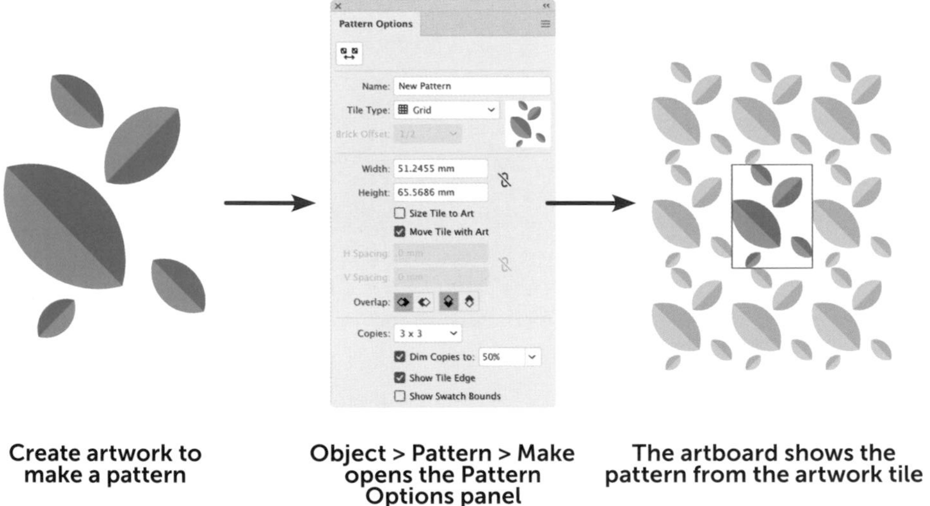

Create artwork to make a pattern | **Object > Pattern > Make opens the Pattern Options panel** | **The artboard shows the pattern from the artwork tile**

The center bounding box (the tile) in the middle of the pattern displays the artwork. This can be edited, moved, rotated, and scaled to create the pattern. The grayed back area around the tile shows how the pattern will look when repeated.

To set the attributes and size of the pattern, there are many option in the Patterns panel.

- **Tile Type:** This sets how the pattern will be repeated. There are many choices for how the pattern can repeat: Grid, Brick by Row, Brick by Column, Hex by Column, and Hex by Row. The Brick by Row and the Brick by Column options also allow different ways the pattern can be offset. Choosing any of the Tile Types will dynamically update the pattern.
- **Width/Height:** This sets the overall height and width of the tile.
- **Size Tile to Art:** Sets the size of the tile to shrink to the size of the artwork.
- **Move Tile with Art:** This allows the artwork and the tile to move together.
- **H Spacing/V Spacing:** This determines the space between adjacent tiles.
- **Overlap:** When adjacent tiles overlap, this option sets which tile will appear in front.
- **Copies:** The dropdown menu of the Pattern Options panel will show the number of copies so you can get a better idea of how the pattern interacts and repeats with itself.
- **Dim Copies to:** This sets the opacity of the copies.
- **Show Tile Edge:** This displays a bounding box around the tile.
- **Show Swatch Bounds:** This displays the part of the pattern that is repeated.

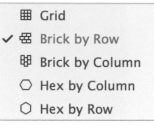

Save a Pattern (Done)

When you are done editing the pattern options, you are ready to save the pattern and apply it to your object. Patterns are automatically saved to the Swatches panel when they are created.

Once the edits are made to your pattern, choose **Done** in the upper-left bar at the top of the open document window. Done simply means that you are done with the edits and those edits will now be applied to the currently saved pattern in the Swatches panel. If you click **Save a Copy** when you are done with your pattern, it will create two of the same patterns in the Swatches panel. Save a Copy is for when you are editing an existing swatch and you may also want to keep the original. It uses the same name as the original pattern and lists it as a copy.

Editing and Scaling Patterns

To edit an existing pattern, double-click it in the pattern swatch in the Pattern Options panel. Select an object containing the pattern and choose Object > Pattern > Edit Pattern, or double-click on the pattern in the Swatches panel.

The Pattern Options panel will appear and you can edit the pattern. Click Done when editing to save the edits or click Save a Copy to create a new pattern based on the one you edited.

You may notice some interesting behaviors with objects that have a pattern fill. When moving the object, the object will move but the pattern will stay in the same location, much like moving a window and having the scenery (the pattern) stay in the same place. This is the default setting for patterns. They stay in the same place when you move your object. When scaling the object, the default setting is that the object will scale but the pattern will not. The pattern stays the same size regardless of the size of the object.

To change the preferences from the defaults and have the pattern scale with the object, choose Preferences > General > Scale Pattern Tiles or in the Transform panel dropdown menu choose Transform Both (this will transform both the object and the pattern together).

Transform Object Only
Transform Pattern Only
✓ Transform Both

If you want to return the pattern to its original size after you have scaled an object, select the object and click once on the pattern in the Swatches panel.

Pattern applied to the object

Object scaled smaller with the pattern

Pattern returned to original size by clicking on it in the Swatches panel

Alignment and Distribution

Guides and Grids

Setting up guides in an Illustrator file helps you align objects and text. Guides can be pulled from rulers for vertical and horizontal alignment or you can make custom guides at angles. Smart Guides are guides that show the alignment between the edges or centers of objects or the spacing between objects. There are several methods of aligning and distributing spacing and creating guides in Illustrator.

Rulers

Rulers can help you place and measure objects as well as the spacing between them. To turn the rulers on choose View > Rulers > Show Rulers. Rulers appear at the top and left of the document window. The rulers measure from the origin, which is on the upper-left side.

Rulers can be set up to measure each active artboard or to globally measure the entire document. When the ruler is set to measure the artboard, the ruler origin will snap to the upper-left corner of the active artboard. When the rulers are set to be global, they will start at the first artboard in the document. To change the rulers from artboard rulers to global rulers choose View > Rulers > Change to Global Rulers. To hide the rulers choose View > Rulers > Hide Rulers.

To change the origin point of the rulers to measure from a different location, click on the small area where the rulers meet and drag the guideline to the location you choose. The crosshairs will indicate the new ruler origin point. To reset the ruler origin point back to the default, double-click at the ruler intersection on the ruler origin point.

Change Units of Measurement

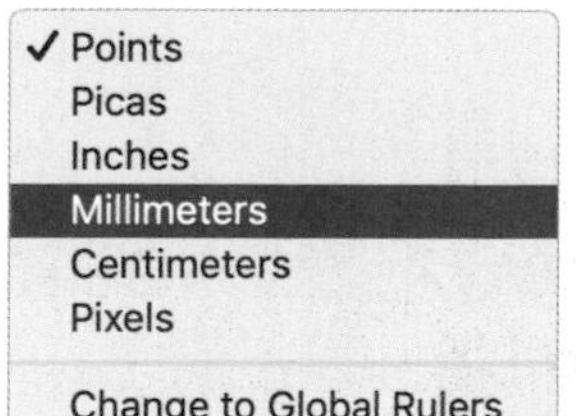

The default unit of measurement in Illustrator is points. You can set a different unit of measure when you create a new document. After you create a document, turn on the rulers. Right-click on the ruler and choose from a list of units. You can also find these options under Illustrator > Preferences > Units (Mac) / Edit > Preferences > Units (PC). Set the units for General (rulers, size, spacing of objects, and general measurements), Stroke, and Type. Each of the three settings can be different units of measure.

Guides

Guides can help you align text, objects, and paths. You can use manually created guides or Smart Guides.

Create Guides

Guides can be pulled from the vertical or horizontal rulers and positioned on the artboard. If the rulers are not turned on, choose View > Rulers > Show Rulers. Drag a guide from a ruler and release it on the artboard. The guide will stop at the edge of the artboard. If you drag a guide outside the artboard, the guide will extend the entire distance of the document. The default color of guides is cyan, but that can be changed in the Preferences panel.

Guides can also be created by selecting a line or path on the artboard and choosing Guides > Make Guides. The path or shape selected will now be a guide and no longer a path or object. Guides can be created on a layer, which then can be turned on or off via layer visibility.

Show/Hide Guides

Guides are turned on by default and can be turned off by choosing View > Guide > Hide Guides. To show or hide guides, choose View > Guides > Show Guides or View > Guides > Hide Guides.

Unlock, Lock, Delete, Move, or Release Guides

Guides are locked by default (Illustrator CC 2019; in older versions the default is unlocked) and can be moved, copied, and deleted from the artboard once they are unlocked.

To **unlock guides**, choose View > Guides > Unlock Guides. This will unlock all the guides.

To **lock guides**, select View > Guides > Lock Guides.

To **delete guides**, unlock the guides then click on the guide and hit the Delete key or choose Guides > Clear Guides to delete all the guides.

To **move guides**, unlock the guide, click on it and move it with the Selection tool.

To **release guides**, unlock the guides and choose View > Guides > Release Guides. This function is only available when the original vector object or path was turned into a guide. A guide pulled from a ruler cannot be released back to a path.

Snap to Point

The Snap to Point function makes points adhere to their origination point or snap to a an existing point. To use this feature, choose View > Snap To Point, select the object with the Selection tool, and move the object. Position the cursor on the exact point you want to align the object with; to an anchor point or another object or guide. The cursor will change from black to white when the anchor point or guide has been snapped to. The Snap to Point alignment depends on the position of the cursor and not the edges or center of the object being moved. The cursor will Snap to Point when it comes within two pixels of an anchor or guide. This snap distance can be set in the Preferences > Selections & Anchor Display > Snap to Point. Snap to Point will not snap to a line or path, it will only snap to an anchor or a path or a guide.

Smart Guides

Smart Guides are guides that appear when you are moving, editing, or drawing an object or path. These guides help align or constrain the object or path being moved or drawn to align to other objects, paths, or to the artboard.

Unlike the Snap to Point feature, Smart Guides will align to any anchor point, guide, center point, or edge of any object, path, or artboard. Smart Guides are on by default and can be turned on and off under View > Smart Guides.

Smart Guide Labels

Smart Guides can be very helpful when creating and editing in Illustrator. When aligning, spacing, selecting, drawing, moving, and editing anything, Smart Guides will display helpful labels. These labels show how things align on the top, bottom, left, right, and center of the object you are moving or editing with any other object on the artboard. When drawing, Smart Guides will show when a circle or a square is constrained. Spacing is easy when you have three or more objects that you want to evenly space the last two objects and the third object you are moving. When selecting objects or paths, the Smart Guides will show the path, anchor, or center being selected.

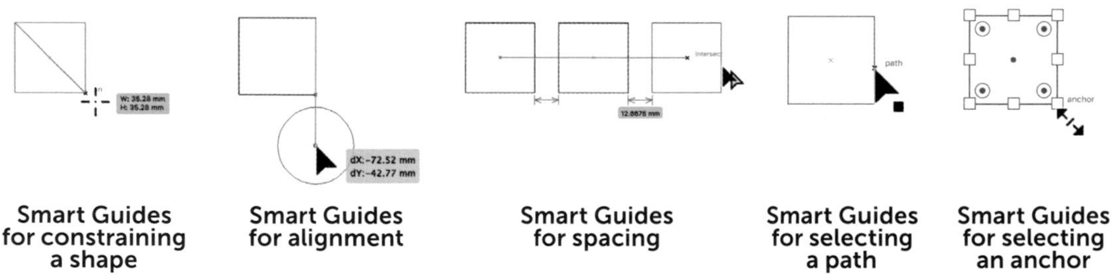

Smart Guides for constraining a shape | Smart Guides for alignment | Smart Guides for spacing | Smart Guides for selecting a path | Smart Guides for selecting an anchor

Smart Guide preferences can be set to show the specific smart guides and helpful hints such as labels for measurement, anchor points and path, spacing guides, rotation angles, and object highlighting. Choose Preferences > Smart Guides to edit the Smart Guide preferences. The default color of Smart Guides is magenta.

Using Smart Guides

Since Smart Guides are on by default, you will see them appear when you begin to draw or move anything in Illustrator.

- **Constrain:** When you draw a square, a Smart Guide will appear at a 45-degree angle when the square is constrained. When drawing a circle, the Smart Guide will appear vertical and horizontal to show the circle is constrained.
- **Measurement:** A small box in the lower-right area of the shape or path being drawn will show the measurements of that object or path. When rotating, the angle of rotation will be displayed. When moving, the X and Y offset will be displayed.

- **Spacing:** When there are three or more objects being spaced apart, the third object being moved will show Smart Guide distance spacers between it and the first two objects and between it and the second object. This is very helpful for keeping spacing consistent without using the Align panel.
- **Selection:** When you select an object or path, Smart Guides will show the path, anchor, center point, or pull handle so you know exactly what you are selecting. These Smart Guides show when you are selecting with the Selection tool, Direct Selection tool, Group Selection tool, Pen tool, Curvature tool, and many other tools.
- **Intersection:** When objects or paths are being moved, edited, or transformed, the intersection of anchors, centers of objects, or paths will be displayed.
- **Moving:** When moving an object or path, a Smart Guide will appear vertically or horizontally in relation to where the object or path was initially located.
- **Extension:** When a straight path has been drawn and you want to extend the line, Smart Guides will appear to keep the line on the same trajectory.

Grids

Illustrator has a base grid that can be turned on under View > Show Grid and turned off under View > Hide Grid. This grid is for precise creation and alignment of shapes and paths. When using the grid, you can choose to have objects snap to the grid for very precise drawing, editing, and moving of objects and paths. When creating icons or working on UI (User Interface)/ UX (User Experience) buttons, this grid can be of great help.

Snap to Grid

To turn on Snap to Grid, choose View > Snap to Grid. This Snap to Grid feature will snap any object, path, anchor point, or handle to the grid, even if the grid is not showing. Smart Guides are disabled when the Snap to Grid feature is turned on because the snap overrides all the Smart Guide features.

Grid measurements and preferences can be set up under Preferences > Guides & Grid. You can set the color of the grid and set the increments of the cells of the grid. Main grid lines can be set by using the Gridline every field; smaller subdivisions of the main grid are set in the Subdivisions field. Grids can be set in back of the artwork so all objects cover it.

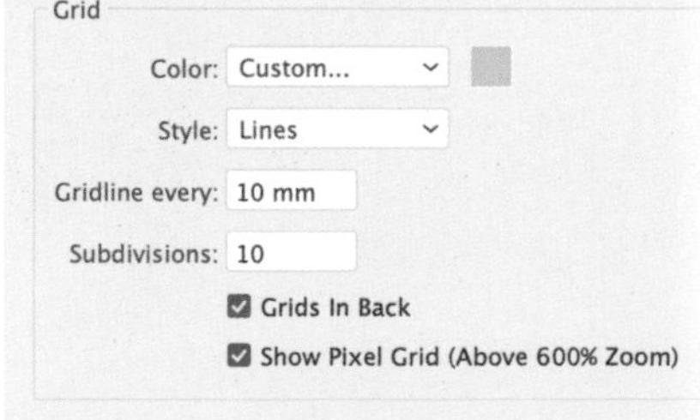

Align Panel

The Align panel allows alignment, distribution, and spacing of objects or paths. Choose Window > Align to open the Align panel. The Align panel shows Align Objects, Distribute Objects, and Distribute Spacing. If the Distribute Spacing section is not showing, choose Show Options from the Align panel dropdown menu to see all the align and distribute options.

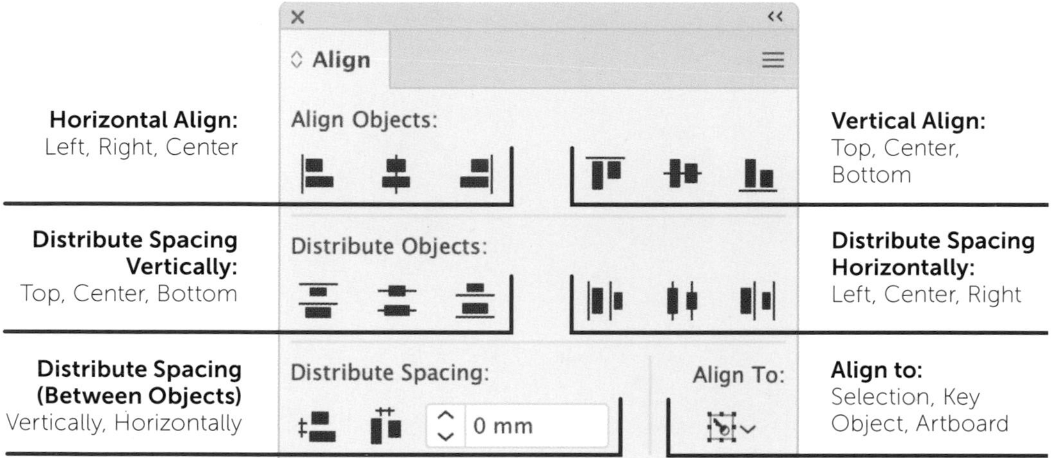

To use the align features, choose two or more objects and click on the desired align method. When aligning objects that are grouped, individual objects in the group will not align—the entire group will align.

Distribute

Using the Align panel, you can distribute objects based on common reference points. When objects are the same size, the spacing between them appears the same. Spacing objects and distributing objects is not the same, though the function appears to give similar results when the objects are the same size. When the objects are different sizes and the Distribute Spacing feature is used, the spacing between the objects is not the same. The Distribute Objects only controls the space between common points on the objects, in this example, the centers. To get the spacing between the edges of objects the same, use the Distribute Spacing option.

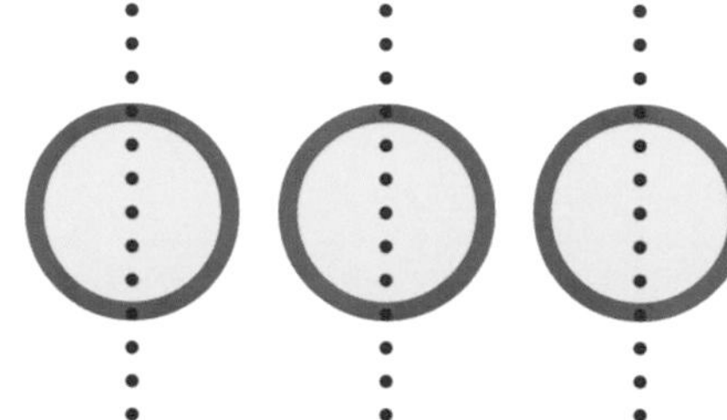

Distribute Objects when Distribute Between centers is applied. When the objects are the same size, the spacing between the object appears the same.

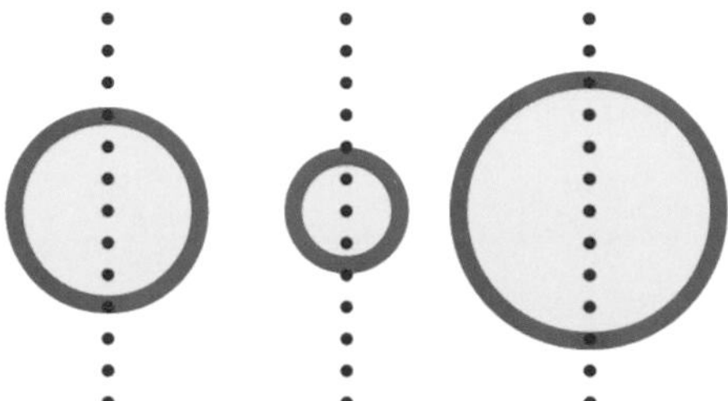

Distribute Objects when Between centers is applied. When the objects are different sizes, the spacing between the objects appears to be different yet the centers are equally spaced.

Spacing

The Distribute Spacing section of the Align panel is hidden by default. Choose Show Options from the Align panel dropdown menu to see the Spacing and Align to selections.

Choose two or more objects to define the space between them. Then choose the Align to function. The spacing field in which you enter a value will be grayed out and unaccessible unless you choose the Align to Key object or Align to Artboard.

- **Align to Key Object:** A Key is an object in the selected set of objects that you specify as the alignment point for all the other objects. Select all the objects, then click on one object to specify it as the Key Object. A thicker bounding box will appear around the selected Key Object. It will remain in its current position and all the other objects will move based on the alignment or distribution setting you specify.

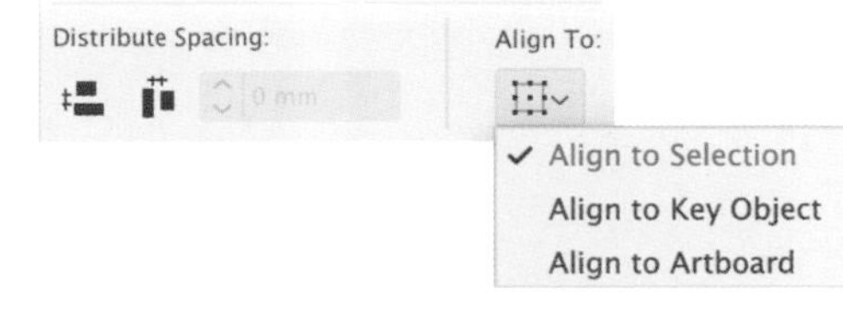

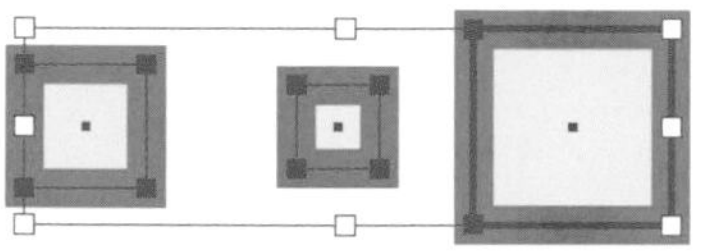

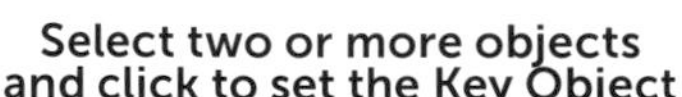

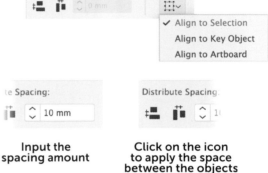

Select two or more objects and click to set the Key Object

Choose Align to Key Object

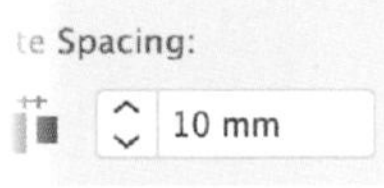

Input the spacing amount

Click on the icon to apply the space between the objects

- **Align To Artboard:** When aligning or distributing objects, you may want to align or distribute items to the artboard. When you select two or more objects and choose any of the Align options, it will align all the objects to that location on the artboard. When choosing Distribute Object or Distribute Spacing, it will distribute the objects to the edges of the artboard and distribute or space them evenly between the artboard edges. When using Distribute Spacing and Align to Artboard, the spacing field will be grayed out because this is not available when Align to Artboard option is chosen.

Precise Positioning and Sizing

When precise positioning and sizing is a requirement, you can snap your objects and paths to a grid. However, if the grid increments are not what you need to make each object a precise size, snapping to a grid can be frustrating.

Using the Transform panel, you can enter in measurements and sizes up to 3 decimal places. After 3 decimal places it will round to the nearest whole number. Precise positioning along the X and Y axis can be controlled, referencing the position of the object from the Ruler Origins point, and the Reference Point of the object. The size and rotation angle can be precisely controlled as well.

Using Math

Any field in the Transform panel uses common math equations to calculate values in the fields. By using add, subtract, multiply, divide, and percentage math calculations, you can precisely control the size, placement, and rotation of an object or path. All calculations will be performed based on the reference point of the object indicated in the Transform panel. In this example the reference point is the upper-right corner of the object.

- **Add (+):** Add any number(s) or percentages after the value in the field and press Return/Enter, and the value will be calculated.
- **Subtract (-):** Subtract any number(s) or percentages after the value in the field and hit Return/Enter.
- **Multiply (*):** Multiply any number(s) or per-cent-ages after the value in the field and hit Return/Enter.
- **Divide (/):** Divide any number(s) or percentages after the value in the field and hit Return/Enter.
- **Percentage (%):** Add, subtract, multiply, or divide percentages after the value in the field and hit Return/Enter.

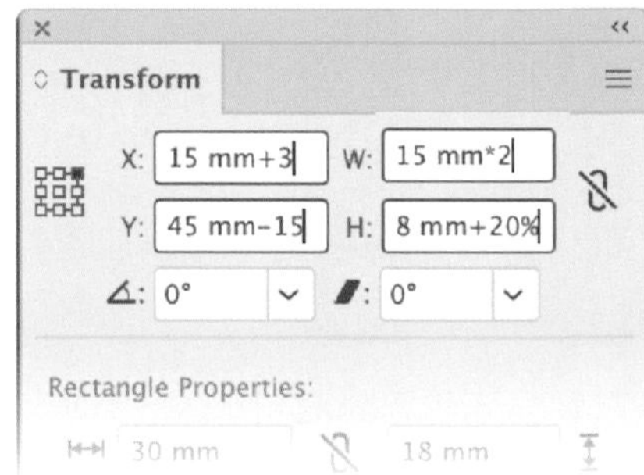

The values being entered into the equation can be different units. If the width is 50 mm, you can add 3 points to the value and it will convert the points to millimeters.

Duplicating

- **Option:** You can duplicate an object by holding Option/Alt, and dragging the object. Hold Shift while using the Option+click / Alt+click and drag to constrain the movement of object vertically, horizontally or at a 45-degre angle depending on which way the object is dragged.

  When moving or editing an object's location or size using the values in the Transform panel, you can also duplicate objects with precision. When entering a math equation into a field, hold Option/Alt before hitting the Return/Enter key will duplicate the object with that equation. For example, for an object that is 30 mm wide, I enter the following equation into the width field: 30 mm * 2. If I hold Option/Alt and hit Return/Enter, it will duplicate the object based on the reference point, and the duplicate object will be 60 mm wide.
- **Repeat Duplicate:** After a duplication, move, or transformation has been done, either using the manual Option/Alt click and drag method or using the Option/Alt math equations in the Transform panel, you can repeat the duplication or transformation by choosing Object > Transform > Transform Again.

> **SMART SHORTCUT**
> Command + D (Mac) or Ctrl + D (PC) will duplicate the last action you performed. Object > Transform > Transform Again.

Copy/Paste can also be used in a precise way. If you copy an object, then choose paste, it will paste the object in the middle of the screen with no regard to the artboard or the location from which you copied the object.

- **Paste in Place:** Choose Edit > Paste in Place to paste an object in the exact location it was copied from.
- **Paste on all Artboards:** To paste an object in the same location on multiple artboards choose Edit > Paste on all artboards.
- **Paste in Front:** This functions in the same way as Paste in Place. It pastes the object on top of or in front of the copied object.
- **Paste in Back:** This functions in the same way as Paste in Place. It pastes the object behind the copied object.

Moving

- **Shift Key:** Moving objects with precision can be done by using Smart Guides and moving objects along the horizontal or vertical axis to keep items in line with each other. You can also hold Shift when moving an object to keep it in line along the horizontal, vertical, or 45-degree angle axis.
- **Keyboard Arrows:** Moving objects by using the keyboard's up, down, left, and right arrows can also be done precisely. Under Preferences > General, you can set the Keyboard Increment, which is how much a selected object, anchor, or path will move using the keyboard arrows. To increase the Keyboard Increment amount by a factor of 10 when using the keyboard arrows, hold the Shift key.

General

Keyboard Increment: 0.1 mm
Constrain Angle: 0°
Corner Radius: 10 mm

Scaling

Scaling objects can be done using the fields in the Transform panel. Setting the reference point to the location you want to scale the object or path to is very important. Select the object and click on the reference point to scale the object from that point.

Hold shift down when scaling objects to keep them proportional. Manually scaling means the object is not governed by the selected reference point in the Transform panel. To manually scale an object from the middle, hold Option/Alt as you scale. Shift can be added to the Option/Alt command to scale from the middle of the shape and keep the object in proportion.

Make It Pixel Perfect

Creating artwork that is going to be used on the web, a mobile device, video, or digital display should be built pixel perfect so there are no fuzzy edges on the artwork, icons, or type. Illustrator shapes and paths are all vector-based, so why are we talking about pixels? Aren't pixels for photos and images? Many icons, buttons, and UI/UX items are built in Illustrator as vector then exported into a pixel-based image, so creating artwork that is pixel perfect is an absolute necessity. Pixels are squares of color and are displayed as whole pixels: and therefore should be built as whole pixels. Any pixel that is a partial pixel creates fuzzy edges. Creating artwork with partial pixels will leave it looking soft and unclear, no matter how large or small it is scaled.

- **Set up:** When creating any pixel-based creation, set your Units to pixels for General, Stroke and Type in Preference > General.
- **Pixel Preview:** Turn on the Pixel Preview under View > Pixel Preview.
- **Snap to Pixel:** Turn on Snap to Pixel under View > Snap to Pixel. When Pixel Preview is turned on, you can't use Smart Guides because the objects and paths will be snapped to the pixel grid.

When creating pixel-perfect artwork, a grid will appear in the background that you can snap objects to. This gird by default is not visible if you are below 600% zoom. Under Preferences > Guides & Grids, you can turn off the pixel grid. Creating artwork that snaps to the pixel grid may not give a clean effect on screen. Always create lines that are whole pixels. Any path or object that is on an angle or curve will look fuzzy; since pixels are squares, their rendering cannot create a smooth curve or angle. The pixels on the preview of objects or paths on an angle or curve will be blended to make them appear smoother. This is called anti-aliasing and it is how all pixel-based artwork is made to look smooth.

Even though you create objects and paths that snap to the pixel grid, and you use whole pixel weights, many of the vertical or horizontal lines may still look fuzzy. To fix this, select the objects and choose Object > Make Pixel Perfect. This will make any object snap perfectly to the pixel grid to create the cleanest object possible.

Artwork that is not pixel-perfect will have edges that are fuzzy

Artwork made pixel-perfect will have edges that are clean

Type that is not pixel-perfect will have horizontal and vertical edges that are fuzzy

Type made pixel-perfect will have vertical and horizontal edges that are clean

5 Effects and Graphic Styles

Opacity and Blending Modes

Transparency can be applied to objects or artwork to allow other objects to show through them. Any artwork or objects can be made transparent, as can gradients, patterns, and placed images. Open the Transparency panel under Window > Transparency to set the opacity and blending mode of objects and create opacity masks. You can also check the Knockout Group option to knock out sections of one object with a section of an overlapping transparent object. The Transparency panel does not show all the options by default; choose Show Options from the dropdown menu to display all the options.

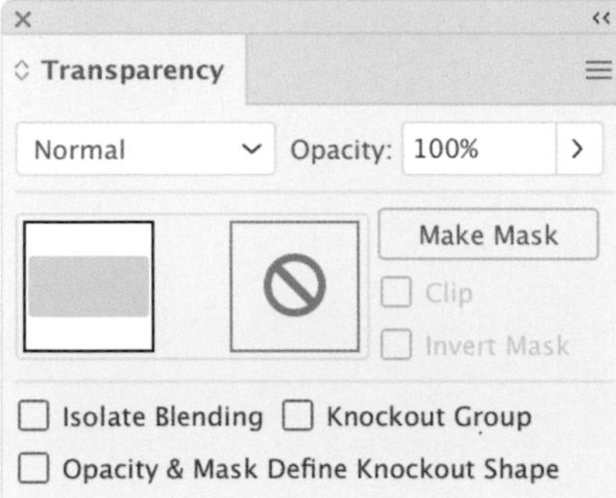

Show Transparency

When creating artwork and applying transparency, the transparent areas may look lighter or may not be easily identifiable since the artboard color is white. To see the transparency of objects, choose View > Show Transparency Grid. A checkerboard grid of white and gray will appear across the document. Any artwork that has transparency applied will show the grid through it.

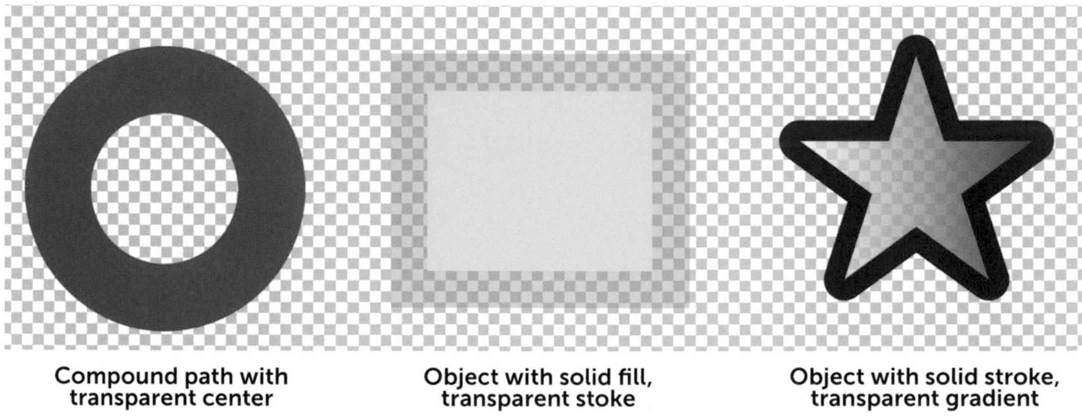

Compound path with transparent center | **Object with solid fill, transparent stoke** | **Object with solid stroke, transparent gradient**

The Transparency Grid preferences can be edited to change the color of the checkerboard pattern under File > Document Setup. You can also set the color of the artboard to mimic the color of the paper you want to print on in the Grid Options section, as well.

Change Opacity

You can change the opacity of a single object, grouped objects or layers, or an object's fill or stroke. The fill opacity and stroke opacity can be controlled separately or together. This can be done in the Appearance panel. Setting the opacity on the entire object or group of objects can be done in the Transparency panel or the Properties panel.

Setting Opacity

Open the Appearance panel (Window > Appearance), select an object, and then select the fill or stroke opacity in the Appearance panel. Click on the twirly to the left of the stroke or fill icon to open the options and show the Opacity link.

- **Stroke:** Set the opacity of the object's stroke in the Stroke entry field on the Opacity link.
- **Fill:** Set the opacity of the object's fill in the Fill entry field on the Opacity link.
- **Object:** Set the opacity of the entire object in the Opacity entry field on the Opacity link.

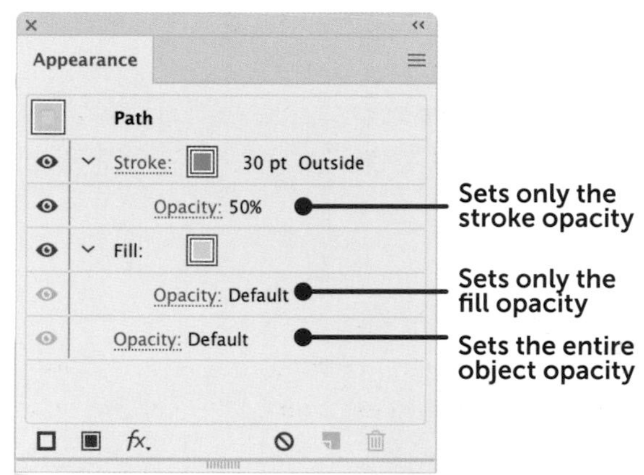

Knockout Group

A transparency Knockout Group is a group of objects that don't show through each other, but that will show through any object that is placed behind it. Create two or more objects and apply a transparency to those objects. Group the objects together under Object > Group. Select the Knockout Group checkbox in the Transparency panel. The grouped objects will no longer show through each other. Place the grouped objects over another object; the group will be transparent but the individual objects in the group will not be transparent with each other.

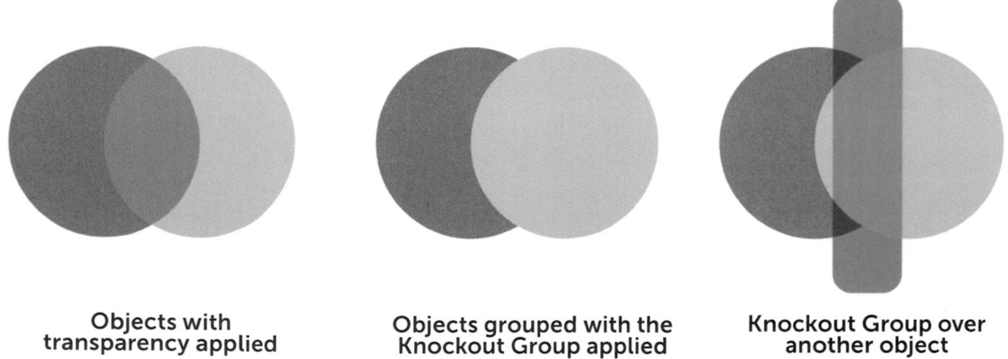

Objects with transparency applied | Objects grouped with the Knockout Group applied | Knockout Group over another object

Selecting the Knockout Group option cycles it through three states: on (check mark), off (no mark), and neutral (square with a line through it). Use the neutral option when you want to group artwork without interfering with the Knockout behavior determined by the enclosing layer or group. Use the off option when you want to ensure that a layer or a group of transparent objects will never knock each other out.

Opacity Masks

Masks are used to show and hide portions of artwork or images in Illustrator. Clipping masks and opacity masks differ from each other in that clipping masks will mask an object based on the shape or outline of the mask and opacity masks will mask an object based on the opacity of the fill of the shape.

A very simple way to understand a mask is: black conceals, white reveals. Any dark-colored object will conceal or mask more of the artwork; any light-colored mask will reveal more of the object. Illustrator bases the mask's opacity on the grayscale equivalent of the corresponding color in the mask. Black hides everything, white reveals everything, and shades of gray will partially hide the artwork.

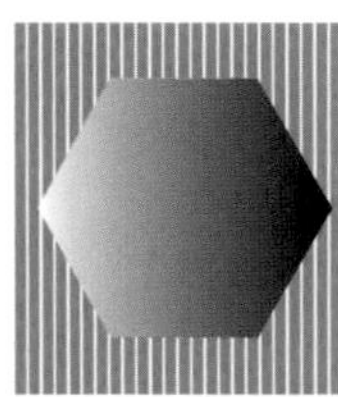

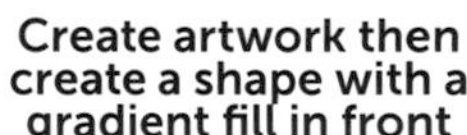

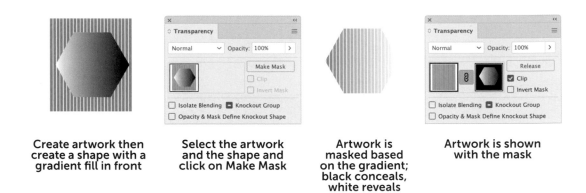

Create artwork then create a shape with a gradient fill in front

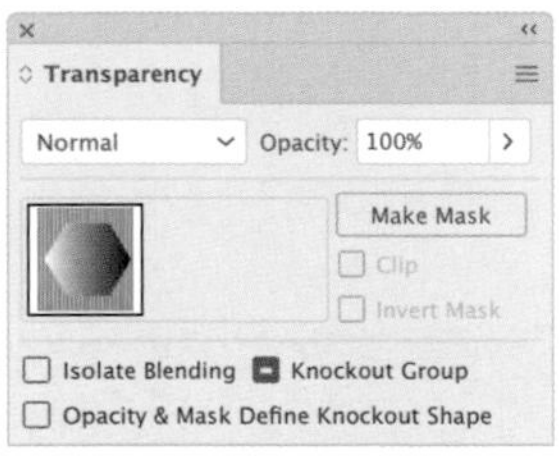

Select the artwork and the shape and click on Make Mask

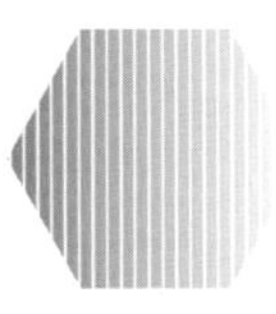

Artwork is masked based on the gradient; black conceals, white reveals

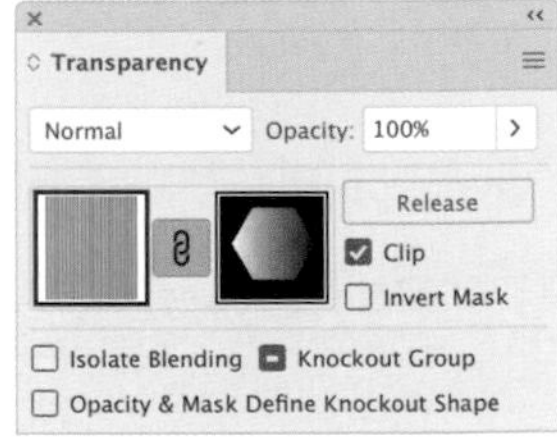

Artwork is shown with the mask

Create an Opacity Mask

To create the opacity mask, select the artwork or image you want to mask. Draw a shape over the artwork and fill it with a gradient. Darker colors will hide the artwork; lighter colors will show the artwork. Select both the shape and the artwork and click the Make Mask button in the Transparency panel. A thumbnail of the masking object appears to the right of the thumbnail of the masked artwork. By default, the artwork and the masks are linked together.

Editing an Opacity Mask

When you click the artwork thumbnail, the artwork and the mask move together. However, when you click on the mask thumbnail (the link between both is grayed out) you can move the mask without the artwork. You can unlink the mask and artwork by clicking in between the thumbnails and moving the mask or artwork independently. If you are used to working with clipping masks and editing clipping masks via Isolation mode, note that you can't enter Isolation mode to edit an opacity mask.

Blending Modes

Blending modes let you vary the ways that the colors of objects blend with the colors of underlying objects. When you apply a blending mode to an object, the effect is seen on any objects that lie beneath the object's layer or group.

My go-to blending modes are **Multiply** and **Screen**. Multiply takes anything that is white and makes it transparent; anything that is not white is made translucent. Screen does the same thing that Multiply does but it makes black transparent and everything else translucent. These blend modes are time savers not only for your vector artwork, but for images as well.

Due to the number of blending modes, and the very specific interactions that one color has with another, I went right to the Adobe Help website and copied the list of each blend mode and description to clarify the way the different blending modes work. Trying to put these into my words would not be of any help.

Here are a few terms to understand before jumping into each blending mode:

- **Blend color** is the original color of the selected object, group, or layer.
- **Base color** is the underlying color in the artwork.
- **Resulting color** is the color resulting from the blend.

Normal: This paints the selection with the blend color, without interaction with the base color. This is the default mode.

Darken: This selects the base or blend color—whichever is darker—as the resulting color. Areas lighter than the blend color are replaced. Areas darker than the blend color do not change.

Multiply: This multiplies the base color by the blend color. The resulting color is always a darker color. Multiplying any color with black produces black. Multiplying any color with white leaves the color unchanged. The effect is similar to drawing on the page with multiple magic markers.

Color Burn: This darkens the base color to reflect the blend color. Blending with white produces no change.

Lighten: This selects the base or blend color—whichever is lighter—as the resulting color. Areas darker than the blend color are replaced. Areas lighter than the blend color do not change.

Screen: This multiplies the inverse of the blend and base colors. The resulting color is always a lighter color. Screening with black leaves the color unchanged. Screening with white produces white. The effect is similar to projecting multiple slide images on top of each other.

Color Dodge: This brightens the base color to reflect the blend color. Blending with black produces no change.

Overlay: This multiplies or screens the colors, depending on the base color. Patterns or colors overlay the existing artwork, preserving the highlights and shadows of the base color while mixing in the blend color to reflect the lightness or darkness of the original color.

Soft Light: This darkens or lightens the colors, depending on the blend color. The effect is similar to shining a diffused spotlight on the artwork. If the blend color (light source) is lighter than 50% gray, the artwork is lightened, as if it were dodged. If the blend color is darker than 50% gray, the artwork is darkened, as if it were burned in. Painting with pure black or white produces a distinctly darker or lighter area but doesn't result in pure black or white.

Hard Light: This multiplies or screens the colors, depending on the blend color. The effect is similar to shining a harsh spotlight on the artwork. If the blend color (light source) is lighter than 50% gray, the artwork is lightened, as if it were screened. This is useful for adding highlights to artwork. If the blend color is darker than 50% gray, the artwork is darkened, as if it were multiplied. This is useful for adding shadows to artwork. Painting with pure black or white results in pure black or white.

Difference: This subtracts either the blend color from the base color or the base color from the blend color, depending on which has the greater brightness value. Blending with white inverts the base-color values. Blending with black produces no change.

Exclusion: This creates an effect similar to but lower in contrast than the Difference mode. Blending with white inverts the base-color components. Blending with black produces no change.

Hue: This creates a resulting color with the luminance and saturation of the base color and the hue of the blend color.

Saturation: This creates a resulting color with the luminance and hue of the base color and the saturation of the blend color. Painting with this mode in an area with no saturation (gray) causes no change.

Color: This creates a resulting color with the luminance of the base color and the hue and saturation of the blend color. This preserves the gray levels in the artwork and is useful for coloring monochrome artwork and tinting color artwork.

Luminosity: This creates a resulting color with the hue and saturation of the base color and the luminance of the blend color. It creates an inverse effect from that of the Color mode.

Note: The Difference, Exclusion, Hue, Saturation, Color, and Luminosity modes don't blend spot colors—and with most blending modes, a black designated as 100% K knocks out the color on the underlying layer. Instead of 100% black, specify a rich black using CMYK values.

Opacity

To select all objects that use a specific opacity, select an object with that opacity, or deselect everything and enter the opacity value in the Transparency panel. Then choose Select > Same > Opacity.

If you select multiple objects in a layer and change the opacity setting, the transparency of overlapping areas of the selected objects will change relative to the other objects and show an accumulated opacity. In contrast, if you target a layer or group and then change the opacity, the objects in the layer or group are treated as a single object. Only objects outside and below the layer or group are visible through the transparent objects. If an object is moved into the layer or group, it takes on the layer's or group's opacity, and if an object is moved outside, it doesn't retain the opacity.

Effects Menu

Raster and Vector Effects

The Effects menu in Illustrator has a list of effects that can be a applied to text, objects, or layers. The Effects menu is divided into two sections. The first section is labeled Illustrator Effects. These are effects that are applied to vector artwork such as objects, paths, strokes, and fills. The second set is labeled Photoshop Effects. These effects can be applied to either vector or raster objects. Not all the effects listed under Illustrator Effects are strictly for vector objects; 3D effects, SVG Filters, Warp effects, Transform effects, Drop Shadow, Feather, Inner Glow, and Outer Glow can be applied to both vector and raster objects.

Apply an Effect

Select an object and choose an effect from the Effects menu. Once the effect is applied it will appear in the Appearance panel. Effects can also be applied from the Appearance panel by clicking on the *fx.* button and choosing from the list of effects.

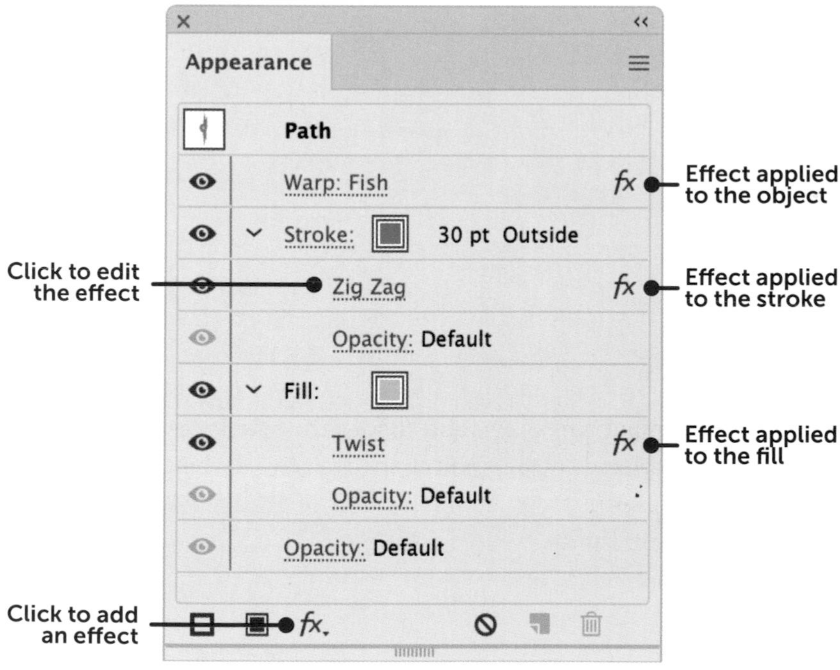

Modify or Delete an Effect

To edit the effect, click on the Effect link in the Appearance panel.

If you select the same effect from the Effect menu with the intent to edit the original, a dialog box will appear warning you that this method will create the same effect twice on the same object. By clicking on Apply New Effect, you will see another instance of the same effect appear in the Appearance panel.

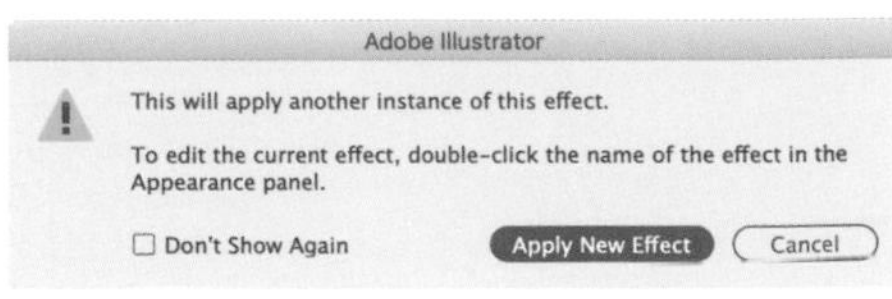

To delete an effect, select the effect in the Appearance panel and click the Delete button.

To turn off the effect but not delete it, click on the eye to the left of the Effect.

Applying Multiple Effects

Multiple effects can be applied to an object, stroke, or fill in the Appearance panel. Select the object name at the top of the Appearance panel to apply an effect to the entire object. Click on the path or fill to apply the effect.

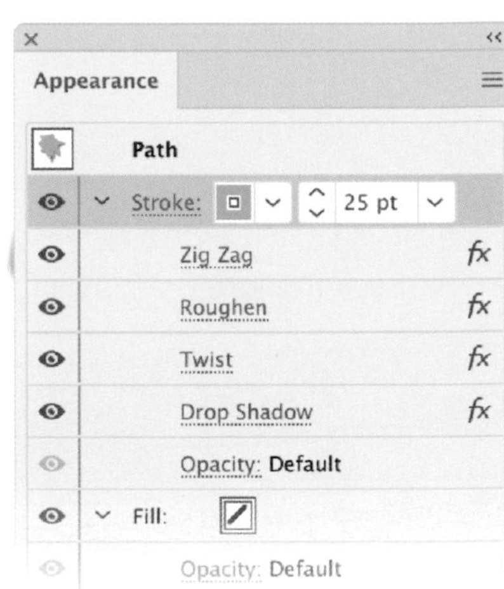

When effects are applied, they build on each other. Each effect will affect the previous effect. In this example, the Zig Zag Effect was applies first, the Roughen Effect second, the Twist Effect third, and the Drop Shadow Effect last.

The order of the effects can be changed by dragging them up or down the list, which will change the look of the object.

Scaling Effects

When effects are applied to an object, the default in Illustrator is that the effects do not scaled with the object. If an object is scaled with an effect, the effect will be distributed over the entire object, but it will not scale with the object. To turn on the Scale Stroke & Effects as needed, click the box in the Transform panel. To change the default settings of this feature, choose Illustrator > Preferences > General (Mac) and select Scale Strokes & Effects or choose Edit > Preference > General (PC).

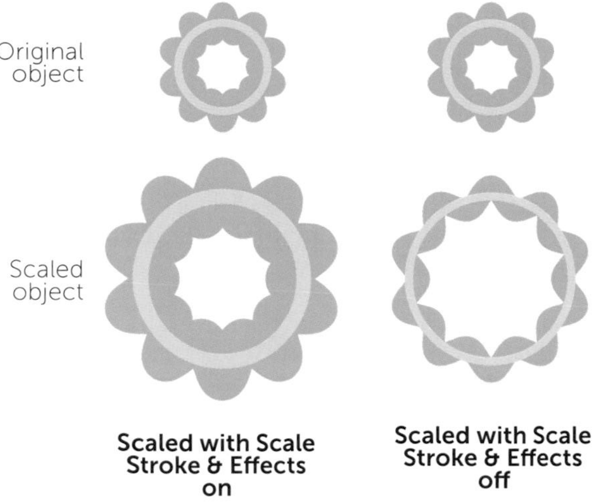

When the Scale Stroke & Effects is turned off, the object can be scaled. The stroke weight will remain the same and the effects will be distributed over the entire shape. The appearance of the effect may change as the object is scaled larger or smaller and may look quite different from the original shape. With Scale Stroke & Effects turned on, the effects and stroke will scale in proportion to the original shape. This will keep the same look of the effect when the object is scaled, keeping the object consistent.

Expand / Expand Appearance

The Expand function converts a path into a shape that matches the appearance of the path. Expand Appearance converts an objects effect to an object that matches the appearance of the effect. The Zig Zag (wavy line) Effect will be used to explain how these functions work.

The left side of this example is shown in Preview mode and the right half of this example is shown in Outline mode (View > Outline [Command + Y]).

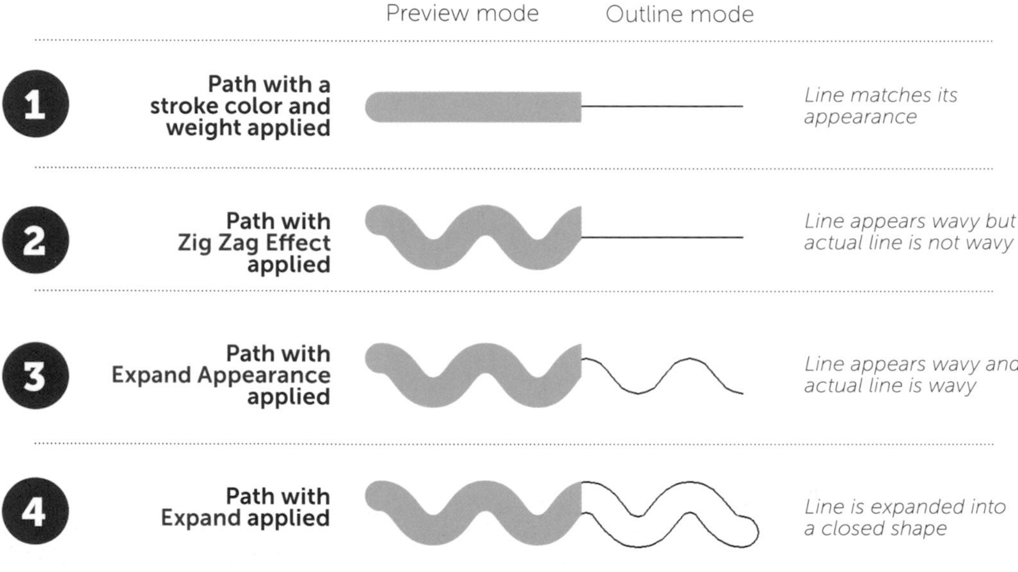
Preview mode
Outline mode
1
Path with a stroke color and weight applied
Line matches its appearance
2
Path with Zig Zag Effect applied
Line appears wavy but actual line is not wavy
3
Path with Expand Appearance applied
Line appears wavy and actual line is wavy
4
Path with Expand applied
Line is expanded into a closed shape

1. When you draw a line, the preview of the line is how it appears in Outline mode. In Outline mode, it displays the wire frame rendering of the object or path but does not show line weight, end caps, or stroke corners.
2. When an effect is added to the line (Effect > Distort & Transform > Zig Zag), you see the straight line with the effect of the Zig Zag Effect applied. In Outline mode, it appears as a line. The effect is just that, an effect. The wavy line isn't wavy, it just appears to be wavy.
3. Select the line with the Effect applied, and choose Object > Expand Appearance. This will convert the line into the actual shape of the effect. By Expanding Appearance, it is no longer a straight line, it matches the wavy appearance, thus the term Expand Appearance. This is still a line. The stroke weight and color can be edited, and end caps applied.
4. Expand Appearance in the Object menu is now grayed-out and no longer an option. Since the Effect or "Appearance" has been expanded, the line, in Preview mode, matches what you see, the only option left is Expand. Using Expand will convert the line to a shape. It will longer be editable as a line. The Expand feature works in a similar way as Outline Path.

Outline Path

Object > Path > Outline Path appears to work the same as Expand. Outline Path will take a path and create a shape based on its appearance.

If Outline Path is applied to step **1** in the example, the result will be a closed shape that looks like the original line.

If Outline Path is applied to step **2**, the result will be step **4**. It takes the line, expands the appearance, and expands the path to create a closed shape. It essentially combines Expand Appearance and Expand in one step.

Graphic Styles

A graphic style is a set of appearance attributes that can be saved and applied to other objects or paths. The Graphics Style panel is under Window > Graphic Styles. Illustrator has hundreds of preset styles to use, or you can create your own. To use a preset style, choose Open Graphic Styles Library in the Graphic Styles dropdown menu and choose from the list of Libraries. You can also click on the Library icon at the lower-left corner of the Graphic Styles panel to access the Graphic Styles Libraries.

Create a Graphic Style

To create a Graphic Style, create an object or path and apply appearance attributes such as fills, strokes, effects, and transparency in the Appearance panel. Once these attributes are applied to the object, select the object and drag it into the Graphic Styles panel to save the style. The pointer will show a green circle with a + in the center, indicating that the style has been added to the panel. Double-click on the name of the style to highlight and edit it. Hit return to exit and save.

Apply a Graphic Style

Create an object or path and click on the Graphic Style in the Graphic Styles panel to apply it. A Graphic Style stores the attributes applied to an object, however they do not store the shape of the object.

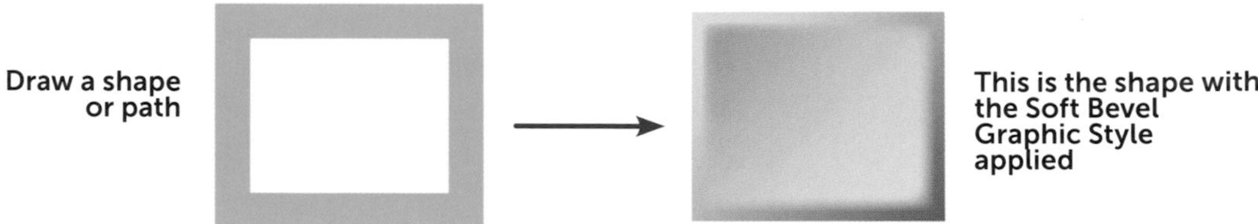
Draw a shape or path
This is the shape with the Soft Bevel Graphic Style applied

You can apply Graphic Styles to single objects, groups of objects, and layers. When you apply a graphic style to a group or layer, every object in the group or layer assumes the attributes of the Graphic Style.

Remove a Graphic Style

To remove a Graphic Style from the object or path, select the object or path and choose Clear Appearance from the Graphic Styles dropdown menu. This will remove all the attributes on the shape and leave it with no stroke, fill, or effects.

Edit/Replace a Graphic Style

In Illustrator, an existing Graphic Style can not be directly edited and resaved, it has to be replaced with another style. Select an object, path, or group that has the new attributes you want to use. Option-drag (Mac) or Alt-drag (PC) it onto the Graphic Style you want it to replace in the Graphic Styles panel. All occurrences with the Graphic Style will have the new style applied to it.

Break the Link to a Graphic Style

When replacing a Graphic Style, all occurrences of it are updated. If you want to have certain object not be affected by the updated Graphic Style, select the object or path and choose Break Link to Graphic Style from the Graphic Style dropdown menu. The object, group, or layer will retain its original appearance attributes, but they will no longer be associated with the original Graphic Style.

Effects Expanding

When an effect is applied to an object, the appearance of the object will reflect that effect. To edit the effect, you will need to go to the Appearance panel and select the shape, and then double-click on the effect to edit the appearance. To turn the effect into an editable object, it needs to be Expanded. The results of Expanding an effect will appear in the Appearance panel, and in some cases, in the Layers panel.

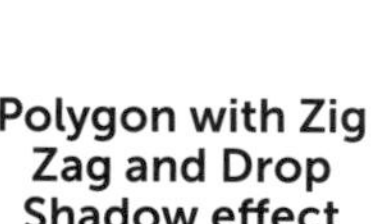

Polygon with Zig Zag and Drop Shadow effect

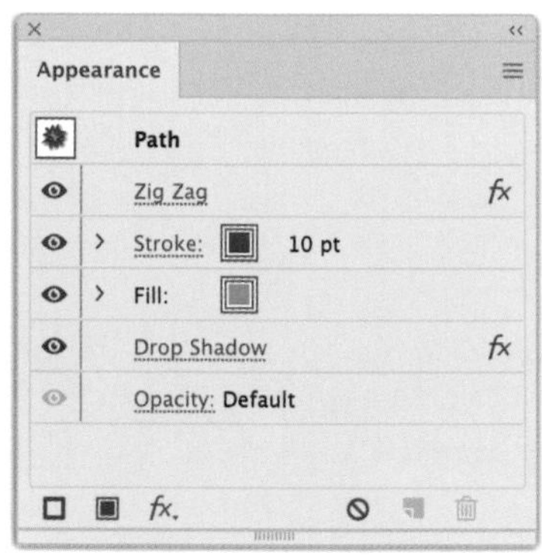

Appearance panel showing the effects applied to the Polygon

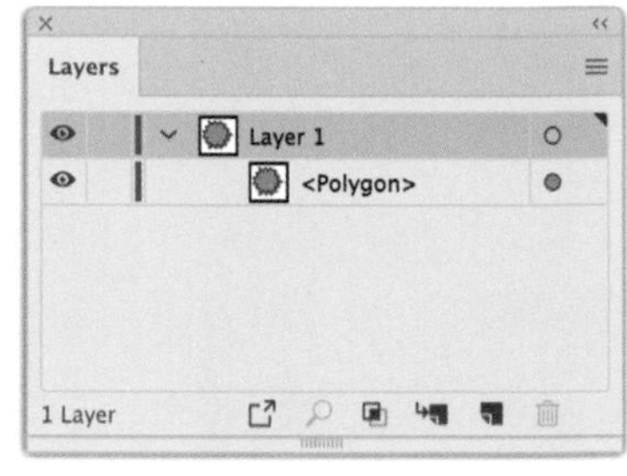

Layers panel showing the Polygon as a single shape and layer

Select the object and choose Expand Appearance from the Object menu. The effects on the shape will be broken out into actual shapes or images that will look like the original object. Vector effects will be expanded into shapes, raster effects (such as bevel and emboss and drop shadow) will be expanded into pixel-based images. These shapes and images will appear in the Layers panel as individual layers, and the Appearance panel will no longer show the editable effects.

Polygon with Expand Appearance applied

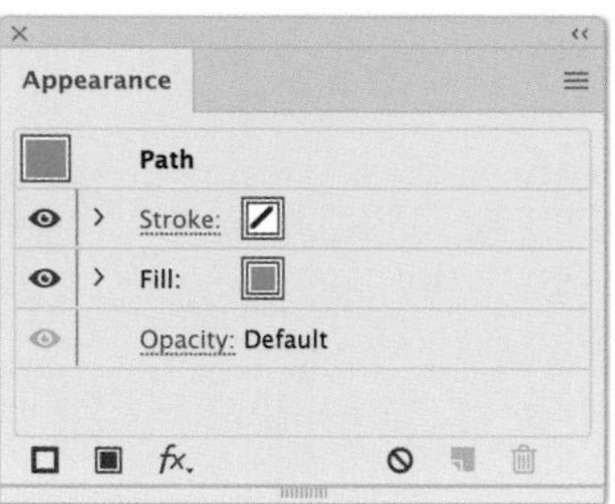

Appearance panel showing no effects on the polygon

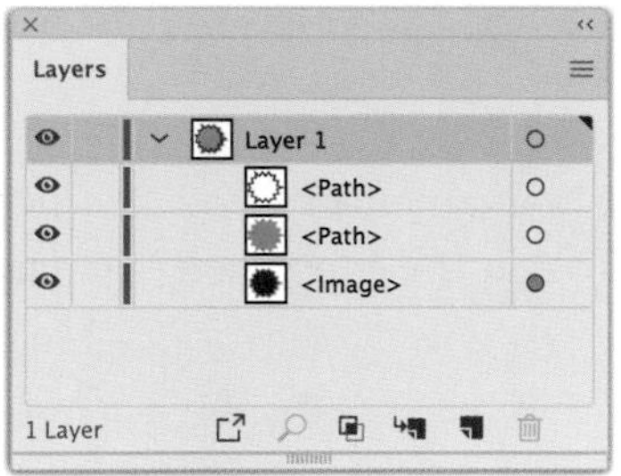

Layers panel showing that the polygon and each effect has been expanded into shapes or images

Clipping Masks / Draw Inside

A clipping mask in Illustrator is a shape that masks other artwork, making only the artwork inside the shape visible. To understand how a clipping masks works, think of a frame and a painting you want to put in that frame. You choose the frame and mount the artwork from behind the frame to make a clipping group. Once a clipping group is created, accessing the frame or the content, or adding or deleting content to a frame requires a few steps to get into editing mode.

Create a Clipping Mask

There are three ways to create a clipping mask with a shape and artwork:

- **Clipping Masks** > **Make**: Create the artwork and group it together. Create a frame on top of the artwork. Select both the group and the artwork and choose Object > Clipping Mask > Make. **Note:** There is one drawback to this method; it removes all the attributes from the frame when the Clipping Mask command is performed.

Artwork created and grouped

Frame created on top of the artwork

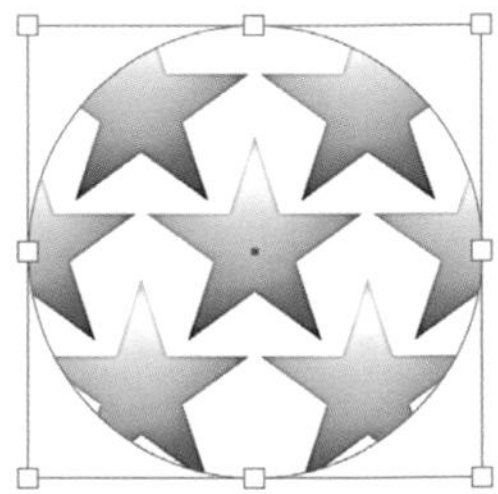

Artwork and frame selected; Object > Clipping Masks > Make

- **Draw Inside Mode**: Create the artwork and group it together. Create a frame with the fill and stroke attributes you want on the frame. Select the grouped artwork and choose Edit > Cut. Select the frame and click on the Draw Inside Button at the bottom of the toolbar. The frame will be highlighted with a dashed line. Choose Edit > Paste to paste the grouped shape into the frame. Move and scale the grouped artwork and then click on the Draw Normal mode to exit the Draw Inside mode.

  The benefit to using this method is that the frame retains its attributes when a Clipping Mask (Draw Inside) is performed. Another benefit is that once the frame is set to Draw Inside, you can create shapes or lines inside the object. With Draw Inside mode you have to draw or raster content inside the frame; you cannot drag any content into the frame.

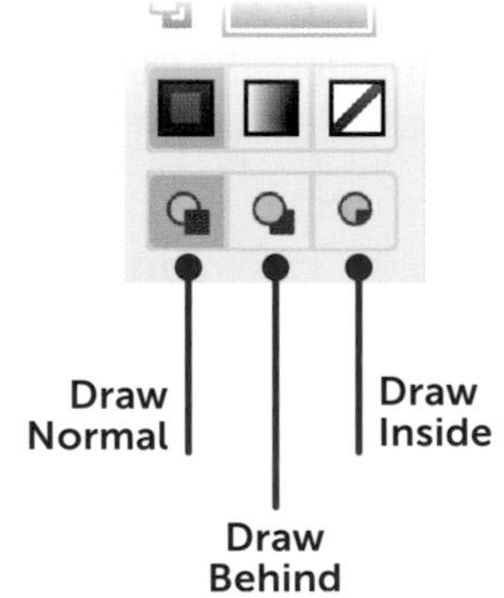

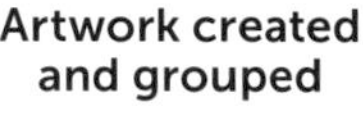

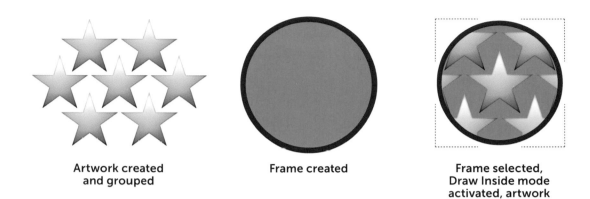

Artwork created and grouped

Frame created

Frame selected, Draw Inside mode activated, artwork pasted inside

- **Layer Clipping Mask**: Create a frame with the fill and stroke attributes you want on the frame. Open the Layers panel and choose Make Clipping Mask from the Layers dropdown menu. The name of the object on the layer will show as underlined, indicating it is a clipping mask. All the objects drawn on that layer will appear inside the frame or clipping mask. The frame will lose its attributes once it is made into a clipping mask. The frame attributes can be reapplied once the clipping mask has been applied.

Edit a Clipping Mask

When a clipping mask is made, the frame and the content act as one object. To edit a clipping mask and its content, you have to access either the content or the frame. There are four ways to edit the frame and content of a clipping mask.

- **Isolation Mode**: Select the clipping group and double-click on the frame or content to isolate one from another. At the top of the document window a gray bar will appear showing the isolation of the objects. Select the frame or content to edit them. To exit Isolation mode click on the arrow left of the isolation bar or use the Escape key.

Layer 1 <Clip Group>

- **Properties Panel:** Select the clipping group and choose the Edit Clipping Path icon or the Edit Contents icon in the Properties panel. Edit the clipping path or the content once it is selected. An alternate way to select the Clipping Path is the Isolate Mask button at the bottom of the Properties panel in the Quick Actions section.

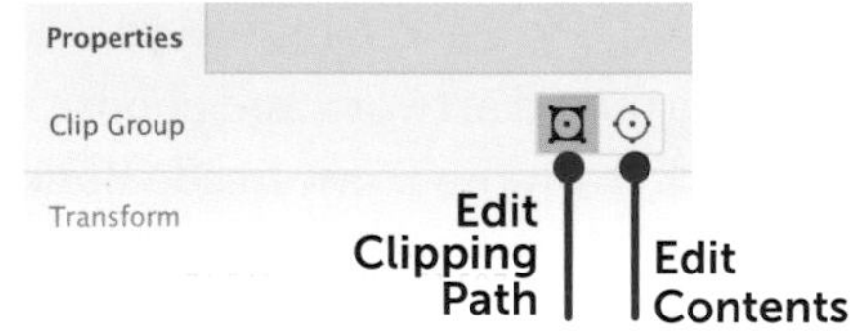

- **Object > Edit Contents:** Select the clipping group and choose Object > Clipping Mask > Edit Contents. This will allow you to isolate and edit the contents of the clipping group. To release the content from the mask, choose Object > Clipping Mask > Release.

- **Layers Panel**: Use the Layers panel to select and edit the clipping group and all of its components. Select the clipping group in the document or in the Layers panel. Expand the clipping group by clicking on the twirly to the left of the clipping group, revealing the frame and all the clipped content. To select the frame, click on the target dot to the right of the layer name. To select the content inside the clipping group, click on the target dot to the right of the item in the group (or the entire group). Using the Layers panel to edit, add, or remove content from a clipping group streamlines the editing process; a big advantage over the three previous methods of editing a clipping group.

  To add an object to a clipping group, create the artwork and position it where you want it to appear on the existing artwork. Expand the clipping group in the Layers panel by clicking on the twirly to the left of the clipping group. This will show the frame and the contents. Drag the layer that the artwork is on into the clipping group and rearrange the layer order if necessary. The advantage of using the Layers panel to add or remove content from a clipping group is you can do so without isolating the frame or the content. Another advantage is the frame attributes can be edited directly by selecting the frame in the Layers panel and editing it in the document. Contents of the clipping group can be selected directly in the Layers panel and edited without the need to isolate or release the clipping mask.

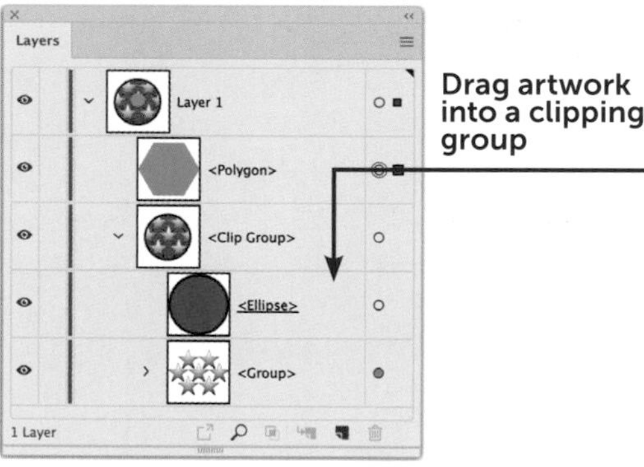

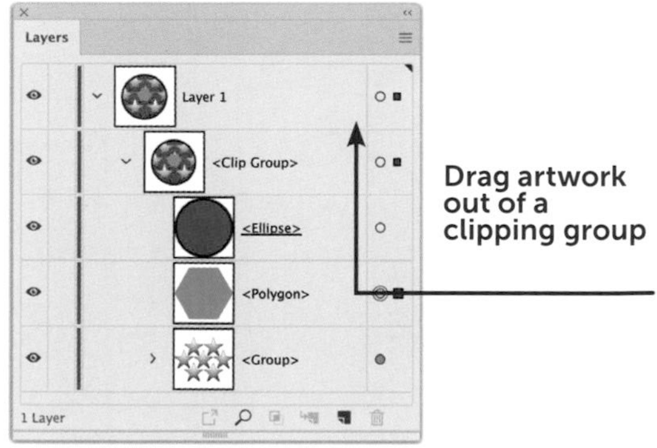

To add content to an existing clipping group, create content and position it on the artwork. Locate the layer in the Layers panel that the content it is on and drag the layer into the clipping group.

Release a Clipping Mask

To release a clipping group and return the frame and artwork to its original, non-clipped state, select the clipping group and choose Object > Clipping Mask > Release. Release Mask is also in the Quick Actions section at the bottom of the Properties panel. This allows you to release the clipping mask and its content back to the original frame and content, taking it out of clipping mode. To release the clipping group, select the clipping group and click on Release Clipping Mask in the Layers panel dropdown menu.

6 Type and Text

It's never too early to talk to your child about typography.

Working with Type

The world of typography is wonderful: making logos, designs, and artwork expresses voice and feeling in a small yet impactful way.

Whether you're an aspiring designer, typography enthusiast, or you just enjoy working with type, understanding basic terminology of type characteristics, spacing, layout, and attributes is essential. Here is the basic terminology regarding type and typography to help you understand how type is measured and expressed.

Typefaces, Categories, and Styles

Font/Typeface:

Back in the days of metal type and printing presses, fonts and typefaces were two different things—the typeface was the specific design of the letters, say Times New Roman or Baskerville; while the font referred to the particular size or style of that typeface, say 10 point regular or 24 point italic (each created as its own collection of cast metal letters and other characters). Today, however, many designers use the terms more or less interchangeably.

"A collection of letters, numbers, punctuation, and other symbols used to set text (or related) matter. Although font and typeface are often used interchangeably, font refers to the physical embodiment (whether it's a case of metal pieces or a computer file) while typeface refers to the design (the way it looks). A font is what you use, and a typeface is what you see."

Character:

An individual symbol, letter, number, punctuation mark, etc., which is a part of the overall set in a typeface.

Alternate Character/Glyph:

Character variations that may be decorative, and can be an optional character within the set of characters. Not all fonts contain alternate characters but many contain optional characters that are not readily accessible through the keyboard keys. The Glyphs panel (Type > Glyphs) shows the additional characters in a font.

Serif:

A serif is a small line or stroke attached to the open end of a larger stroke in a letter or symbol within a particular font family. Serif can also be used to describe an entire font set.

Sans Serif/Sans:

Sans Serif means "without" serif, sometimes simply shortened to Sans. This is a character that has no small line or stroke added to the open ends of the stroke.

Sans Serif
Myriad

Serif
Minion

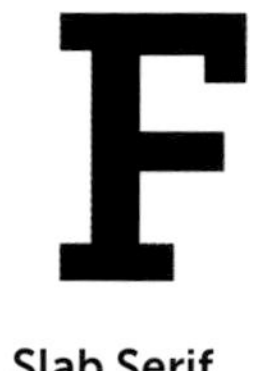

Slab Serif
Museo Slab

Hand Written
Motion OT

Ornamental
Edwardian Script

Preferences & Workspaces
Shape Creation
Advanced Construction
Editing & Transformation
Effects & Graphic Styles
Type & Text

Positioning, Spacing, and Measuring

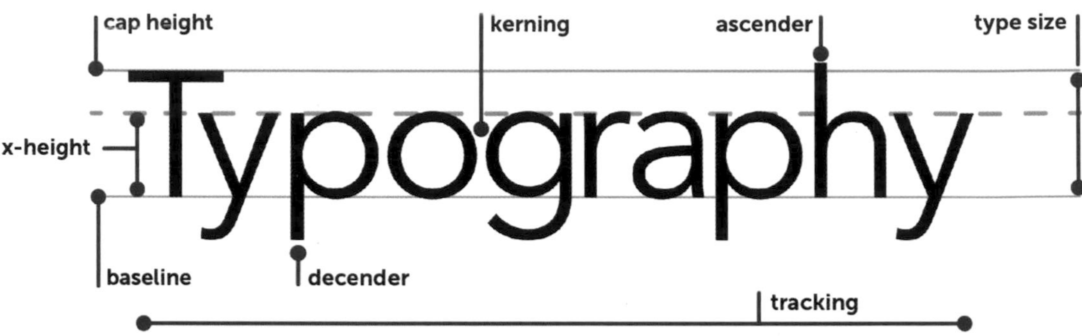

Measuring

The two units of measurement commonly used for measuring type are points and picas. Points are used to indicate the size of type or the leading (spacing between lines). A point is 1/72 on an inch. Picas are larger units of measure, as 6 picas make one inch. Quick conversion is 12 points make up 1 pica, 6 picas make up an inch, and 72 points make up an inch.

Baseline

This is the invisible line that letters and other characters sit on. Rounded characters sit slightly below the baseline; if the rounded part of the character sat exactly on the line, it would appear to not line up with other characters.

Kerning

The horizontal spacing between two characters. Adjusting the kerning between characters helps to create the appearance of consistent spacing between certain letter combinations. The spacing between each letter combination is different so kerning is not an exact spacing amount—it's about overall visual consistency. A rule of thumb is to kern any type larger than text because the letter spacing is more noticeable in headlines and subheads.

Spectacular — Kerning with visually consistent spacing

Spectacular — Kerning with visually inconsistent spacing

Tracking

Kerning and Tracking are in essence referring to the same attributes, the only difference being that kerning is the space that is adjusted individually between two characters (normally in a headline), while tracking is the spacing between characters that is adjusted over all the text. Tighter tracking reduces the space between each character; looser tracking increases the space between each character. Tracking that is too loose or tight is harder to read; the right amount of spacing between each letter helps to define each character.

A spectacular time was had by all, culminating in an event like no other.	Positive (more space) amount of tracking applied to text over multiple lines
A spectacular time was had by all, culminating in an event like no other.	Default amount of tracking applied to text over multiple lines
A spectacular time was had by all, culminating in an event like no other.	Negative (less space) amount of tracking applied to text over multiple lines

Leading/Line Spacing

The vertical spacing between lines of text is measured from baseline to baseline. The term leading comes from the day when type characters were made of lead. The lead type was placed into a jig, which was then mounted on a press. Ink was applied to the type and pressed into paper. Leading referred to actual strips of lead that gave spacing between the lines of the characters. Leading is measured in points.

The Anatomy of a Letter

X-Height

This is the height of a typeface measured from the baseline to the top of lowercase letters disregarding ascenders and descenders.

Descender

This is a portion of a letter that extends below the baseline.

Ascender

This is a portion of a lowercase letter that rises above the x-height.

Text Creation

Creating text in Illustrator requires the Type tool. There are seven type tools depending on how you want to create and edit type. The main type creation and editing tool is the Type tool (shortcut T). Once you select the Type tool, choose between point text and paragraph text.

Point Type/Paragraph Type

Point type is text that is created by selecting the Type tool and clicking on the artboard to begin typing. The type starts at the point of clicking and continues in a horizontal line. **Paragraph type** is text that is inside a text container. To create paragraph type select the Type tool and click and drag on the artboard to create a text container. Once the container is created, you can type inside it and the text will stay within the bounds of the container.

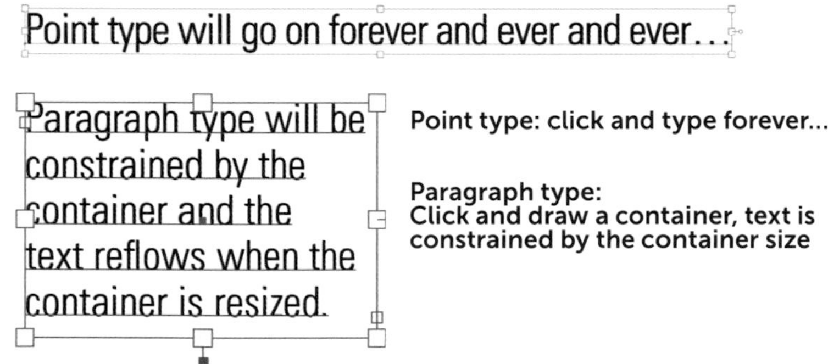

Point type is generally used for short headlines, labeling a map, numbers on an infographic, or short bits of type. Paragraph type is for when you are typing something with several lines of text that flow within the text container.

Start Typing

Select the Type tool and click on the artboard to begin typing using Point text. Click and drag using the Type tool to create a container for paragraph text. Either method will create placeholder text when you begin. The placeholder text shows the font and the size of the type; once you begin typing the placeholder text disappears. You can disable the default Illustrator setting of filling new type containers with placeholder text. Choose Preferences > Type > Fill New Type Objects With Placeholder Text and click that option off.

Clicking on the artboard and typing with the Type tool will result in a single line of text that will go on forever. Point type can be broken into more than one line by inserting the type cursor and hitting the return key to break the line.

Creating a Text Box

Clicking and drawing a container with the Type tool will result in a text container (text box) that will hold the type. This is referred to as paragraph type. Scaling or editing the container with the Selection tool will cause the type to adjust to the container as it is resized.

Text in a Container

Text can be typed into any shape of container. Draw any container with the Shape tools, select the Area Type tool and click on the path of (not inside) the container with the arrow of the area type tool to activate the container to receive type. The type cursor will show a dotted line around it, indicating the text will go inside the container. Once the container is activated with the area type tool, the shape will fill with placeholder text that you can replace. If the object has stroke or fill attributes, all the attributes will be removed when the shape is turned into a text container. Click on the Selection tool when you are done typing.

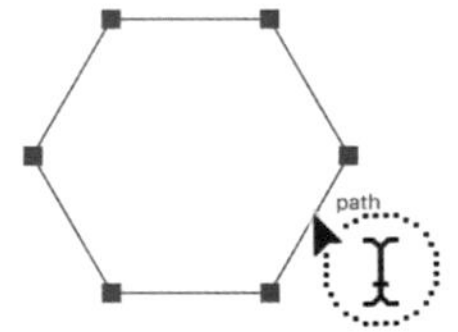

Type on a Path

To create type on a path, draw a line or a shape. The path can be an open path or closed shape. Select the Type on a Path tool. Click on the path at the location where the “arms” are coming out of the Type on a Path tool to convert the path to accept type. Once the path can accept type, it will show placeholder text.

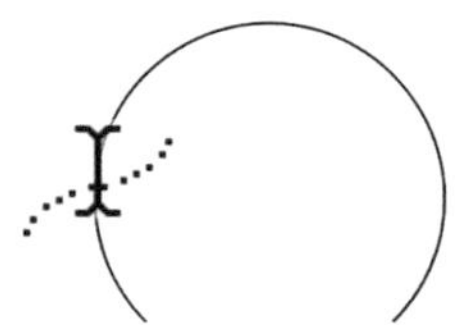

Setting Character Attributes

Setting the character attributes can be done in the Properties panel. Select the Type tool and click on and select the type you want to edit, then choose from the Character attributes. The Properties panel displays the basic Character formatting. By clicking on the three dots in the lower-right portion of the Character section will open the full Character panel. The Character panel can also be accessed under Window > Type > Character (shortcut Command + T). When the Character panel is open, the default display does not show all the options. Click on the dropdown menu and choose Show Options to see the entire Character panel as seen here.

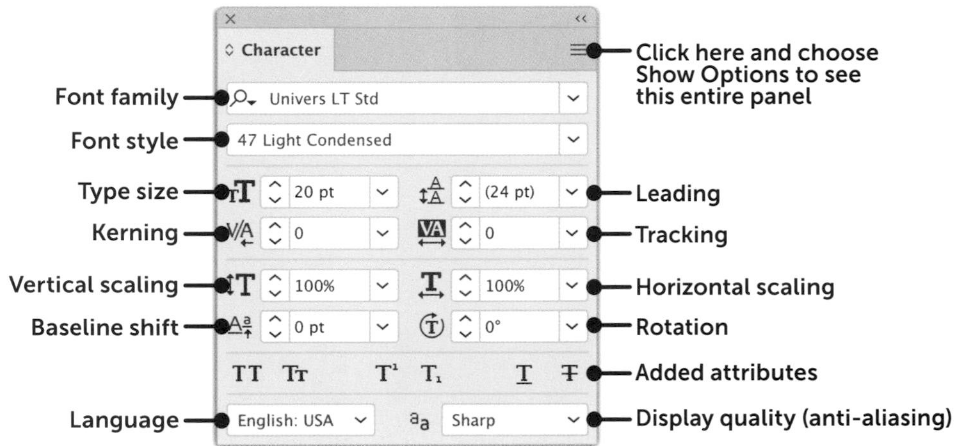

Fonts

Select a font from the dropdown menu. To see how the selected type looks in the selected font, click on the Selected text dropdown and choose from a list of display options. To increase and decrease the size of the display, click on one of the three display options. Filters can filter the fonts by classification (serif, sans serif, decorative, etc.), favorite fonts, recently used fonts, and activated fonts respectively. Font styles are different styles of the font family: light, book, bold, italic, heavy, condensed, or extended options. Not all type faces will have additional styles of the chosen font; some fonts have numerous styles.

Size

1. To set the type size, select the type you want to resize and choose one of the following options: Click on the type size dropdown menu and choose from a list of preset sizes.
2. Click on the up and down arrows to the left of the size field.
3. Click inside the type size field and use the up and down arrows on the keyboard.

4. Select the value in the field, enter in a new value and hit return.
5. Use the shortcut to resize the type. Command + Shift + > (Mac), Ctrl + Shift + > (PC) to increase the type size, Command + Shift + < (Mac), Ctrl + Shift + < (PC) will decrease the type size. When using the keyboard shortcuts, the point size will increase and decrease in preset increments. To change these increments choose InDesign Preferences > Type (Mac) or Edit > Preferences > Type (PC). At the top of the type preferences section you can set the increments for adjusting type. The defaults are 2 points and 20/1000 em for Tracking and 2 points for baseline shift. I like to reduce the defaults to smaller increments for finer adjustments when using the shortcuts. I use 1 point for Size/Leading, 10/1000 em for Tracking and 1 point for Baseline Shift. When you change the settings with a file open, you are only changing the preferences for that file. The Tracking setting also sets the Kerning increments as well.

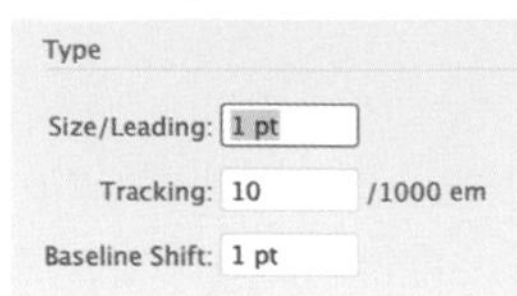

Leading

To set the leading amount (space between lines of text), select the type you want to edit the leading on and choose one of the following options:

1. Click on the Leading dropdown menu and choose from a list of preset sizes.
2. Click on the up or down arrows to the left of the size field.
3. Click inside the Leading field and use the up or down arrows on the keyboard.
4. Select the value in the field, enter in a new value and hit return.
5. Use the shortcut to adjust the leading. Holding Option + the up arrow will decrease the leading amount (move the lines of type up), Option + down arrow to increase the Leading amount (move the lines of type down).

6: Choose Auto Leading, which will increase/decrease the Leading amount automatically along with the type size. Auto adjusts the leading 20% more than the type size; if the type is 12 pt, the leading is automatically set to 14.4 pt.

Kerning

To set the kerning amount (space between two characters), click the type cursor and the type you want to edit and choose one of the following options:

1. Click on the kerning dropdown menu and choose from a list of preset sizes. Positive numbers will offer more space; negative numbers offer less space.
2. Click on the up or down arrows to the left of the size field.
3. Click inside the kerning field and use the up or down arrows on the keyboard.
4. Select the value in the field, enter in a new value, and hit return.
5. Use the shortcut to adjust the kerning. Press Option + the right arrow to increase the kerning amount, or Option + left arrow to decrease the kerning amount.

6: Auto (Metric) Kerning was built into the font when it was created. Illustrator uses this auto kerning to create tighter and more visually pleasing spacing between letters. Optical Kerning is also built into a font. It uses an algorithm that calculates the space between letters based on their shape.

Tracking

Tracking and kerning are the same feature. The only difference is that kerning is adjusted when you click the cursor between two characters and adjust the individual spacing; tracking is used when lines or paragraphs of text are selected and the letter spacing is adjusted all at one time. The shortcuts are the same for both kerning and tracking.

Scaling

Scaling text vertically or horizontally can be done, however, it is not generally advised. From a graphic design point of view, choose a font that is extended or condensed instead of stretching the type. Stretching the type is considered a typographic sin in the eyes of designers and type aficionados because it distorts the original letter form design.

Baseline Shift

Baseline shift is used for footnotes and the ® and ™ symbols. This shifts the type off the baseline and positions it generally above the x-height of the type. Baseline shift can be done one character at a time and is not meant to be a substitute for leading. In other words, do not use baseline shift for an entire line of copy! To set the baseline shift, select the type you want to edit and choose one of the following options:

1. Click on the baseline shift dropdown menu and choose from a list of preset sizes. Positive numbers will move the type up, negative numbers will move the type down.
2. Click on the up or down arrows to the left of the size field.
3. Click inside the baseline shift field and use the up or down arrows on the keyboard.
4. Select the value in the field, enter in a new value and hit return.
5. Use the shortcut to adjust the kerning. Shift + Option + the up arrow will shift the type up, Shift + Option + the down arrow will shift the type down.

Character Formatting Styles

Setting type to be all caps, lowercase (initial caps), super script, subscript, underlined, or strike through can be selected at the bottom of the Character panel.

You may notice that there are no Bold or Italic options (called fake bold or fake italic). If you want to make type bold or italic, choose the bold or italic style of the font in the font dropdown menu. If the font you are using does not have bold or italic, choose another font that does. Adobe took out the fake bold and fake italic because it caused problems with printing. Applying a fake bold or italic would display one way but would not reproduce correctly.

Language

The language dropdown menu does not convert your text into the language selected (that would be awesome though). This is used by the built-in dictionary for spell checking and hyphenation options.

Display

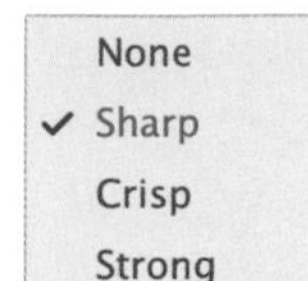

The Display dropdown menu is for rendering type when it is to be used for web-based export. When an Illustrator file is saved for the web, in many cases the type will be rasterized (converted into pixels). These options are for you to choose the look that the type will render as. In order to see the text as it would appear on a website, choose View > Pixel Preview and then choose the None, Sharp, Crisp, or Strong option. If an Illustrator file is printed, saved as a PDF or an SVG file, these options are not relevant. This is only for saving the file for the web.

Setting Paragraph Attributes

Setting the character attributes can be done in the Properties panel or in the Paragraph panel under Window > Type > Paragraph (the shortcut is Option + Command + T) select the Type tool, and highlight the type you want to format.

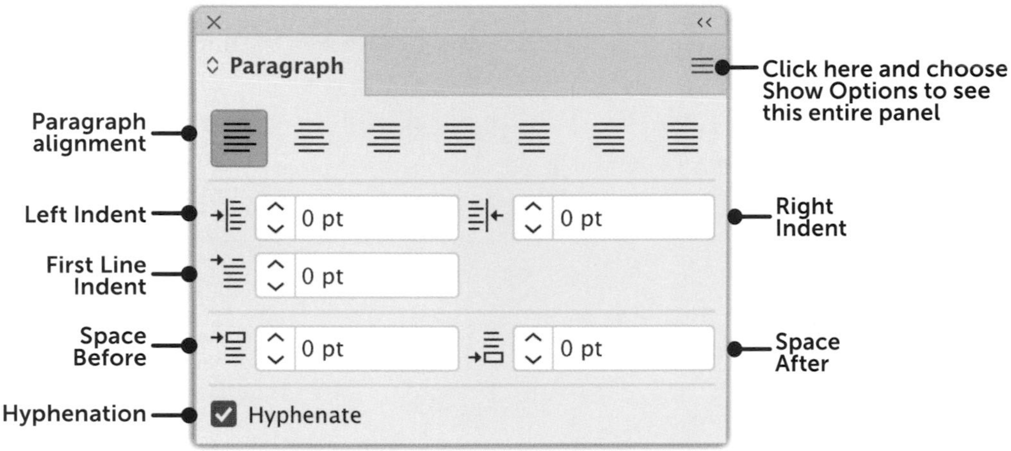

To see all the panel options, click on the dropdown menu and choose Show Options. Alignment can be set to make the type flush left, centered, flush right, force-justified, force-justified, with the last line centered, force-justified with the last line right-justified and force-justified for all lines. Left and right indents are used when formatting type when you want to have the selected type not be at the left or right margins. Paragraph formatting is controlled by hitting the return key. When using the first-line indent alignment, the first line of a paragraph following a paragraph return will be indented. Space before and space after set how much space there is between paragraphs. Hyphenation controls the breaking of the line of text. Hyphenation can be applied to the selected type or to the entire text container. To set the Hyphenation controls, click the Paragraph panel dropdown window and choose Hyphenate....

Import Text

Text can be imported from supported files in the .txt, .rtf, .doc, and .docx formats. To import a text file, choose file > Place. Select the text file you want to place and click Place.

The text file will show a loaded cursor with the text. Click and draw a container to flow the text into. If you have drawn a shapc to flow tlie text into before placing the text, choose Place and click on the edge of the existing container to activate it as a text container and the text will flow in. Any stroke or fill attributes on the existing container will be removed once the text is flowed in.

Manage Text

Resize a Text Area

Resizing the text area depends on what text format was created: point type, area type, or text along a path.

Point type. When resizing point type, choose the Selection tool and click on the point type. The type will activate and show pull handles to resize the text. Since point type starts a at point and can continue indefinitely, resizing with the pull handles can scale or distort the type. To resize point type and keep the correct proportions, use Shift when scaling to prevent the type from being stretched. If point type has been stretched, you can restore the scaling back to 100% in the Character panel. To break lines of text in point type, hit the return key.

Point type will go on forever and ever and ever...

Point type can be scaled with the pull handles

Point type scaled without shift, stretching the type

Area (paragraph) type: With area type, changing the size of the container using the pull handles resizes the container and will reflow the text. Holding Shift while scaling will resize the container proportionately, but not the type. If the container is too small to contain all the type or the type size is larger than the container can hold, a red plus will show in the lower-right corner of the container This is the text overflow indicator. To solve this, pull the window shade down to open the container until it is large enough to display all the text. You can also double-click on the window shade to automatically resize the container to the size of the text. This window shade feature will also work when the container is larger than the type and can be double-clicked to fit the container to the text.

Paragraph type will be constrained by the container and the text reflows when the container is resized.

Paragraph type will be constrained by the container and the text reflows when the container is resized.

Area type can be scaled with the pull handles, reflowing the text

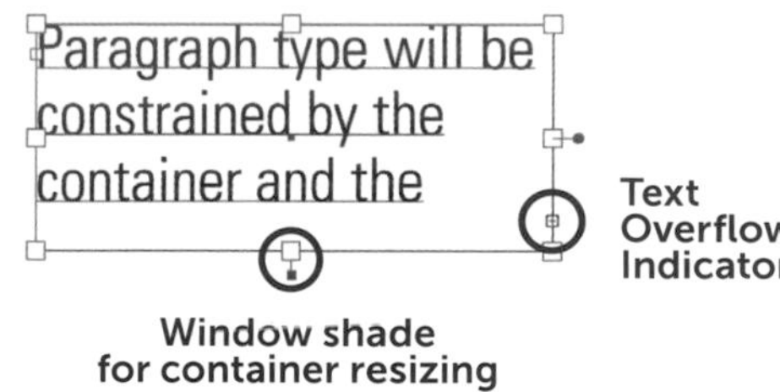
Paragraph type will be
constrained by the
container and the
Text Overflow Indicator
Window shade for container resizing

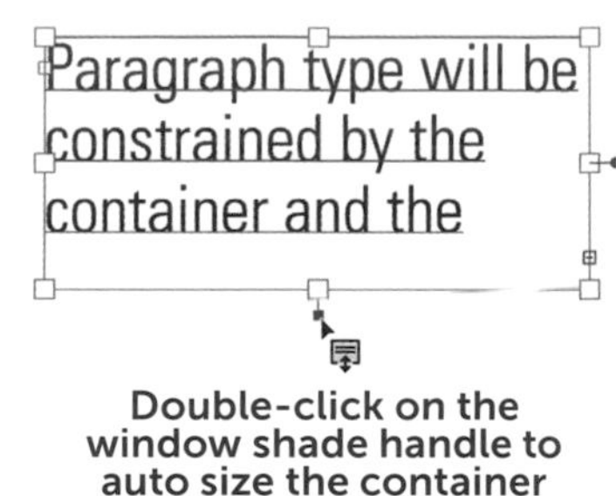
Paragraph type will be
constrained by the
container and the
Double-click on the window shade handle to auto size the container

To change the shape of the area type, you can use the Direct Selection tool to move corners or edges, or the Pen tool to edit the shape. The type will flow inside the shape regardless of how it is resized and reshaped.

Type on a Path: When resizing an open line or closed shape of Type on a Path, the type will respond as it does with point type and area type. If type is on an open path, scaling the line with the pull handles will cause the type to distort as it would with point type. Holding Shift to scale an open line with type will scale the type and line proportionally. When Type on a Path is applied to a closed shape, the resizing works as it does with area type. The overall shape will resize but the type will remain the same size. If the shape is scaled down, the text overflow indicator may appear, meaning there is not enough space for the type. There is no auto-resize with Type on a Path. To change the open line or closed shape of the Type on a Path, you can use the Direct Selection tool to move corners or edges or the Pen tool to edit the line or shape.

Selecting Text

Selecting text to change its attributes can be done by selecting the Type tool, clicking on the text, and dragging the cursor to select the text. A shortcut to activate the text tool is to double-click on the text with the Selection tool. This double-click will automatically activate the Type tool. To exit out of type editing, click Escape or select the Selection tool.

To edit type attributes on an entire set of text, there is no need to use the Type tool to select the text. If you select the type with the Selection tool, you can apply text attributes to the entire selected text. If you want to apply text attributes to a portion of the text, use the Type tool to select the text you want to affect.

Text Options

Convert to Point Type/Area Type: Point type can be converted to area type and vice versa. The advantage of being able to convert one to another is you may have started with point type and found you have more text and would like to use the container to reflow the text instead of manually breaking the lines using Returns. To convert text, choose Type > Convert to Area Type (if you have point type) or Type > Convert to Point Type if you have area type. This can also be done directly on the bounding box of either area or point type. The round handle that appears on the right side of the selected text indicates what kind of type it is. A solid handle indicates area type, and a non-filled handle indicates point type. Double-click on that handle and it will convert from point type to area type and vice versa.

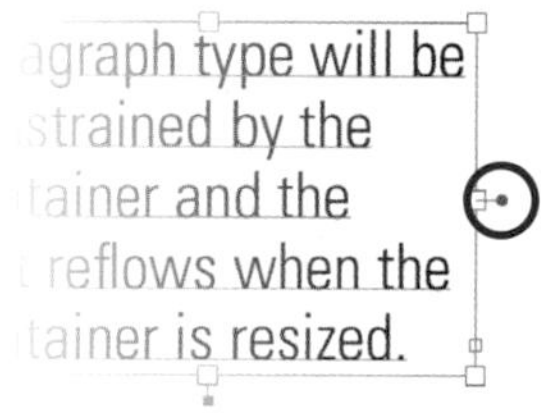

Area type: Double-click on the handle to convert to point type

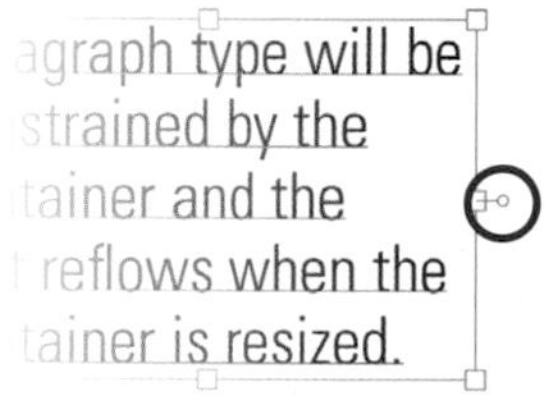

Point type: Double-click on the handle to convert to area type

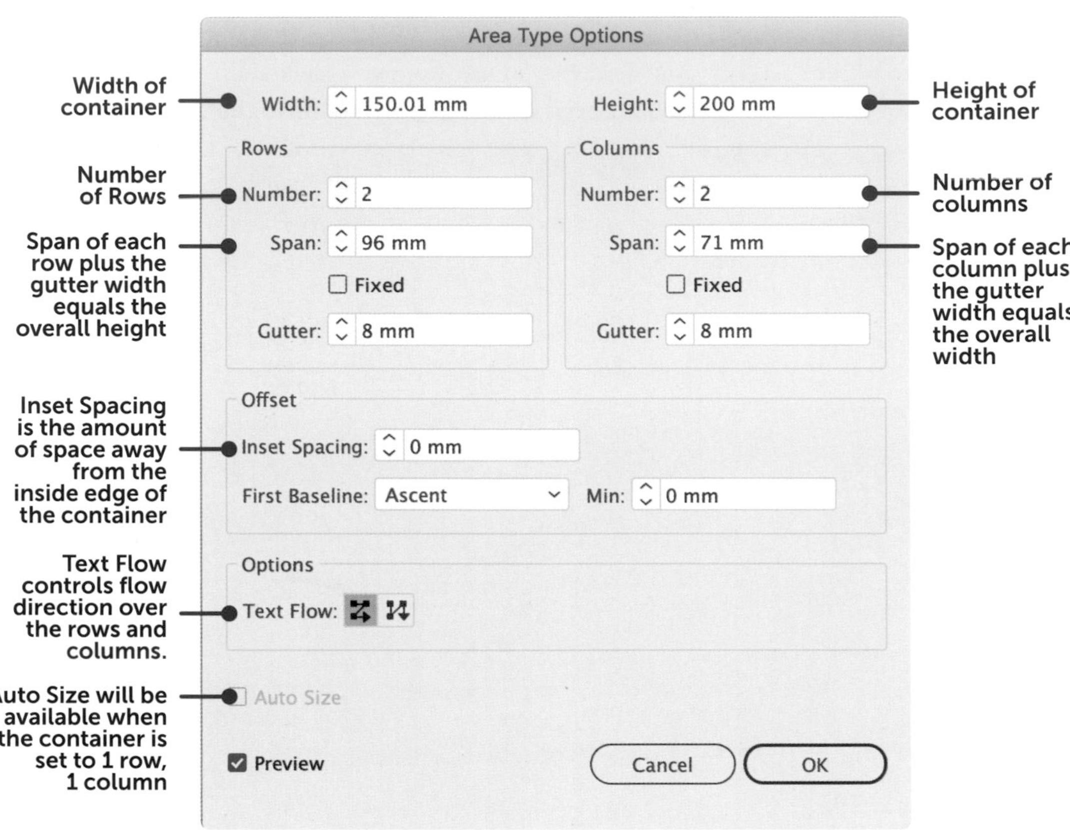

Area Type Options

Select the type container with the Selection tool and choose Type > Area Type Options. Width and height control the overall size of the container or the size of the bounding box if the shape is not a rectangle.

Rows and columns can be set, dividing the container into sections.

Gutter is the spacing between each section.

Span is the size of each row and column.

Offset is the buffer zone between the inside of the container and the text. The default is 0 which allows the text to sit against the edge of the container. Add a value to inset the text from the edge of the container.

Text Flow is how the text will flow over multiple rows and columns; flow can be left to right or down each column.

Auto size allows the container to resize based on the amount of text in the container and is only available when the container is set to one row and one column.

Type on a Path Options

Editing Type on a Path can be done with the Selection tool or the Direct Selection tool. Select the path and click on the stops at either end of the text. The stops are vertical lines at the beginning and end of the text that appear at the point where you clicked on the Path with the Type on a Path tool. The stops can be moved along the path to move the text to a different location. The center-line handle on the path can be selected and used to move the text along the path (like a joystick) and can also be used to flip the text to the other side of the path. The center handle is halfway between the start and stop point of the type.

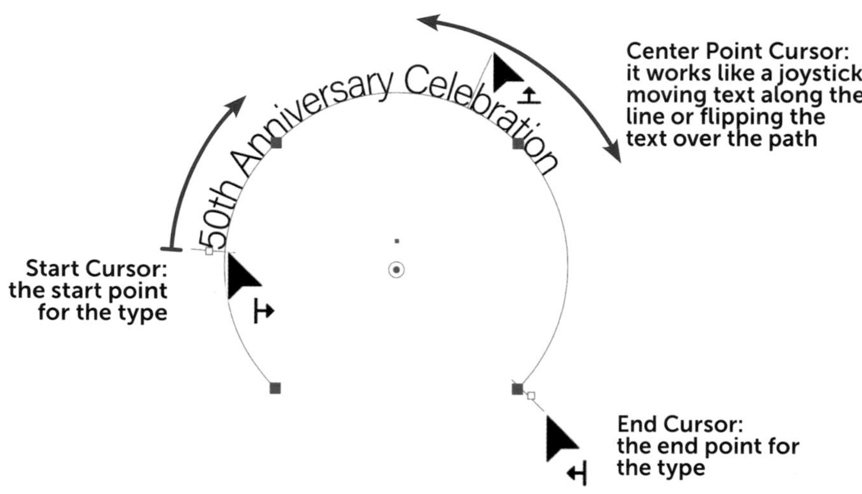

Type on a Path has several options that can be edited. Choose Type > Type on a Path > Options.

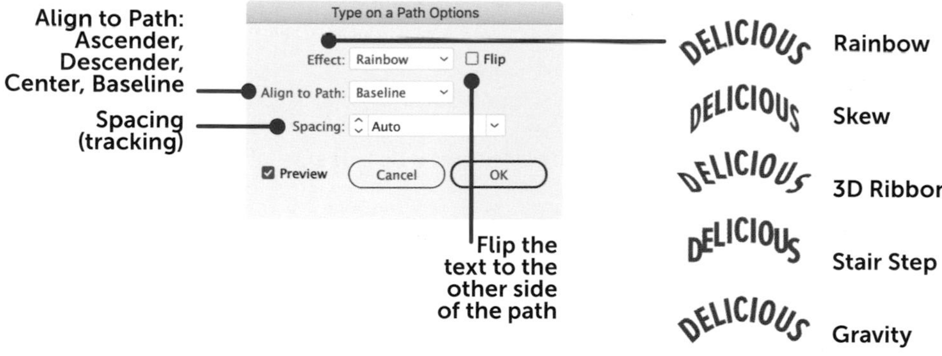

Thread Text/Text Linking

To link text containers together so text can flow from one section to another or one artboard to another, the containers can be linked together. Select an area type container with the Selection tool and click the outgoing port on the lower-right side. If there is more text than the container can hold, there will be a red plus; if there is no text overflow, the port will not show a red plus. Once you click on the port, the cursor changes to the loaded text cursor and there are two options: you can draw another container with the loaded cursor and the text will link and flow into the newly created container, or you can click on a existing shape or container and have the text flow into it.

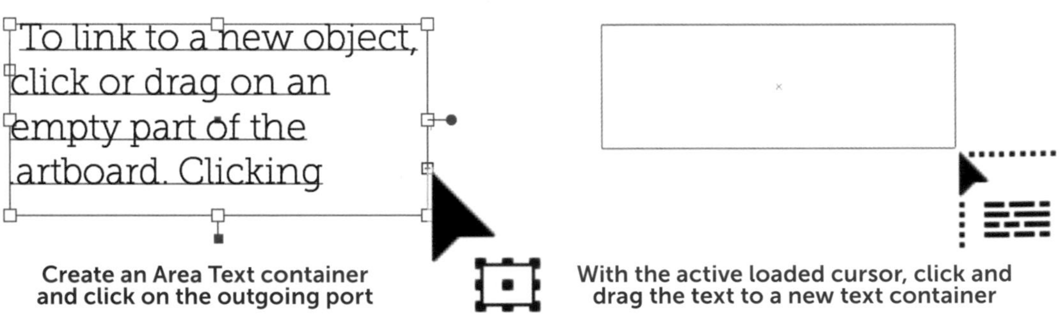

Create an Area Text container and click on the outgoing port

With the active loaded cursor, click and drag the text to a new text container

Another method for linking text containers and threading the text is to select two or more shapes or existing containers and choose Type > Threaded Text > Create.

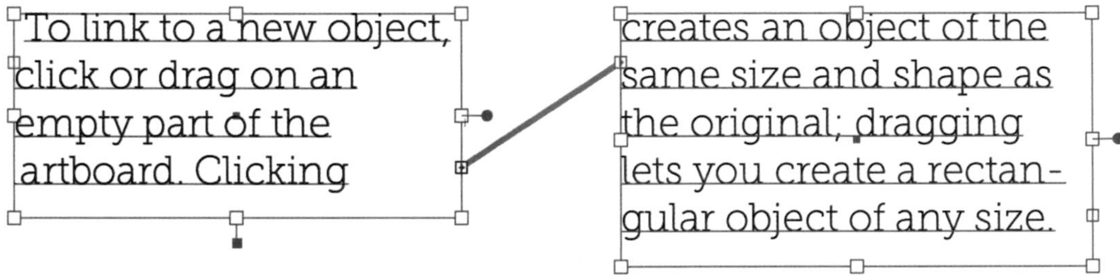

These Area Text containers are linked and the text threads show from one container to another

Remove or Break Threads

To break the thread between two objects, double-click the port on either the outgoing end of the thread or the incoming port on the linked text container. The text will not be lost, it will flow back into the first object. To break container links and release the text thread, choose Type > Threaded Text > Release Selection and the text will flow into the next or previous object depending on the location of the container in the linked chain. To break the containers at the threads, leaving the text in each container yet removing all the links, choose Type > Threaded Text > Remove Threading.

Wrap Text Around an Object

Text wrap is a setting that creates a force field around an object so the text in an area type container wraps around the object. You can wrap area text around any object, such as other type containers, images, shapes, and lines you create in Illustrator. If the object is an embedded bitmap image, Illustrator wraps the text around opaque or partially opaque pixels and ignores fully transparent pixels.

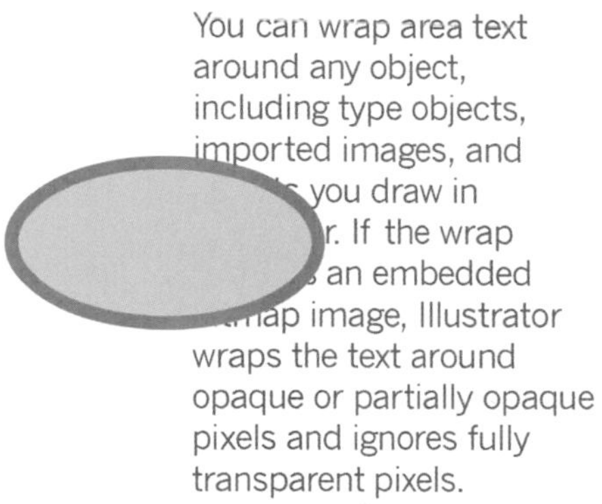

Area type with a shape in front of the type

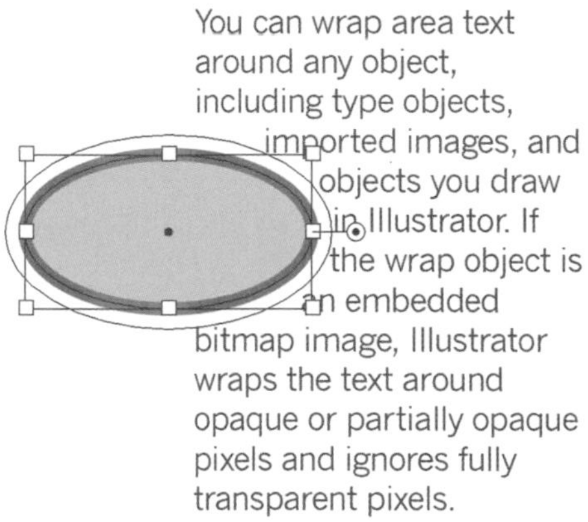

Text wrap applied to the shape, forcing the type to wrap around

Wrapping around a shape is determined by the stacking order of objects. The object with the text wrap applied has to be in front (on top) of the type container in order for the text wrap to be effective. To bring an object to the front, choose Object > Arrange > Bring to Front. Move the object in front of the text to make the wrap effective. You can also move the object with the text wrap applied in front of the text using the Layers panel, selecting the object and dragging it above the text container in the list of layers. The object and type container have to be on the same layer in order to make the text wrap work.

You can set wrap options by selecting the shape and choosing Choose Object > Text Wrap > Text Wrap Options to set the offset or the amount of distance between the object and text. This can be set before or after the text wrap is applied to the object. To remove the text wrap from the shape, Choose Object > Text Wrap > Release.

Touch Type Tool

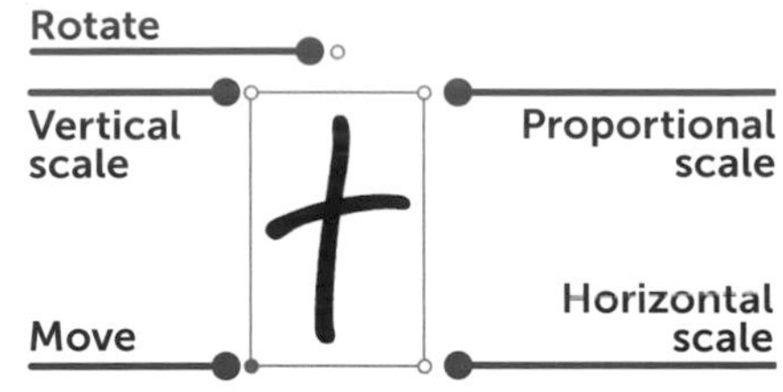

The Touch Type tool is used to individually adjust the scale, baseline position, and rotation of individual characters in active text, not text that has been outlined. To edit using the Touch Type tool, create text using any of the type tools. Select the Touch Type tool and click on any letter in the set text and a bounding box will appear. Using one of the five handles on the selected type, you can rotate the letter, scale it vertically or horizontally, or move the letter—all while keeping the type active. Once you are done, you can adjust the point size of the text and it will keep the proportions and spacing set by the Touch Type tool.

Begin with active text, and use the Touch Type tool to scale, rotate, and position each letter

The end results with different letter sizes, positions, and nesting

Outline Fonts

Converting active type to outlines gives you the ability to adjust each letter as a shape using any of the shape editing tools. Outlining type is also a way to convert active fonts in a logo or artwork so the fonts don't need to accompany the file. An example is providing a final logo to a client. Having to make sure each end user has the fonts would not be realistic, so converting the type to outlines is the solution.

To convert type to outlines, select the type you want to convert using the Selection tool and choose Type > Create Outlines. The active type will now be shapes that look the same as the original type. One way you an tell if the type is active or outlined is to view it in Outline mode. Choose View > Outline. Active type will appear as solid black and outlined type will show as a black outline. Once the type is outlined, choose Object > Ungroup to ungroup the letter shapes into individual editable shapes. Once the font has been outlined, there is no command to convert it back to active text unless you choose Undo Create Outlines under the Edit menu.

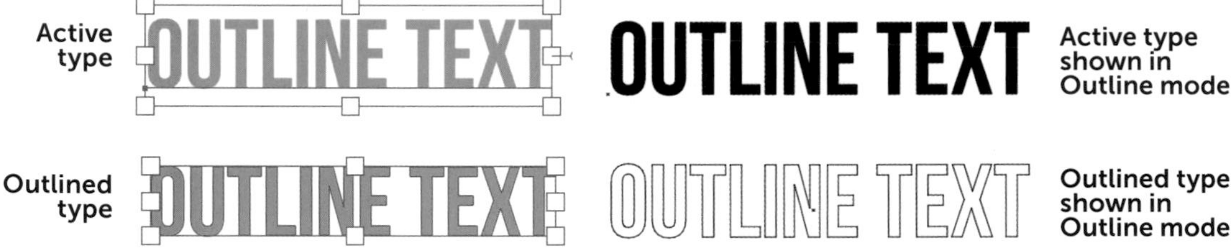

Exporting Text

To export active text from Illustrator, use the Type tool to select the type, and choose File > Export > Export As.... In the Export dialog box, choose Text Format (TXT) from the dropdown list of items, select a location to save the file, enter a filename, and click OK. Set the format options (Mac or PC) and click Export.

Adobe Fonts

Adobe gives you access to use hundreds of fonts through their Creative Cloud applications. These fonts are included in your Creative Cloud subscription and are available through the fonts menus of the Adobe applications or through the web at www.fonts.adobe.com. The Adobe Fonts website allows easy browsing through different classifications of typefaces, including serif, sans serif, handwritten, ornamental, slab serif, monoblock, script, and hand-drawn styles. Fonts can be displayed for easy comparison and selecting. Activating the fonts requires that you sign in to your Creative Cloud account in the upper right of the website. Once singed in, you can activate as many fonts as you choose simply by sliding the Activate Fonts slider at the bottom of each font display.

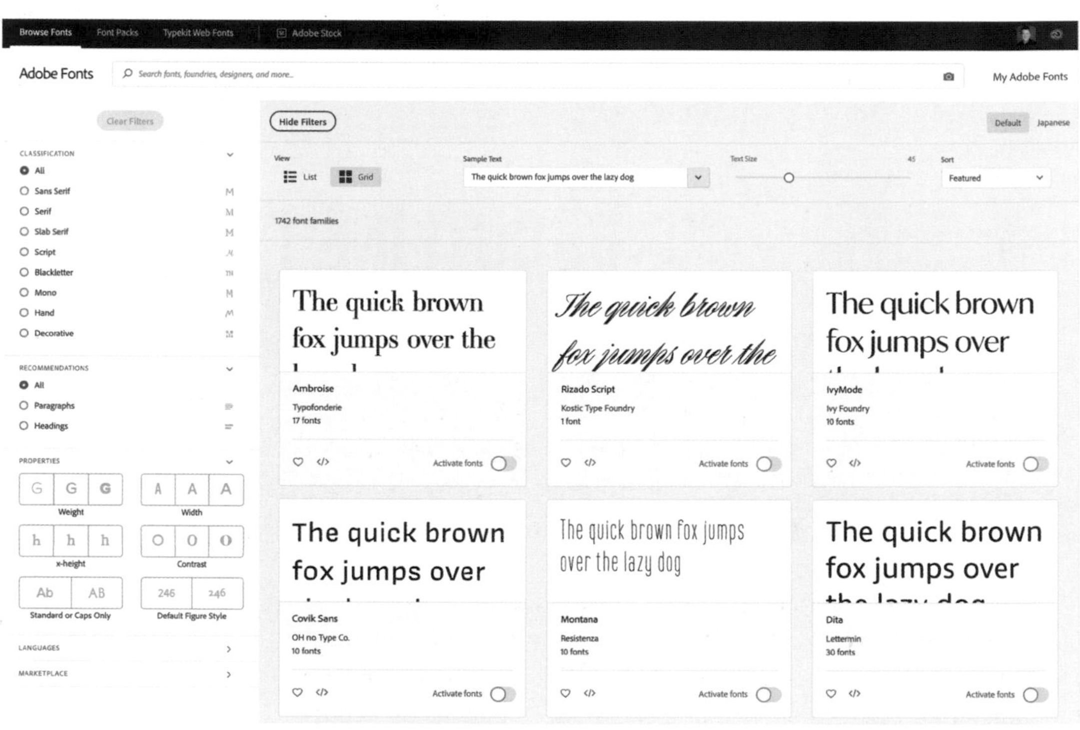

Activating Fonts

You can browse and activate Adobe Fonts from within the Illustrator font menu. Click on the font menu dropdown next to the font name and choose Find More at the top of the menu. The list will display of all the available Adobe fonts. Find the fonts you would like to activate and click on the cloud icon at the far right of the font. Once activated, these fonts are available for use in all your Creative Cloud applications.

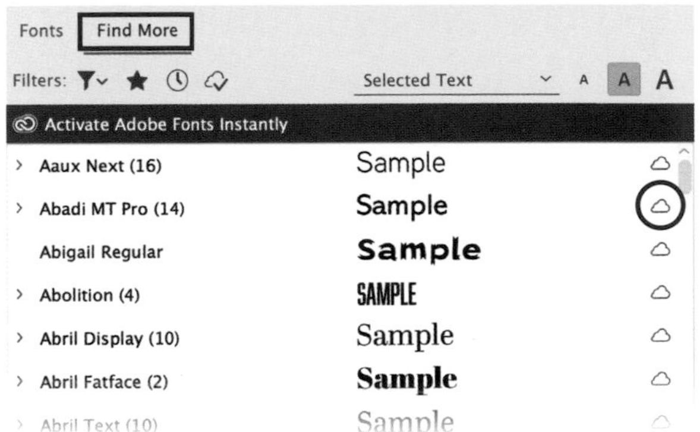

Find/Replace Missing Fonts

When opening a document that uses fonts not installed on your system, an alert dialog box appears indicating which fonts are missing. If the missing fonts are included in the Adobe fonts catalog, you can click on the Activate checkbox to the right of the font. If this checkbox is not checked, the missing font will not be part of the Adobe Fonts library and will need to be installed or substituted for another. Click on the missing fonts and then click on the Find Fonts... button.

If you choose to close this dialog box and continue on to the file, the missing type will display with the closest available font or it will be highlighted in pink. You can find and replace fonts after closing this dialog box by choosing Type > Find Fonts or call this same dialog box by choosing Type > Resolve Missing Fonts.... To substitute missing fonts with a different font, select the text that uses the missing font and apply an available font.

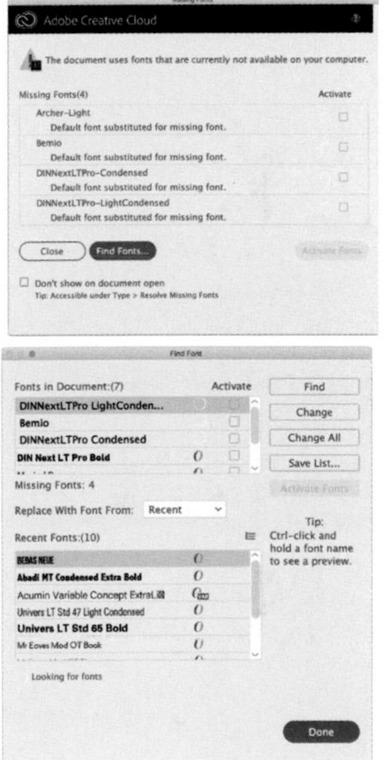

To find a font that is used in the file, select the name of a font in the top section of the dialog box. The first occurrence of the font is highlighted on your artboard. Select a replacement font from the bottom section of the dialog box and click the Change button to change that one occurrence or Change All to change all occurrences of the selected font.

You can replace fonts from the Replace With Font From list of recently used fonts, document fonts, or system fonts to list all fonts on your computer. Document fonts are only the fonts used in that artwork file; system fonts are all the installed fonts available in Illustrator.

Packaging Fonts

When sending your Illustrator file to another person, you can collect the fonts that have been used and any linked graphics, and package the file using File > Package.... A packaged file will create a folder that contains the Illustrator document, fonts, linked graphics, and a package report.

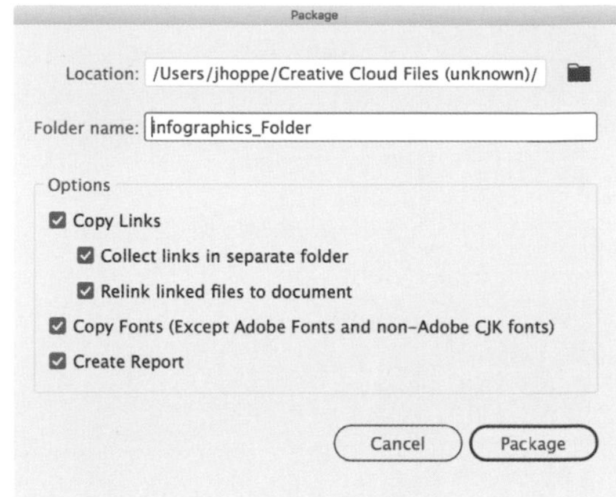

When you choose File > Package you can select the location for the package folder and the folder name.

Copy Links copies any linked graphics (embedded graphics will not be copied) so they can be relinked to the new links folder. This way, when the file is opened by the end user, they will not have to relink the files from the links folder.

Copy Fonts Used in the Documents copies the fonts files used in the document, but may not copy the entire font family. CJK refers to Chinese, Japanese, and Korean characters. If any are used in the file, they will not be packaged. Fonts that are activated from Adobe Fonts are not collected in the package. When the end user opens the file the missing fonts dialog box will appear. The missing Adobe fonts can be activated by selecting the missing font from the list and clicking the Activate checkbox or the Activate Fonts... button to activate all the fonts in the list.

Paragraph and Character Styles

A paragraph style is a set of character and paragraph formatting attributes that can be applied to a paragraph or range of paragraphs in a single click. The benefit of using paragraph styles is consistency of text formatting and efficiency; when you change the formatting of a style, all text to which the style has been applied will be updated with the new format.

Paragraph Style Attributes

Paragraph styles can include attributes for formatting such as font size and face, color, leading, paragraph indents, paragraph spacing, underlining, drop caps, nested styles, bullets and numbering, keep options, and more. Paragraph styles are applied to an entire paragraph as defined by a paragraph return.

Character Style Attributes

Unlike paragraph styles, character styles do not include all the paragraph formatting attributes of selected text. Character styles only include character attributes such as font style, size, and color. Character styles are for specific characters or words that need to be styled differently within a paragraph style.

Create Paragraph Styles

Open the Paragraph Styles panel under Window Type > Paragraph Styles. To create a paragraph style, format text with the font, size, color, and paragraph formatting options to create a style for that text. Select the text and choose New Paragraph Style... from the Paragraph panel drop-down menu.

Create a name for the style. Click on the Add to My Library button if you would like to store this style in your Adobe library for use in other Illustrator files or other Adobe applications.

When you create a new paragraph style based on formatted text, the style settings section in the middle of the New Paragraph Style dialog box will reflect those settings. You can edit the formatting using the list of items on the left side of the Paragraph Style dialog box.

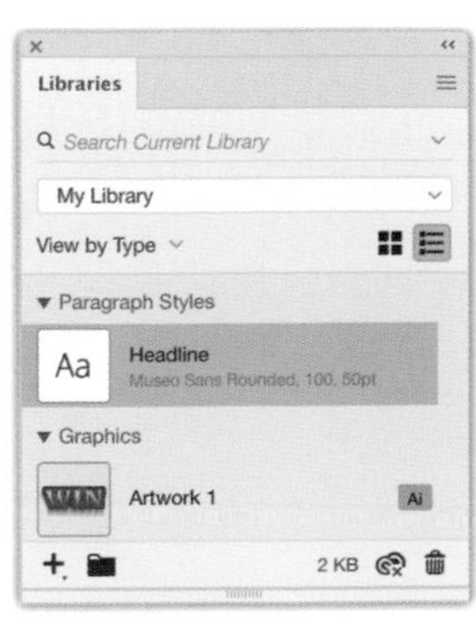

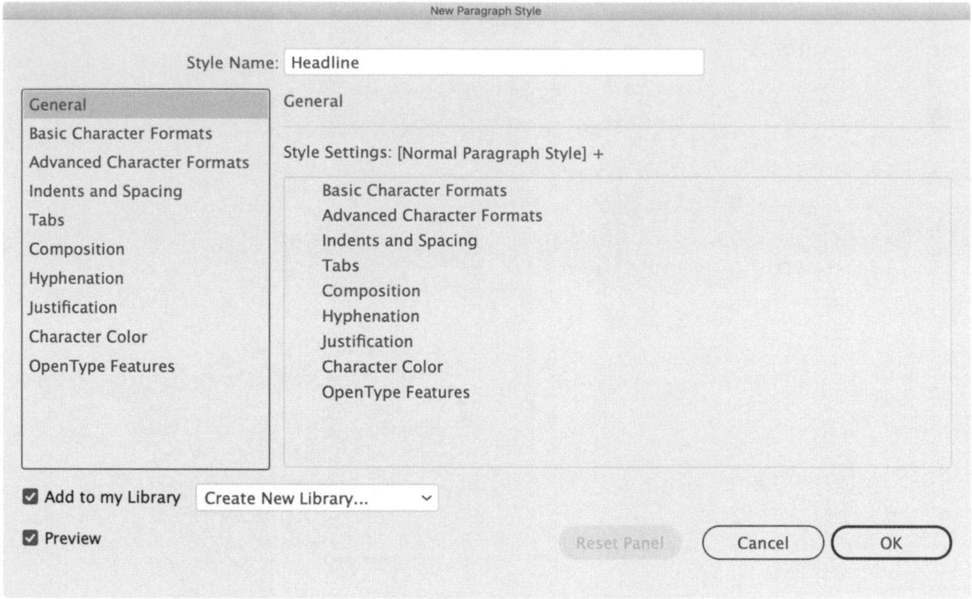

New Paragraph Style
Style Name: Headline
General
Basic Character Formats
Advanced Character Formats
Indents and Spacing
Tabs
Composition
Hyphenation
Justification
Character Color
OpenType Features
General
Style Settings: [Normal Paragraph Style] +
Basic Character Formats
Advanced Character Formats
Indents and Spacing
Tabs
Composition
Hyphenation
Justification
Character Color
OpenType Features
Add to my Library Create New Library...
Preview Reset Panel Cancel OK

General gives an overview of the current settings of the paragraph style.
Basic Character Formats offer settings for the Font Family, Font Style, size, leading, kerning, tracking, case settings (all caps, small caps), position (subscript or superscript), underline, or strike through.
Advanced Character Formats allow you to edit horizontal and vertical scale, baseline shift, character rotation, and dictionary language.
Indents and Spacing edits alignment, left indent, right indent, first-line indent and space before and after paragraph returns.
Tabs: is where to set tab positions and alignment, and leaders.
Composition defines how the text flows using Adobe Single Line Composer or Adobe every Line Composer.
Justification is for editing word, letter, and glyph spacing attributes, and setting auto leading and single word justification.
Character Color is for editing the fill or stroke color of the text, and settings for overprinting of colors.
Open Type Features allow you to edit Open Type additional features.

Applying a Paragraph Style

Once a style has been created you can apply that style to any paragraph. If you want to apply the style to all paragraphs in a text container, select the text container with the Selection tool and click on the style in the Paragraph Styles panel. To apply the style to a single paragraph, select the Type tool and click on any paragraph to apply the style. You do not need to select the entire paragraph in order to apply the style; all that is necessary is to have the text cursor in a paragraph and the style will be applied to the entire paragraph. If you select a single word or phrase and apply the style to those selected items, it will apply the style to the entire paragraph. To apply a style to a single word or phrase, use a character style.

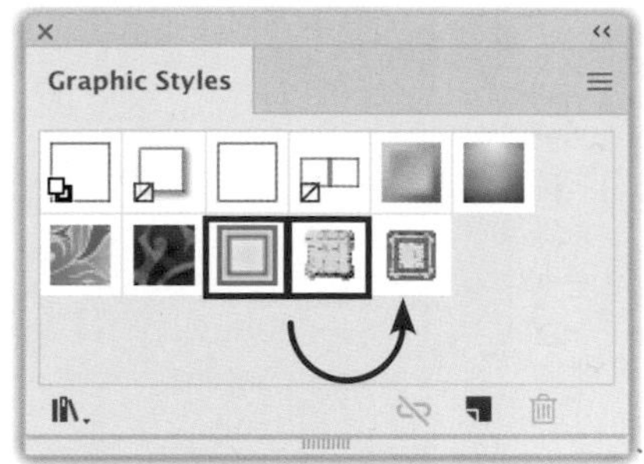

Editing a Paragraph Style

Once a style has been created and applied, it can be edited in the Paragraph Styles panel. Double-click the style name or select the style and choose Paragraph Style Options... in the Paragraph Styles panel dropdown menu. When the edits to the style have been made, they will appear in the document wherever that style is applied.

Style Overrides

Once a style has been created and applied, additional editing done to the text may cause the paragraph style to display a + next to the name. This indicates that some of the text formatting no longer matches the paragraph style. To return the text to the formatting created in the paragraph style, click on the dropdown menu and choose Clear Overrides. This will return the formatted type back to the paragraph style.

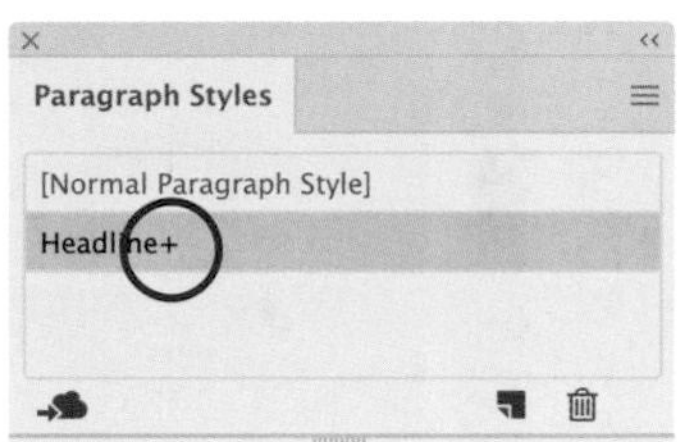

Redefine a Style

There are times when the text in the document is changed manually after a style is applied, and those changes need to be updated in the paragraph style. To redefine a paragraph style after it has been applied, select the Type tool, click on the text, and choose Redefine Style under the paragraph style dropdown menu. All instances of the redefined style will be reflected in those new edits.

Delete a Style

To delete a style, click on the style in the Paragraph Style panel and click on the trash can icon in the lower right of the panel. Deleting a style removes it from the list but keeps the formatting on the text where it was applied.

Load Styles

To bring styles in from an existing Illustrator file, click on the Paragraph or Character Style dropdown menu and choose Load Paragraph Styles (when in the Paragraph Styles panel) or Load Character Styles (when in the Character Styles panel). To load all paragraph and character styles, choose Load All Styles... and navigate to the Illustrator file that has the styles you want to import. Click OK.

Character Styles

A character style is a set of formatting attributes that can be applied to a letter, word, or character in a single click. Unlike paragraph styles, which apply to an entire paragraph, character styles apply only to portions of the paragraph. Character styles are used when a word needs to be made bold, or colored independently from the style applied to the paragraph. Character styles do not have as many options as paragraph styles since they do not include many of the paragraph formatting options such as indents. Character styles simply apply character attributes such as fonts, font size, kerning, and font color.

Open the Character Styles panel under Window Type > Character Styles. To create a character style, format text with a font, size, and color and create a style from that text. Select the text and choose New Character Style... from the Character panel dropdown menu.

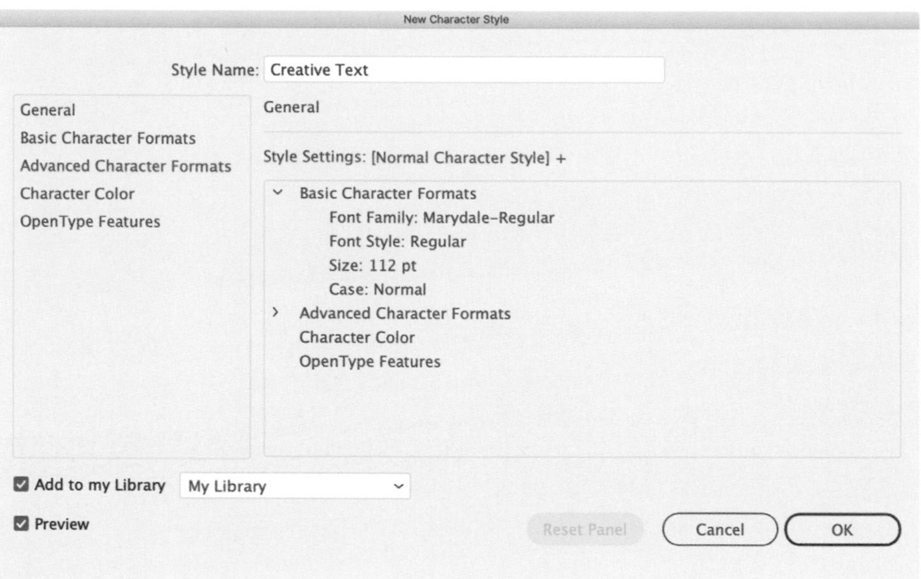

General gives an overview of the current settings of the character style.

Basic Character Formats allow you to edit settings for the Font Family, Font Style, size, leading, kerning, tracking, case settings (all caps, small caps), position (subscript or superscript), underline, or strike through.

Advanced Character Formats is where you can edit horizontal and vertical scale, baseline shift, character rotation, and Dictionary Language.

Character Color is where you can edit the fill or stroke color of the text, and settings for overprint.

Open Type Features is where you can edit Open Type additional features such as ligatures and contextual alternates.

Creating, Applying, and Editing

The creation and editing of character styles is the same for paragraph styles. Applying a character style requires the letter or word(s) to be selected with the Type tool. It is not good practice to apply a character style to entire text container. If the entire text needs to be formatted, use a paragraph style.

Graphic Styles

A graphic style is a set of appearance attributes such as stroke weight, color, fill color, and effects, applied to an object. The style can be saved and applied to other objects. You can apply graphic styles to objects, groups, and layers. When you apply a graphic style to a group or layer, every object in the group or layer takes on the attributes of the applied graphic style.

Create a Graphic Style

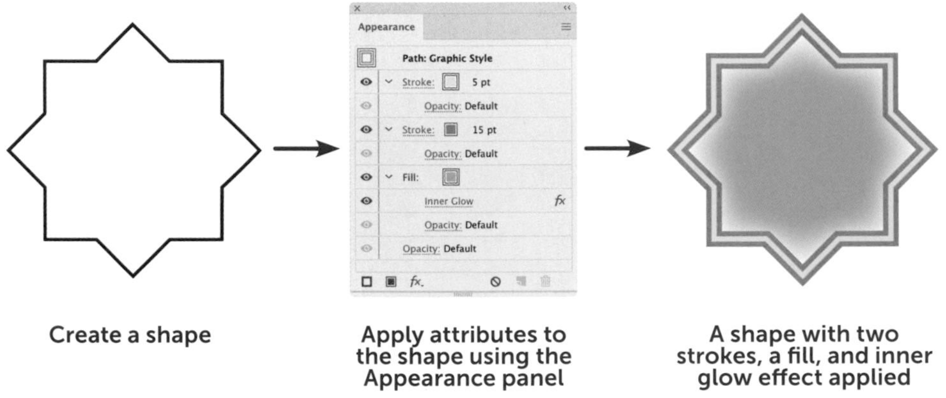

Create a shape

Apply attributes to the shape using the Appearance panel

A shape with two strokes, a fill, and inner glow effect applied

To create a graphic style, start with a shape or line and apply appearance attributes to it by using the Properties panel or the Appearance panel. Add a stroke, set the color and stroke weight, set the opacity, and add a fill color and any effects to the shape.

The Appearance panel is helpful in showing all the attributes applied to a shape. It also makes it easy to adjust and order the appearance attributes, create multiple fills and strokes, and edit effects. All the attributes on the shape make up the graphic style of the shape.

Save a Graphic Style

To save a graphic style to be used on other shapes, open the Graphic Styles panel under the Window menu. Select the shape you have applied the attributes to, then click the New Graphic Style button in the lower-right corner next to the trash can. Name the style and click OK. Other ways to save the style is to drag the object with the attributes applied to it into the Graphic Styles panel. An arrow with a green circle will appear and you can add the style to the panel. You can also click on the thumbnail at the top of the Appearance panel and drag it into the panel to add the style.

Merge Two or More Existing Graphic Styles

To create a graphic style using two or more existing styles, you can select the styles you want to combine together in the Graphic Styles panel. Click on a style and Command + Click on another style you want to merge it with. Choose Merge Graphic Styles from the panel dropdown menu. The new graphic style will contain all the attributes of the selected graphic styles and is added to the panel at the end of the Graphic Styles panel. Certain attributes may hide others such as a fill hiding a stroke set to the inside of a shape. You can edit the attributes of the merged styles in the Appearance panel and update the style by choosing Redefine Graphic Style in the panel dropdown menu.

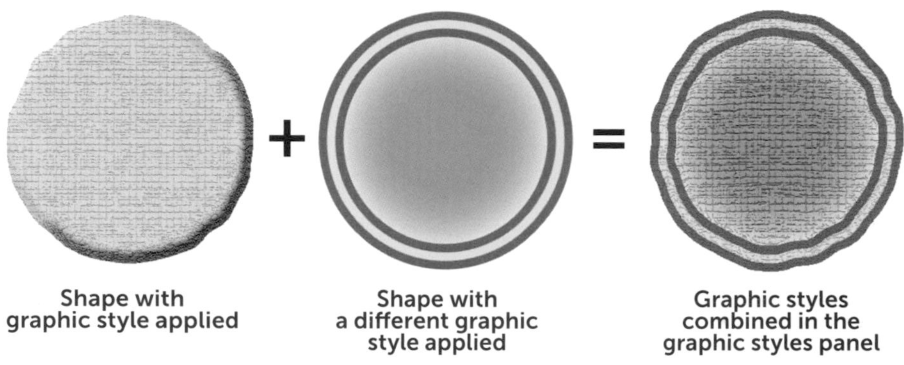

Shape with graphic style applied | **Shape with a different graphic style applied** | **Graphic styles combined in the graphic styles panel**

Styles Applied to a Group

When you apply a graphic style to a group, the objects in the group will not have the styles applied individually. In the Layers panel, the group will display the graphic style attributes, yet the individual shapes will not appear to have the style applied. If the objects are ungrouped, the style will be removed and the shapes will return to their original state before the graphic style was applied.

Create shapes, group them together

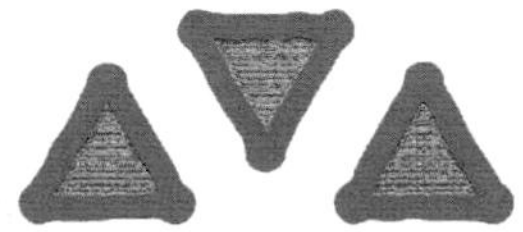

Apply a graphic style to the group

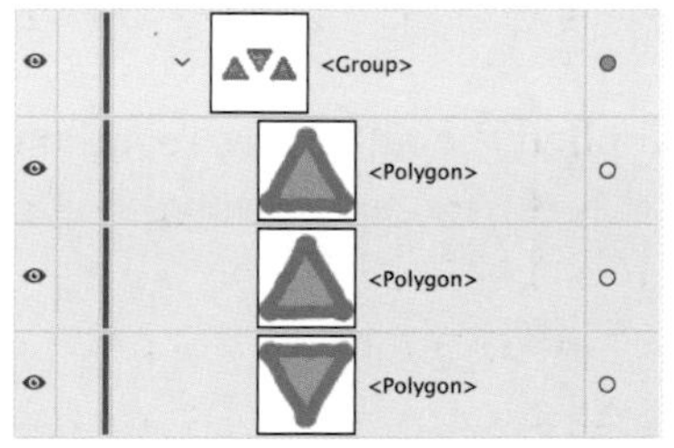

Graphic Styles on Active Type

When you apply a graphic style to active text, the fill color of the style will override the color of the text. To keep the original color of type when applying a graphic style, deselect Override Character Color in the Graphic Styles panel menu.

When you apply a graphic style on type, the effect may look very different from the style as applied to an object. When the graphic style was created, the effects, stroke weights, and fills were based on the object size, so when the style is applied to small type, all the attributes may pile on or overrun each other. This will create different-looking effects as the type is made larger and smaller or when thinner or heavier type is used.

Graphic Styles on Outlined Type

To fix the issue of the scale of the style, outline the text before applying the graphic style. Choose Type > Create Outlines. Once the type is outlined, if you apply the graphic style the fill may stay black. To solve this issue, select all the outlined type and choose Object > Compound Path > Make. Apply the graphic style to the outlined, compound text.

Scale a Style

When a graphic style has been applied to an object or shape, the attributes are fixed. If there is a 10 pt stroke around the object in the graphic style, that stroke will be 10 pts no matter what size shape you create.

A graphic style applied to outlined, compound, heavy/bold type

A graphic style applied to outlined, compound thin/light type

A small shape may look like the stroke is covering up the fill or not showing all the attributes of the graphic style. This is common when a style is applied to small type or thin, lighter type because there is not enough room in the shape to show all the attributes of the style as they were built.

Scale Stroke & Effects

To scale a style with the shape, you can check the Scale Stroke & Effects checkbox at the bottom of the Transform panel. You can also set the Scale Stroke & Effects in the Scale tool and under Preferences > General > Scale Stroke & Effects. When the Scale Stroke & Effects is not checked, the shape can be resized and the graphics style attributes will be fixed; as the shape is scaled larger or smaller, the effect does not scale with it.

To have the graphic style look the same on all shapes regardless of size, check the Scale Stroke & Effects checkbox. Now when the shape scales up or down, the attributes of the graphic style will scale with the shape, keeping the look of the style consistent over all shapes, regardless of size.

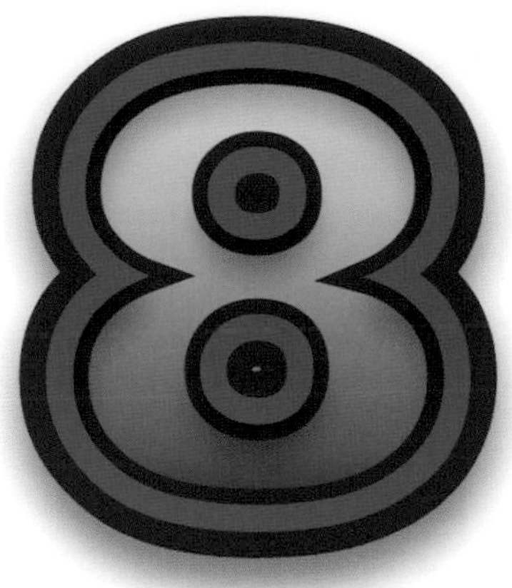

A graphic style applied to an outlined, compound type

A shape with a graphic style applied, scaled down with the Scale Stroke & Effects turned ON

The stroke weight scaled as the shape was scaled

A shape with a graphic style applied, scaled down with the Scale Stroke & Effects turned OFF

The stroke weight stayed the same as it was in the original size

Graphic Style Libraries

Graphic Style Libraries are collections of preset graphic styles. Illustrator has several presets available for use in the Graphic Styles panel. Click on the Graphic Styles panel dropdown and choose from the list or click on the library icon in the lower-left corner of the Graphic Styles panel. When you choose a graphic style from the library, it appears in a new panel separate from the Graphic Styles panel.

Save a Graphic Style Library

To create a library of graphic styles, create styles in the Graphic Styles panel or drag other styles from other Library panels into the Graphic Styles panel. Choose Save Graphic Style Library from the Graphic Styles panel dropdown menu. All the styles in the panel will be saved in the new library. If you save the library file in the default location where the other style libraries reside, the library name will appear in the user-defined submenu of the Graphic Style Libraries and Open Graphic Style Library menu. To load a Graphic Style Library, choose Open Graphic Style Library from the dropdown menu, and choose the library from the list. If it is not listed, choose Other Library and navigate to the saved library, select it, and click open.

Rename a Graphic Style

Choose Graphic Style Options from the panel menu, rename the file, and then click OK.

Delete a Graphic Style

To delete a graphic style from the panel, click on the style in the panel, click on the panel dropdown menu. Choose Delete Graphic Style and click Yes, or drag the style onto the Delete icon. Any objects or layers that the graphic style was applied to will retain the same appearance attributes, but the attributes will no longer be associated with a graphic style.

Break the Link to a Graphic Style

Breaking a link to a graphic style is helpful when you want to edit the attributes of a style on a shape without a style change being made on all the shapes where that style is applied. Select the object that has the graphic style applied to it, and choose Break Link To Graphic Style from the Graphic Styles panel dropdown menu, or click the Break Link To Graphic Style button in the panel.

Once the link to the style is broken, you can change any fill, stroke, and effect attributes of the selected shape. However, these attributes are no longer associated with a graphic style so any edits made to the graphic style will not be reflected in the graphic.

Replace Graphic Style Attributes

To replace a graphic style with another graphic style, you can hold the Option key and drag the graphic style icon you want to use from the Graphic Styles panel onto the graphic style icon you want to replace. This will replace the style with the new style on all the objects that had the old style applied to it. All occurrences of the graphic style in the Illustrator document are updated to use the new attributes when the old graphic style is replaced with the new one.

7 Working with Color

Color is a power
which directly
influences the soul.

wassily kandinsky

The Basics

RGB

RGB is defined as red, green, and blue. These three basic colors make up the visible spectrum of light. Devices that capture and display color use the RGB color model. Computers, televisions, phone screens, all display in RGB color. Cameras and scanners read RGB color.

The RGB color model is what is called an additive color model in which red, green, and blue light are added together in various amounts to create the visible spectrum of color. When all three colors are added together at full strength or intensity, the result is white light. When their intensity of strength is zero, and they are added together, the resulting color is black.

CMYK

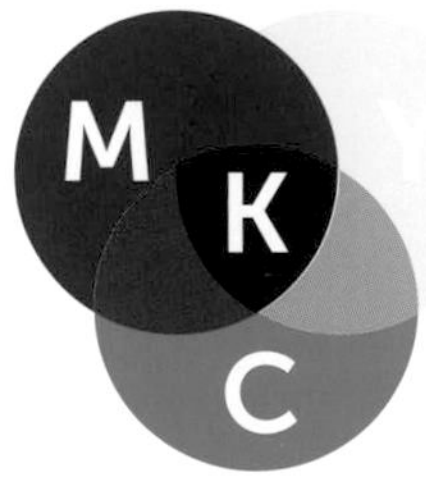

CMYK stands for cyan, magenta, yellow, and key (black), however, in virtually all cases "key" is referenced black. CMYK are the colors defined by the printing process. CMYK printing is also known as 4-color process (4CP) or full color. Inks and toner are made up of CMYK.

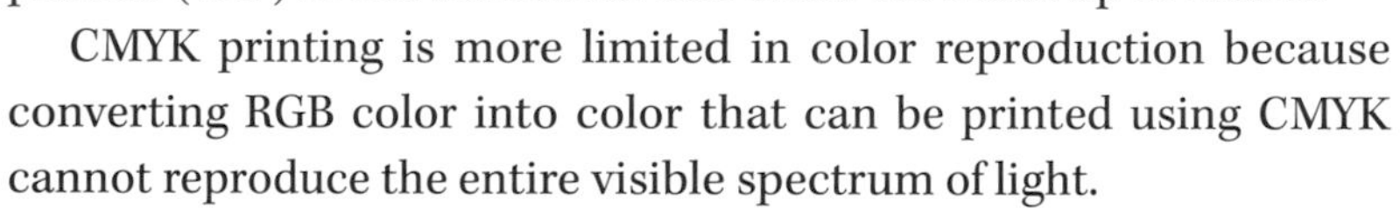

CMYK printing is more limited in color reproduction because converting RGB color into color that can be printed using CMYK cannot reproduce the entire visible spectrum of light.

Process Colors Versus Spot Colors

Spot colors are premixed colors that are not made up of CMYK. Think of spot colors as a single crayon, pencil, or tube of paint; you don't have to combine C, M, Y, and K together to get that single spot color. Spot colors can also be metallics and pastels, which cannot be printed using any combination of CMYK colors. Spot colors cannot be printed on a color printer or color copier since the color of the spot ink is a premixed, single ink. They usually are printed at a print shop or with a professional printer using an offset press that specifies that spot color ink.

There are several companies that produce spot color: Pantone (the most popular spot color printing system in the U.S. and Europe), Toyo, DIC ANPA, GCMI, HKS, and RAL. The Pantone Matching System, usually shortened to PMS, is how spot colors are sometimes referred to. PMS 356 refers to the Pantone system and color 356. In the printing world, the term PMS color and spot color is used interchangeably. In order to correctly see the representation of a spot color, you need to have access to a printed color guide. A spot color guide is a printed book or fan deck accurately representing the color on paper, metal, or vinyl.

Since spot colors cannot be printed using the traditional 4-color process, using a spot color in many cases is not feasible. When printing images, CMYK is used because it can most closely represent the visible range of colors.

Why use spot colors?

Spot colors can give a more vibrant color since it is a single ink instead of an ink color created using percentages of CMYK. Single-color printing is less expensive than 4-color printing. If you want to have a business card printed, printing in one or two spot colors is less expensive than printing in 4-color. Some colors are not possible with any CMYK combination of colors. Florescent colors for instance, can only be made as a spot color. Common uses for spot colors are screen printing, business cards, and packaging.

white is the color of the paper. The colors in nature are often beyond the *gamut* of a printer. It's often thought that RGB devices (like monitors) always show more colors than prints, but that's not necessarily true. A good printer can produce some colors (perhaps yellows and teals) that an average monitor cannot. We say those colors are outside of the monitor's gamut.

Color geeks often use special software to compare devices by the size (and shape) of the devices' color spaces. So when we say that the Epson 4900 printer has a larger gamut or color space than, say, its cousin the 3880, we mean that it can produce more vibrant colors. That is, it produces colors farther out along the **a** and **b** axes in Lab space. Two monitors may both be using 255R 255G 255B to represent white, but one will be brighter than another. This is just like when I turn up the volume of my stereo to 11 and two different loudspeakers produce different volumes from that same signal.

To get identical results from my devices, I may have to dim my brighter monitor, or send different color numbers to the more vibrant printer to match a less vibrant one. As you can imagine, this could make it difficult to produce images or documents for different media.

So What Should I Do?

The old advice when preparing documents was to always and only use RGB for onscreen viewing (for websites, presentations, etc.), and to always and only use CMYK for print. But the puzzle is deciding which numbers to use to achieve a specific color. We crossed our fingers or hoped that someone else would make it work.

Consider the case of making a swatch to best represent the orange of a leaf. To accurately see that color on a computer's display, the numbers *might* be 223R 110G 78B. In fact, that would get me pretty close on most average displays. But on a high-end display with a larger color space (one that produces a wider range of colors), the numbers might be 209R 87G 60B. Since each device requires its own values, there's never one, perfect answer.

The print case is worse. For a press job on good paper, the numbers for that orange could be 9C 67M 70Y 0K. On newsprint (typically a cool-gray paper), we might need 8C 61M 77Y 0K. A different manufacturer's inks may require both of those to change. Different paper stock? Start again. And what happens to the "use CMYK for print" rule if I use a modern inkjet printer with perhaps ten or more different inks that could include red, green, blue, violet, or even orange itself? Can you even imagine the dialog box you'd have to navigate for this situation?

Fortunately, you'll never see a

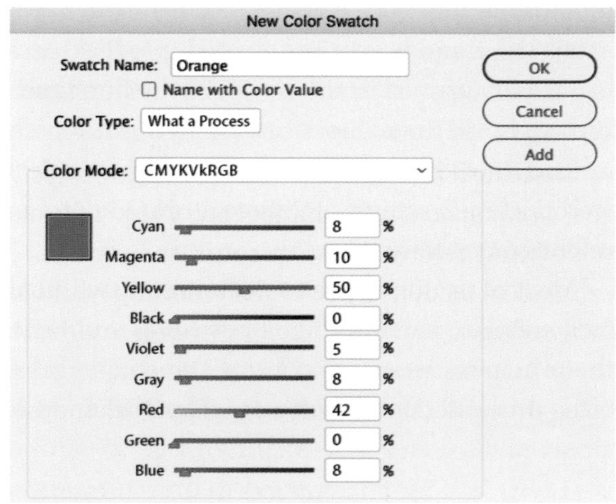

This is not a real dialog box!

It's a good thing, too. If we had to build swatches for every output device, we'd never have time to actually design anything.

fictitious dialog box like that. But honestly, for a color like this, do you really *care* what the numbers need to be? You should care only that the output looks right—that is, it maps to the right point in the Lab color space. We don't have to worry about colors' numbers at final output because Adobe software uses a cool technology to make sure they're the right numbers: color profiles.

Profiles

Long ago, in the dark years of the 1980s and 1990s, we had to manually tune the numbers for each output device, or we paid someone to do it for us. But now, software uses data files called *color profiles* to help compensate for the quirks of our monitors and output devices.

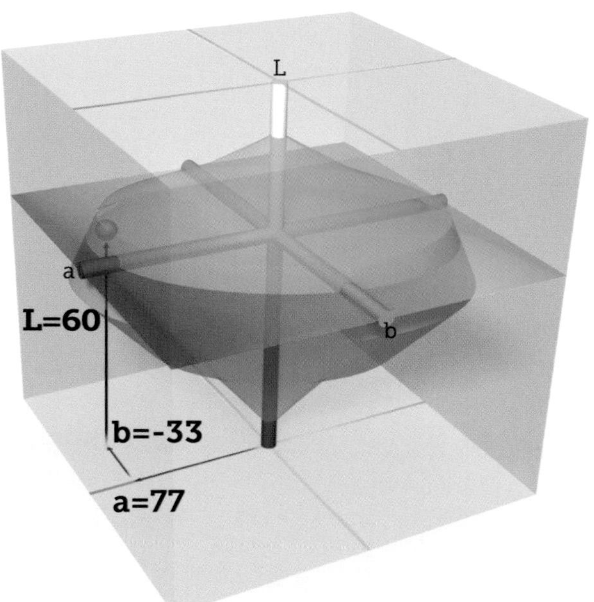

Device Color Space

This volume within Lab space represents the color space of a hypothetical output device (like a printer). The purple under discussion appears to be within **gamut**. That is, we can expect it to output accurately.

All color spaces, and therefore all devices, are compared to Lab (the colors we can see). Some devices produce a greater range of color than others. A **profile** describes a device's color space and lets us compare it to the color space of other devices. Profiles also allow us to convert color spaces across devices.

That is, profiles help us move from an input device, like a scanner or camera, through software like Photoshop, to an output device, while keeping in-gamut colors accurate throughout.

A profile describes a device's color space. In fact, some people use the phrases "color profile" and "color space" interchangeably. The profile captures information about the darkest black, the whitest white, the most saturated versions of each hue, and more. Most importantly, a profile lets us map a perceived color to the RGB or CMYK numbers needed for that specific device. While we work in our documents, the software itself uses a profile for its working color space. That is, our software is treated as yet another device in our workflow. Since it's software, and therefore a virtual device, we can choose the color space it uses—and that color space has a profile to describe it. Photo-editing programs like Photoshop take note of a camera's profile and convert that to Photoshop's color space, and then when we print, the numbers get converted to the printer's profile. The colors are assigned numbers that best represent them at each stage of the process, for each device. Software does the math with the numbers provided by profiles.

The Flexibility of RGB

As always, my main concern is that all colors' appearances are maintained correctly through output (if physics allows it). I may want to borrow a color from an image for other graphic elements so they all match each other. Since all my Adobe software typically uses the same working space, that orange in the example of the leaf, can be represented by 223R 110G 78B in Photoshop, InDesign, and Illustrator.

Imagine we've created an InDesign document with an RGB color swatch with those values. In that publication, we've placed photos from Photoshop and graphics from Illustrator, which also use the same values (you can see this in the example above).

When the job is done and we make a PDF for the public, what form will it take? If it's to be consumed onscreen only, then I'd choose a specific RGB for the PDF's Destination (details to follow). If I'm sending the PDF to a print shop, I'd ask them what they'd like. If they request CMYK, I'd then ask them which press profile they prefer, and I'll use that as the PDF's destination. In either case, the color numbers used by the photo and graphic elements in my layout will change, but they will continue to *match each other*. For a press PDF (using GRACoL, for example), that orange would become 9C 67M 70Y 0K. Since the orange we started with is within the gamut of the destination color space, it will be maintained with high fidelity.

But what do I mean by destination, and where do we specify it?

When saving a document as a PDF from Illustrator, we are given many options (see the Output chapter of this compendium for more). In the Output options, we choose whether and how colors are converted from their current color space. We also choose the destination profile and whether or not that profile will be included in the PDF. The destination will usually be sRGB for those PDFs viewed onscreen, or the CMYK profile that your print vendor desires.

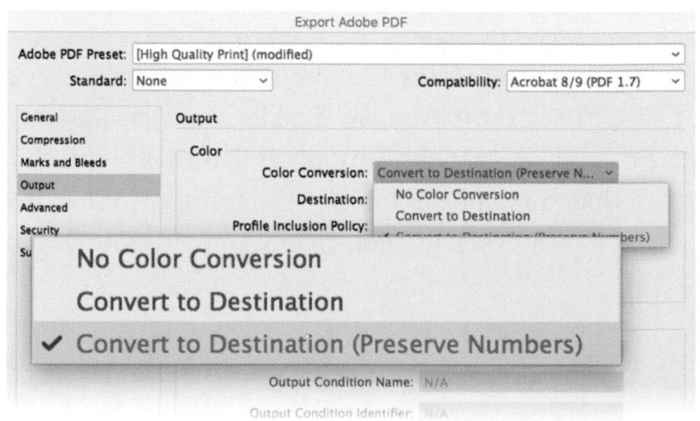

Convert RGB, but preserve CMYK builds

That's what this option does. As we take our flexible RGB content and prepare a PDF for whichever output medium it's intended, this will ensure that things that match each other will continue to do so.

Notice that for Color Conversion, I chose Convert to Destination (Preserve Numbers). There's a lot implied by that short phrase. It means that if I choose an RGB profile for my destination, all my RGB numbers and any CMYK numbers will change to yield the correct perceived colors for the new RGB profile. If I choose a CMYK destination, all my RGB numbers will be converted to that destination, but any native Illustrator content that uses CMYK builds will keep those numbers, even though they'll look different on the destination device. That's helpful for those few CMYK builds that must not change (like pure black for text).

Why preserve some CMYK builds?

We need numbers (builds) to change as we go from one device to another *if we want the colors to remain the same visually*, especially with images. But when something else is more important than color accuracy—like text legibility—we want the build maintained.

Convert to Destination (Preserve Numbers) is made for just that. If we have fine lines or lightweight text, it's best to keep it one ink (usually black). Mis-registration (misalignment of the inks) makes it hard to read. Heavier or larger text *may* withstand the misregistration of two or maybe three inks.

I use RGB for everything using color I care about, and CMYK on pure black and a few specific ink builds.

The Useful Rigidity of CMYK

A clear example of where the CMYK build matters more than the visual appearance is the text you're reading right now. I want it to print as pure black and nothing but black, no matter which press or paper I use. On newsprint, it'll look more gray than black. On a good paper, it'll look darker. On either, it'll be sharp, because with just one ink there's no chance of misalignment. Illustrator's default (black) swatch is perfect for this: a CMYK build of 0 0 0 100K. With the Preserve Numbers option, it'll always be black ink only.

That very special build is more important to me than its final visual appearance. I know that if pure black got converted from one CMYK color space to another, it might become 74C 70M 64Y 77K—a 4-color black! If there's any misregistration in printing, delicate text could become fringed with color, making this conversion a disaster for legibility.

Our example orange is different. As long as I don't apply it to small, delicate text, I don't care what numbers are used to render it. That's why, for this color and most all others, I define InDesign color swatches as RGB. If it's a color I sampled from an RGB photo, using an RGB swatch is easier, too.

If you find that you harbor an inner color geek, you can even use a Lab swatch in Illustrator! If you use Pantone spot colors, you'll notice they're defined with Lab as well, since they're aspiring to a specific perceived color.

Illustrator is color-mulitlingual: it speaks RGB, CMYK, and Lab, and converts between and among them. It can use elements in any of those modes at the same time and output to the one you choose. Since elements of any color mode may end up in an Illustrator file, this is a necessary luxury.

Final Advice

In a nutshell, use RGB for onscreen *and* print documents, and use CMYK swatches for those few kinds of things whose ink-use is more important than color appearance.

Swatches Panel

The Swatches panel is where colors, tints, gradients, and patterns are stored. To open the Swatches panel choose Window > Swatches. A preset list of colors, gradients, and patterns are in the panel by default. These can be edited, and new ones can be created and exported for use in other Illustrator files.

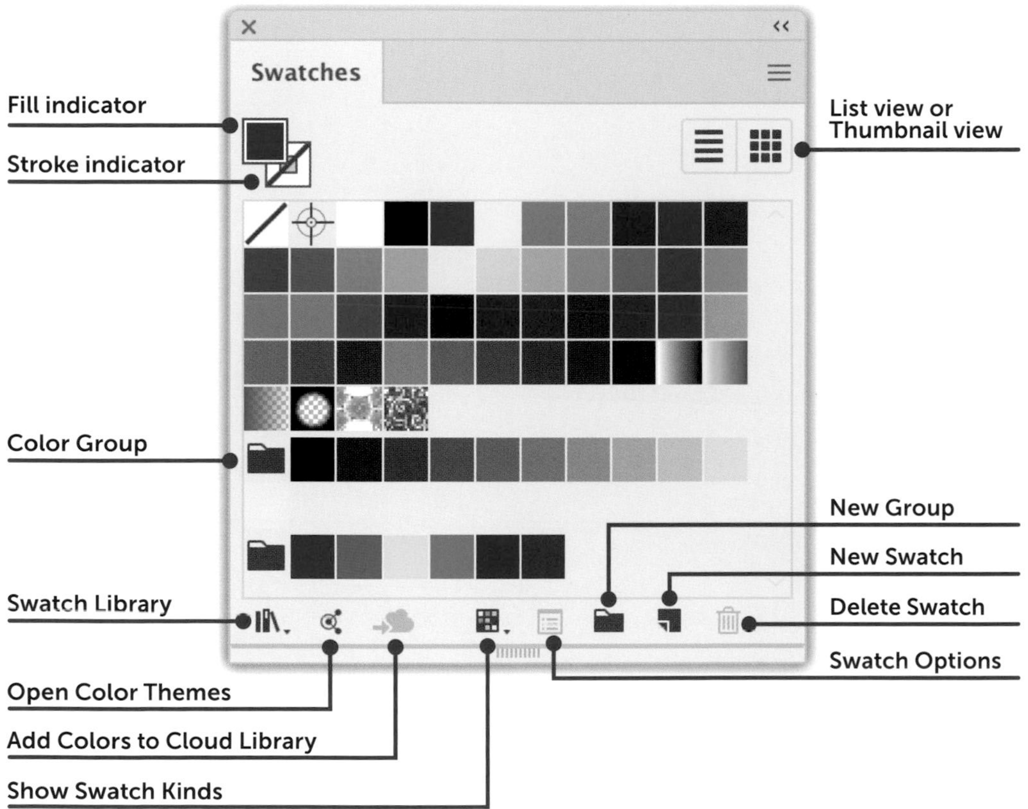

Swatches Panel Options

Change the Swatches Display

To change the view of the swatches from the default thumbnail view, select a view option from the Swatches panel menu: Small Thumbnail view is the default, but you can also choose from Medium Thumbnail view, Large Thumbnail view, Small List view, or Large List view. You can also click on the Thumbnail view/List view icons on the upper right of the Swatches panel to switch between List view and Thumbnail view. The Thumbnail view makes it easy for choosing colors from the Swatches; the List view shows the build of each color.

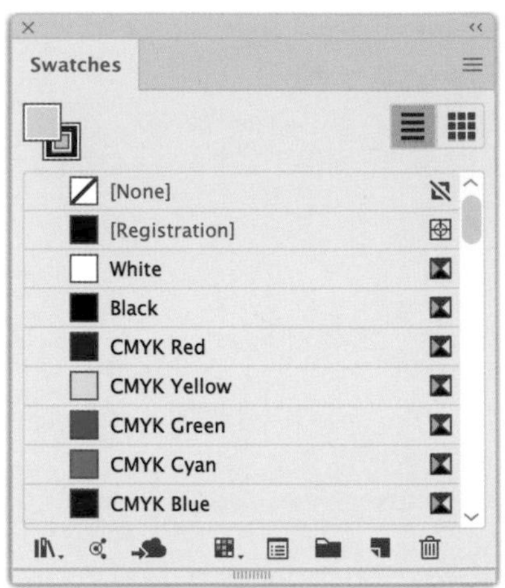

Show a Specific Type of Swatch

In the Swatches panel the display will show color swatches, gradient swatches, and pattern swatches. Click the Show Swatch Kinds button at the bottom of the Swatches panel and choose one of the following: Show All Swatches, Show Color Swatches, Show Gradient Swatches, Show Pattern Swatches, or Show Color Groups. This will show only the selected swatches, hiding all the others.

Select All Unused Swatches

You can delete any unused swatches. If you want to remove colors that are not used in a document, you can choose Select All Unused from the Swatches panel dropdown menu and then delete them.

Create a Color Group

Color groups are useful to keep color together in a folder in the Swatches panel. To create a Color Group click on the Group Folder icon at the bottom of the Swatches panel and drag individual color swatches to the Color Group Folder. You can also select the colors you want in a new color group and click the New Color Group button. The colors will be added when the group is created.

Change the Order of Swatches

You can reorder swatches by clicking and dragging them to a new location in the color list. You can sort them by name or kind by choosing Select Sort By Name or Sort By Kind from the Swatches panel dropdown menu. These commands only work on individual swatches, not swatches in a color group.

Creating New Swatches

Create Process (4-Color) Swatches

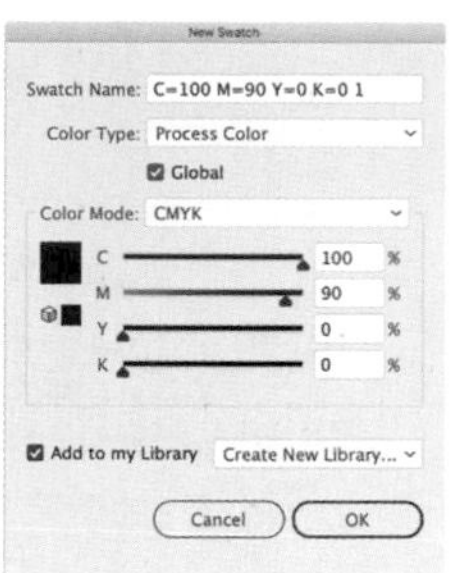

Swatches Panel: Create a new process swatch by clicking on the New Swatch button or select New Swatch from the panel dropdown menu in the Swatches panel. From the Color Type menu, choose Process Color. Check the Global box to make the color a Global color. Slide the sliders for the color values or enter a value in the field.

Color Picker: Select a color using the Color Picker by double-clicking on the color swatch at the bottom of the toolbar. Select the color on the spectrum or enter in the values and click OK. Once you have selected a color from the Color Picker, you have to add it to the Swatches panel. The newly selected color from the Color Picker will appear in the upper left of the Swatches panel. Drag that color swatch into the Swatches panel to add the color to the list of colors.

Double-click on the Color Picker icon

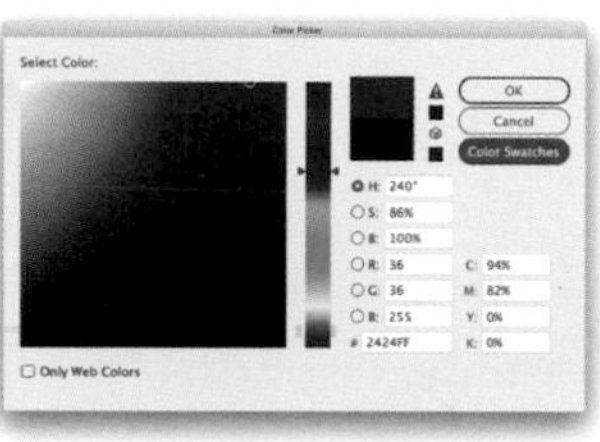

Choose your color from the Spectrum or sliders

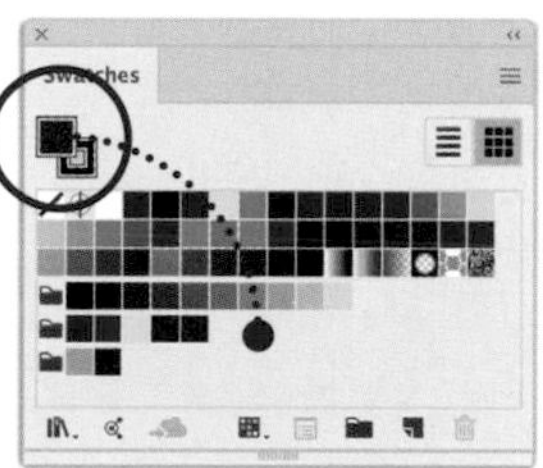

Drag the color swatch into the panel

Color Panel: Select a color using the Color panel by selecting the color on the spectrum or entering the values via the sliders.

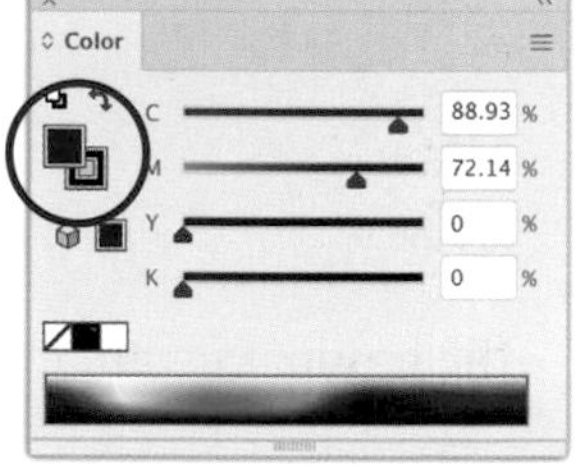

If the Color panel only shows the spectrum, click on the panel dropdown menu and choose Show Options to see the sliders. Once you have selected a color from the Color panel, you have to add it to the Swatches panel. Drag the color swatch into the Swatches panel to add the color to the list of colors or choose Create New Swatch from the Color panel dropdown menu and the selected color will be added to the Swatches panel.

Add Spot-Color Swatches

In the Swatches panel you can create a spot color by choosing Spot Color on the Color Type menu. However, creating spot colors this way creates pseudo spot colors that are not based on a color matching system, and are process colors acting as spot colors. To add a spot color based on a color matching system such as the Pantone Matching System, choose Open Swatch Library from the Swatches panel dropdown menu or click on the Library icon at the bottom of the Swatches panel. Choose Color Books and it will show a list of spot color matching systems.

ANPA Color
DIC Color Guide
FOCOLTONE
HKS E Process
HKS E
HKS K Process
HKS K
HKS N Process
HKS N
HKS Z Process
HKS Z
PANTONE+ CMYK Coated
PANTONE+ CMYK Uncoated
PANTONE+ Color Bridge Coated
PANTONE+ Color Bridge Uncoated
PANTONE+ Metallic Coated
PANTONE+ Pastels & Neons Coated
PANTONE+ Pastels & Neons Uncoated
PANTONE+ Premium Metallics Coated
PANTONE+ Solid Coated
PANTONE+ Solid Uncoated
TOYO 94 COLOR FINDER
TOYO COLOR FINDER
TRUMATCH

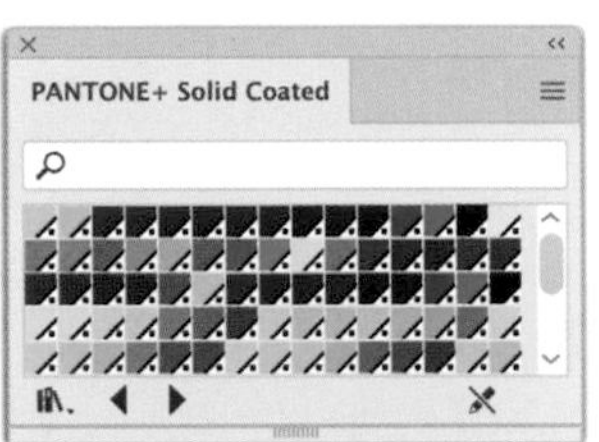

Once you choose the color matching system, a new panel will open with all the color swatches from that color system. Choose from the list of colors and drag the spot color(s) from the panel in the Swatches panel. You can also double-click on the spot color swatch to add it to the Swatches panel.

Add Used Colors

Another way to add colors to your Swatches panel is to create artwork, apply color to the artwork using the Color panel, Color Picker, or Color Theme panel, and then add the colors to the Swatches panel. Click on the Swatches panel dropdown menu and choose Add Used Colors. All the colors used in the document (that are not in the Swatches panel already), will be added to the bottom of the Swatches panel as global colors. This is a nice feature because all the colors come in as global colors and don't require conversion.

Add Selected Colors

If you want to add specific selected colors to the Swatches panel, select the object or artwork that contains the colors, and chose Add Selected Colors from the Swatches panel dropdown menu. This is the same process as the Add Used Colors fucntion, but this limits it to used selected colors. All the selected colors come in as global colors.

Global and Non-Global Colors

A global color is a color that, when edited, automatically updates the color edits in your artwork wherever that global color was used. Global colors have several advantages over non-global colors.

A non-global or local color is a color, that, when edited, does not effect any of the artwork where it was applied. For example: you have an orange swatch that you have applied to your artwork and you edit that orange swatch to make it red. The places where the original orange swatch was applied to your artwork is still there, but the original orange swatch is no longer in your Swatches panel. How do you get that orange swatch back to use it for other artwork? All colors in the Swatches panel are local by default. Another issue with non-global colors is that you cannot create a tint of a non-global color. If you want a lighter value, you have to create a new color swatch and formulate a lighter version using the color sliders.

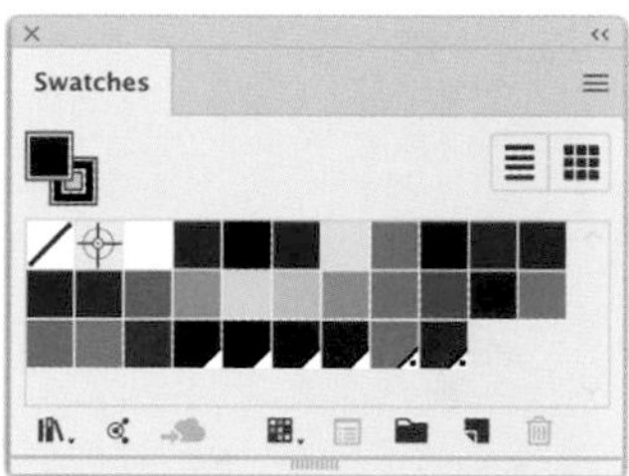

All spot colors are global, yet process colors can be either global or local. A global color is identified by a white triangle in the lower-right corner. A color swatch with a triangle and a dot indicates a spot color. To make a non-global color a global color, double-click on the non-global color swatch to open the Swatch Options panel. Click the Global checkbox and click OK. If you applied a non-global color to your artwork and then made it global, this will not make all the instances of that color a global color. You will have to reapply that global color to all instances to be able to edit the global color and see the changes where that color is applied to your artwork. A best practice is to make all colors global colors before using them in your artwork.

Color Modes

When creating a new document, you can choose between the RGB color mode and the CMYK color mode. The RGB color mode is the default color when you choose a new document preset for Web, Mobile, Film & Video, or Art & Illustration. CMYK color mode is he default color when you choose a new document preset for Print. In RGB mode, the Color panel and Swatch panel will show the colors in RGB values. In CMYK mode, the Color panel and Swatch panel will show the colors in CMYK values.

RGB color mode can reproduce a wider range of color that CMYK color mode. If you choose a color in RGB mode and convert it to CMYK, the color may be out of gamut, meaning the color cannot be reproduced accurately. Converting an RGB color to CMYK limits the color range, which is one of the inherent drawbacks of printing.

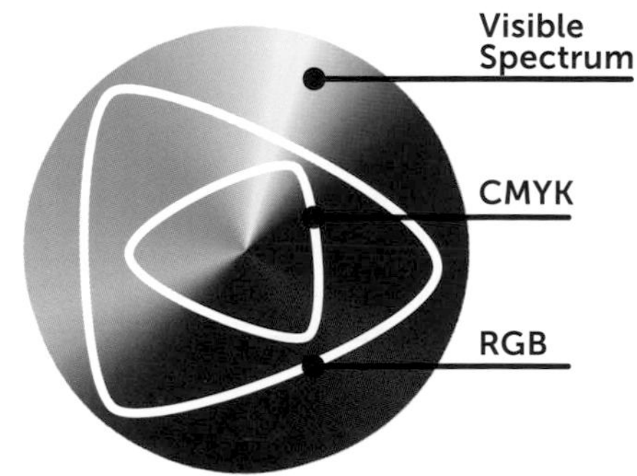

You can also convert the color mode of a document and the colors in the document by choosing File > Document Color Mode and switch between RGB and CMYK.

Spot Colors

A spot color is a premixed ink that is used in place of, or in addition to, process inks, and requires its own printing plate on a printing press. Spot color inks can reproduce colors that cannot be created by the process colors, such as metallic inks, pastels, fluorescents, and varnishes. Since spot colors are not created using process colors, printing the CMYK equivalent is not possible in all cases because the range of spot colors goes beyond what a CMYK ink can create. When you use a spot color, the appearance of the color on your monitor or printed results from your printer can only simulate the color; it can not create a direct match. To see what the spot color will look like, you will need a color book that has the exact color printed on paper that closely matches your printing needs. These color books are produced by the spot color companies that make the colors, and the books are printed on uncoated, coated, and matte papers to best show the results of the spot color as printed on these different surfaces. Uncoated paper absorbs more of the ink and creates a less vibrant representation of the spot color because it reflects less light. Matte papers will provide a moderately vibrant color with little reflection of light off the paper. Coated papers will provide a more vibrant color because the inks sits on the top of the paper and reflects the light.

Spot colors are used in many printed items as a cost saving measure where paying for four inks (CMYK) to create a color is not feasible. A single spot color on a business card makes more sense than four colors, and is less expensive

You can use tints of a spot color to create different tonal ranges while using a single ink. Using opacity of a spot color is not recommended because the spot color will be multiplied with the colors below. This produces a result on screen that does not print the same.

Illustrator includes the popular color-matching system libraries that can be accessed through the Swatches panel library. Confirm which spot matching system your printer uses to choose the accurate matching system for your needs.

Editing Colors

Swatch Panel

Editing a color in the Swatches panel can be done by double-clicking on a color, changing the values using the sliders, and clicking OK. If the color is a non-global color, the color swatch will change but none of the artwork in the document will change. If the color is a global color, all the instances of that color or tints of that color will be changed in the document.

Color Panel

Editing a color in the Color panel can be done by sliding the sliders or changing the color mode from the panel dropdown menu. Any changes that are made in the Color panel are not reflected in the Swatch panel. To add a newly edited swatch to the Swatches panel, choose Create New Swatch from the panel dropdown menu or drag the color square from the Color panel into the Swatches panel.

Color Picker

When you click on the Color Picker, you can choose from an entire spectrum of color in HSB, RGB, CMYK, or HEX color. When you select a color from the Color Picker and click OK, it does not add it to the Swatches panel and there is no direct way to add it from the Color Picker. To add the newly created color, open the Color panel and the color you chose in the Color Picker will show in the Color panel. Then add the color to the Swatches panel from the Color panel.

HEX Colors

HEX colors can be created from the Color Picker by entering in any HEX code in the field, giving you access to the entire range of HEX colors. In the Color panel, choose Web Safe RGB from the panel dropdown menu and enter in the value or adjust the sliders to access only the 216 Web Safe RGB colors. Web safe colors are an informal standard of colors supported by the majority of popular web browsers when computer displays were limited to 256 colors.

Tints

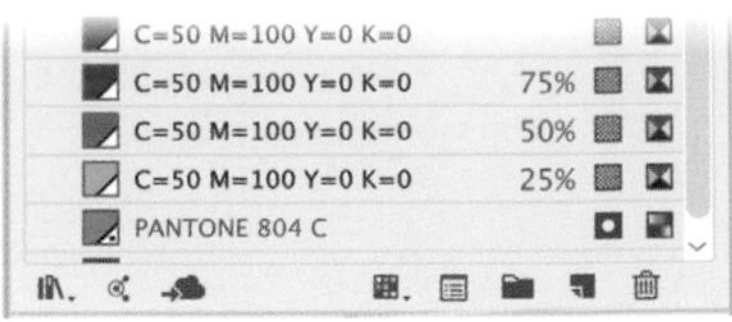

Creating a tint of a color is, by definition, adding white to the color. Tinting can only be done with a global color; non-global colors will not allow tints to be created of them. To create a tint, select the global color in your Swatches panel and open the Color panel. When a global color is selected, the Color panel will show a single color slider that you can slide to create tints of that global color. You can slide the slider, enter a value in the field, or click on the tint ramp to create a tint.

To add the tint to the Swatches panel, drag the tint color swatch into the Swatches panel or choose Create New Swatch from the Color panel dropdown menu. Create as many tints of a global color as you want. Another advantage of global colors and tints is that when the main global color is changed, all the tints of that color are changed in the Swatches panel as well as all the places the global color and its tints are used. Tints show up in the Swatches panel with the same color name and the value of the tint to the right of the name. Since non-global colors cannot have tints created, if you select a non-global color and open the Color panel, the single tint slider will not be available; only sliders to create a new color will be showing, along with the full spectrum of color at the bottom of the panel.

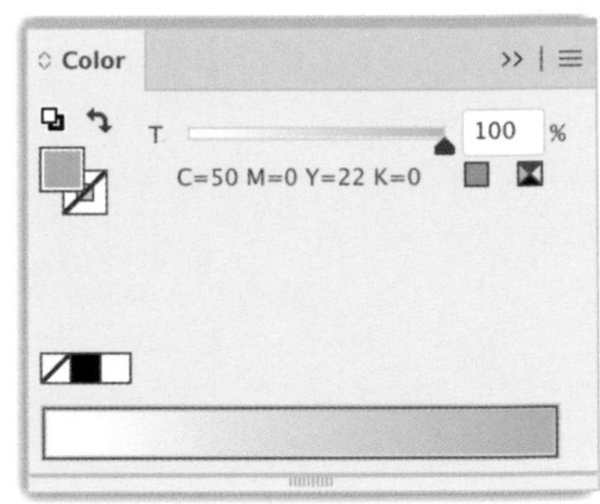

Opacity

The value applied to the Opacity setting defines how strongly the other colors in your document will show through the color or object the opacity is applied to. Tint and opacity may show the same results on screen, but opacity should be used only if you want to have transparency. Opacity should not be used with a spot color because the results may be inaccurate or not reproducible.

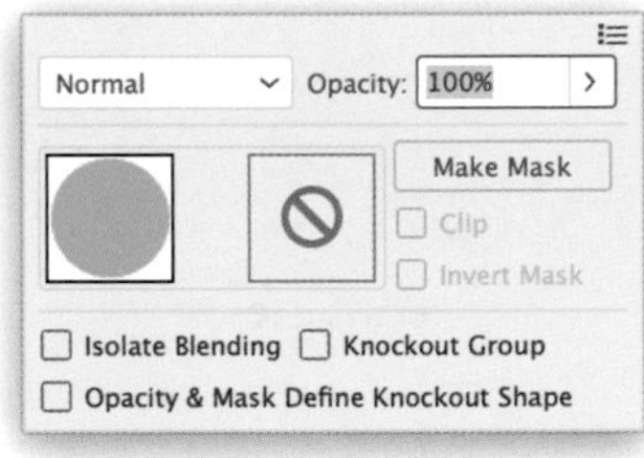

Opacity can be applied to objects, type, gradients, and images, and can be set in the Properties panel or the Appearance panel. In the Properties panel, opacity is set on the entire object. In the Appearance panel, you can independently set the opacity of the stroke, fill, or object overall by clicking on the opacity link for each specific selection. When adjusting the object's opacity, the stroke and fill opacity is adjusted in proportion to the overall opacity.

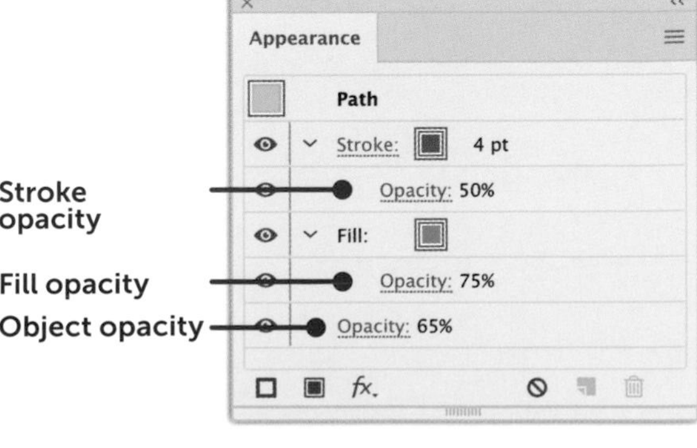

Sampling Color

Eyedropper

The Eyedropper tool allows you to sample a specific color from part of an object, shape, or image inside or outside of Illustrator. You can sample color, stroke weight, opacity, and other attributes and apply them to other objects to keep the appearance of objects consistent. To make this work smoothly, you will need to set the Eyedropper preferences to choose what it will sample and what it will apply to your selected object.

Double-click on the Eyedropper tool in the toolbar to open and set the preferences. In the Eyedropper preferences. You can choose what the Eyedropper will sample (left column) and what it will apply (right column). Type attributes can also be sampled and applied to other type. The Eyedropper tool will not sample and apply shape attributes.

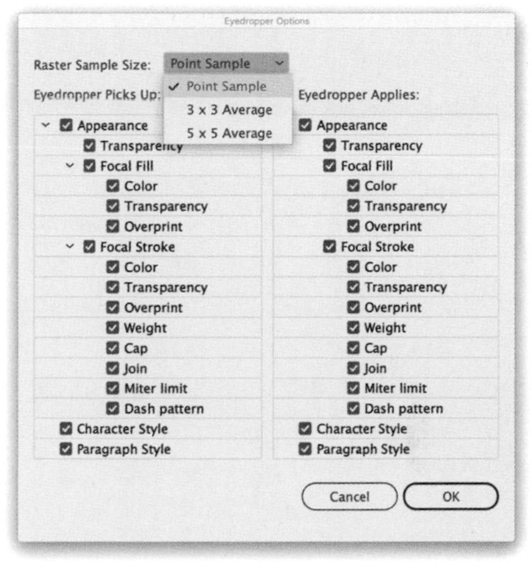

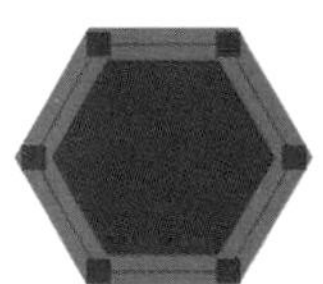

Select the shape to apply the attributes to

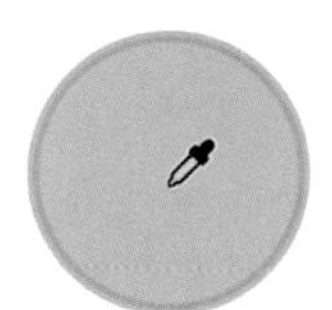

Select the Eyedropper tool and click on a shape to sample the attributes

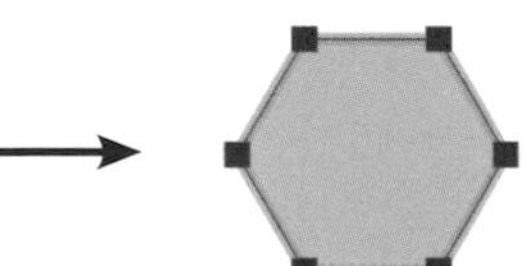

This is the original shape with the selected attributes applied

Sample a Specific Color

Shift+click to sample a color from a specific point on a gradient, pattern, mesh object, or placed image and apply the color to the selected fill or stroke.

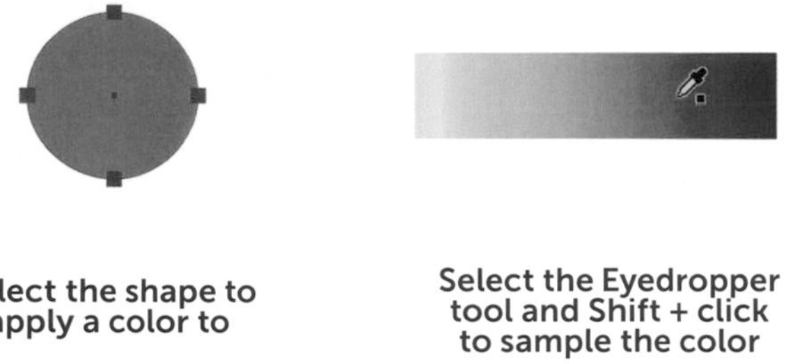

Select the shape to apply a color to

Select the Eyedropper tool and Shift + click to sample the color

Add Additional Attributes

Select a shape you want to add additional attributes to, select the Eyedropper tool, and hold Shift+Option (Mac)/Shift+Alt (PC) while clicking on another object to add the appearance attributes of the object to the selected object. This example has a shape with a purple stroke, no fill. With the shape selected, select the Eyedropper tool and Shift+Option/Alt click on another shape to add the attributes to the original shape. This added the green fill and the orange stroke in addition to the original purple stroke.

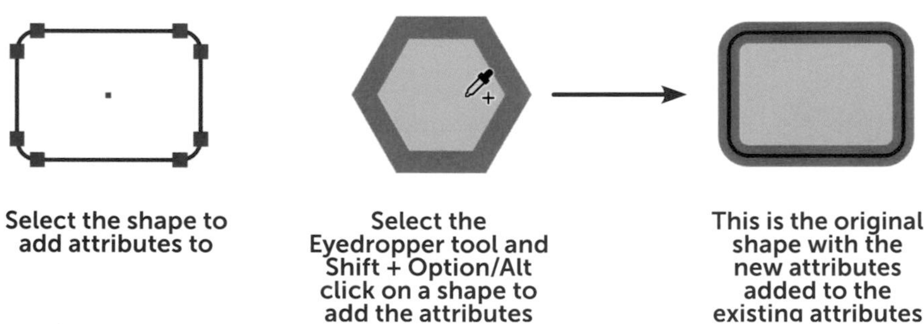

Select the shape to add attributes to

Select the Eyedropper tool and Shift + Option/Alt click on a shape to add the attributes

This is the original shape with the new attributes added to the existing attributes

Raster Sample Size is for sampling color from images in Illustrator or outside Illustrator. Point Sample will sample the color from the pixel you click in the image. If you want more consistent color to be sampled from and image, choose a 3 x 3 Average (this gives you colors from 9 pixels of color) or a 5 x 5 Average (an average of 25 pixels of color). Once the color is sampled, it will appear in the Color panel and can be added to the Swatches panel. Images are sampled in RGB color regardless of the color mode you are using in Illustrator.

To sample a color from outside Illustrator, you can select the Eydropper tool, click on your artboard, and keep the mouse clicked, then move the cursor to any other image or item outside of Illustrator (your desktop or other application window) and sample the object. To see the color you are sampling, have the Color panel open while sampling outside Illustrator and the color window will display the sampled color. Once the cursor is over the item you want to sample, release the mouse to sample the color.

Color Won't Sample

When sampling colors from an image with the Eyedropper tool, you may find that the color will not sample from the image no matter what you click on. This can be very frustrating, especially if the Eyedropper tool will sample other colors not from the image. Double-click on the Eyedropper tool to open the preferences. Uncheck the Appearance box on thc lcft side under Eyedropper Picks Up, leaving all the other boxes checked, and click OK. Now the Eyedropper tool will sample the colors from the image as you would expect.

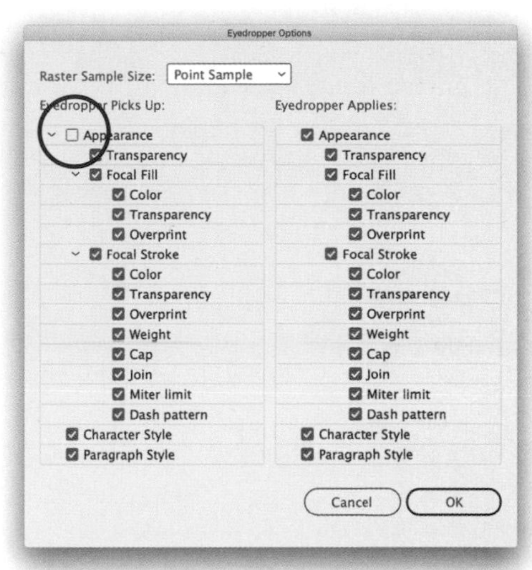

Adding Sampled Colors

To add colors to the Swatches panel from an image, I set up a palette of shapes to add the selected colors to, like a painter's palette. Draw a rectangle (or any shape) for each color you would like to sample from the image. Select a rectangle, then select the Eyedropper tool and click on a portion of the image to sample to the color, and the rectangle will fill with that color. Select the next rectangle and repeat the sample process until you have all the colors you want to use from the image.

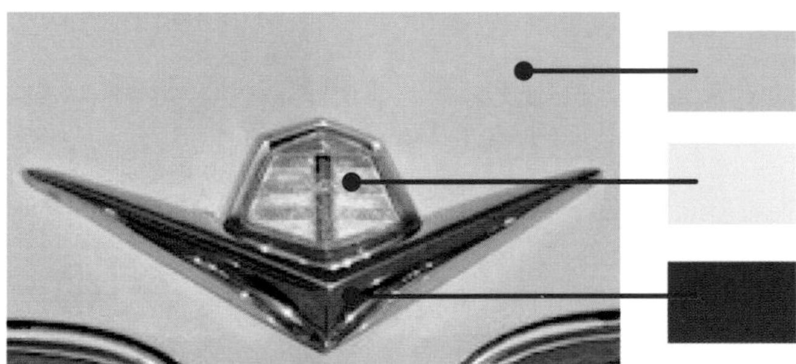

Draw rectangles, one for each sampled color. Select the rectangle, then select the Eyedropper tool and click on the image to fill the rectangle with the selected color.

To add the selected colors that you sampled from the image, select the rectangles with the colors and choose Add Selected Colors from the Swatches panel dropdown menu. This will add all the selected colors to the swatch list, and make them global colors in the process.

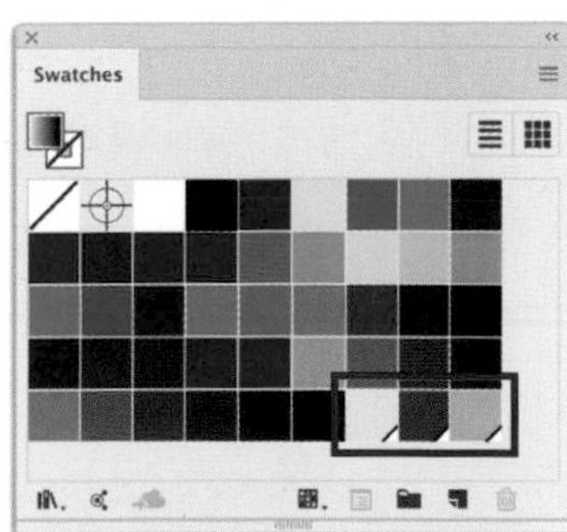

Color Theme Panel

The Color Theme panel is mini version of the Adobe color website at color.adobe.com/create. Open the Color Theme panel under Window > Color Themes.

Create

With the create tab selected, you can create colors using the Color Wheel or the sliders. When using the sliders, you can choose from CMYK, RGB, LAB, HSB, and HEX colors modes. You have five colors that show on the color bar with the base color indicated by the white triangle. The base color is the one that the color harmonies are based on.

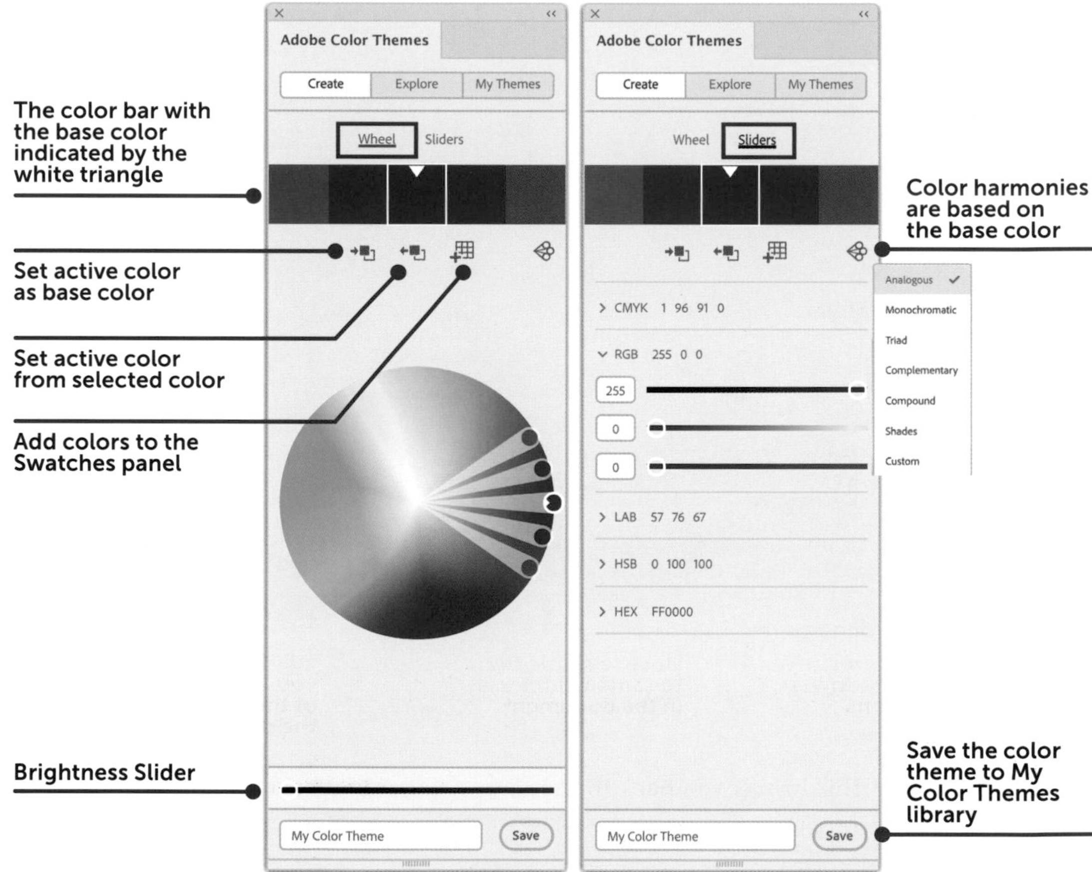

Explore

The Explore tab allows you to explore other color themes. Enter in a search term to find user-generated themes matching that search term. Once you find a color theme, click on the three dots in the lower right of the theme and choose from the options: Edit This Theme, Add to My Favorites, Add to Swatches, or View Online.

Input a search term, then click on ••• in the lower corner for options

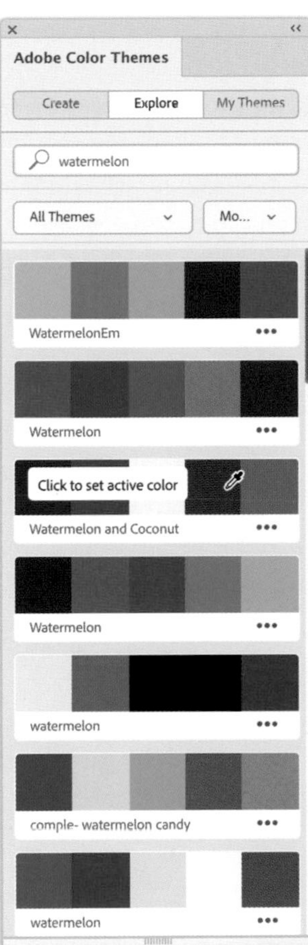

Hover over a color to sample it for use in the document

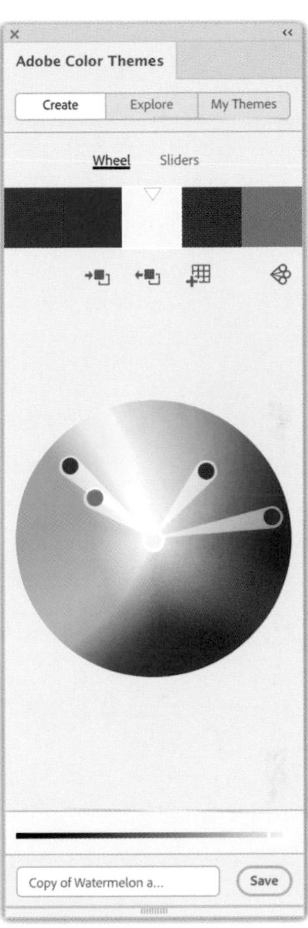

Edit This Theme lets you edit the five colors of the theme from the color wheel

Edit This Theme: This brings you back to the Create panel to edit the colors based on the color wheel. Edit any of the colors in the wheel and click on the Save to Swatches icon.

Add to My Favorites: This saves the colors you searched in the Explore tab. To access My Favorites, click on the dropdown menu located under the Search bar and choose My Appreciations to show your list of Favorites.

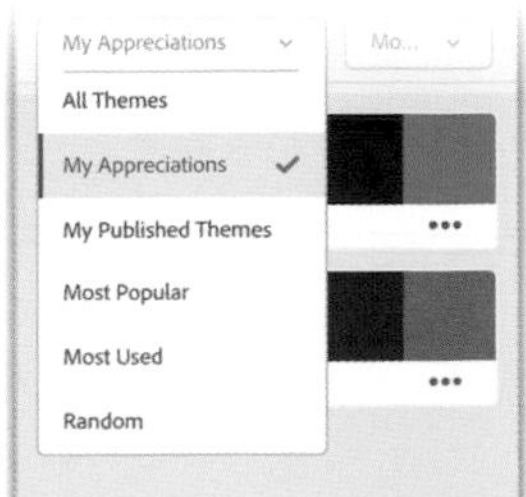

Add to Swatches: This adds the active colors to the Swatches panel in a folder at the end of the list of colors.

View Online: This opens a web browser to the color.adobe.com website.

My Themes

Creating Color Themes using the Color Themes panel can be done under the Create tab, which shows the color wheel. You can also explore a color theme and edit it further. In the Create section, move the color paddles around the color wheel to make a selection. Moving the color paddles out from the center will create a more saturated color, whereas moving them in toward the middle will be a less saturated color. To change the brightness of the color, use the slider at the bottom of the panel to make the overall color darker by sliding it to the left, or brighter by sliding it to the right.

Clicking on the sliders link will convert the color wheel to slider bars so you can edit in the different color modes. In either wheel or slider mode you can create colors. Click on the color harmonies icon to see harmonies associated with your base color selection.

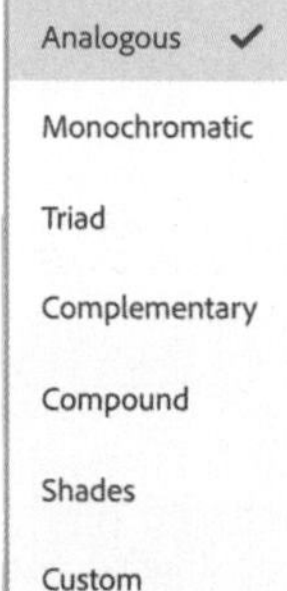

To save a set of colors, type in a name for the theme you want to save at the bottom of the Color Themes panel and click Save. Any saved themes can be accessed by clicking the My Themes tab at the top of the Color Themes panel or online at color.adobe.com.

My Themes Via color.adobe.com

Accessing the Adobe Color Themes online at color.adobe.com/create will give you more options to explore, create, and publish. You can edit color as you would in Illustrator and you can choose a color from an uploaded image to start your color exploration. You can create color themes and publish them so other users can search by keywords in the Adobe Color Themes panel or online. Any themes you create can also be downloaded for use in other Adobe applications as well. Color themes are saved as an adobe swatch exchange (.ase) file, so they can be downloaded for use in other Adobe applications.

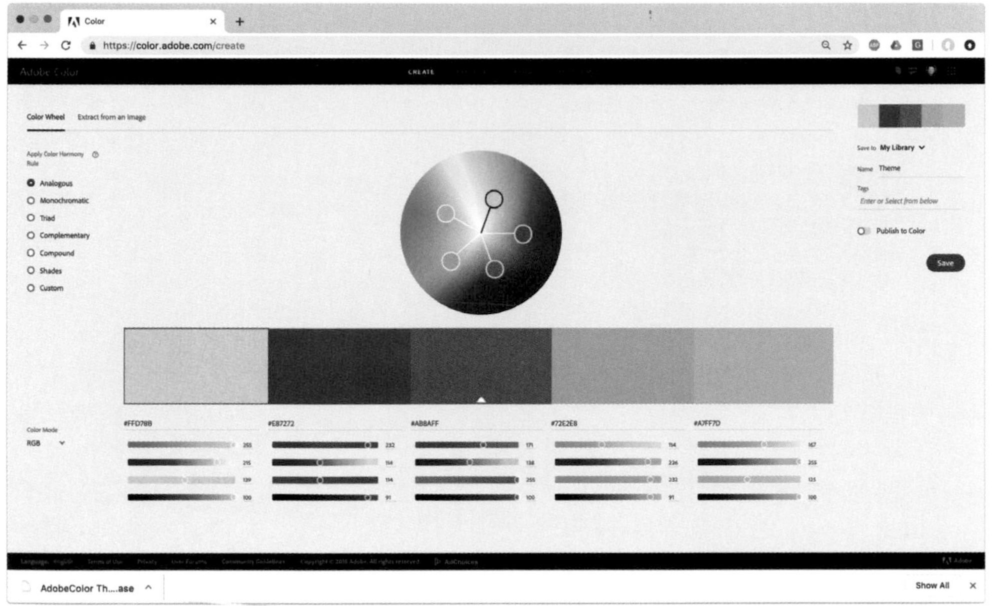

Color Guide Panel

The Color Guide panel is a useful way to create color variations. Open the Color Guide panel under Window > Color Guide. With the Color Guide panel open, click on a color from the Swatches panel, Color Picker, or Color panel to set the base color in the Color Guide panel. Once the base color is set, a list of colors will appear next to it. These colors are a set of color harmonies that are built into the Color Guide panel.

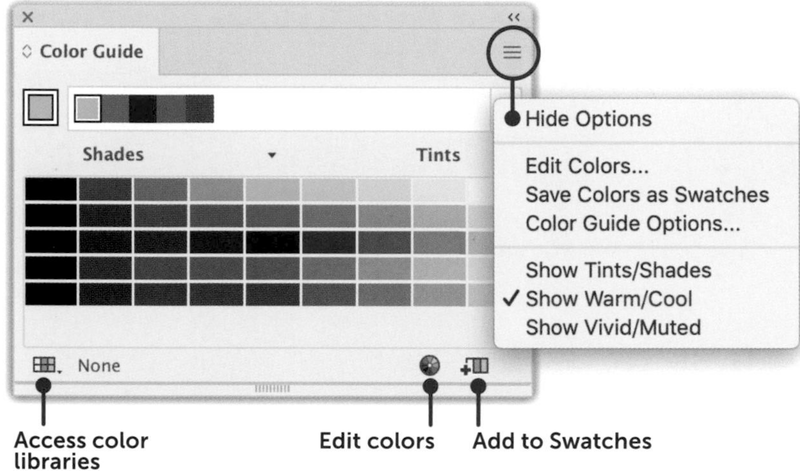

Color Harmonies

A long list of color harmonies are listed at the right of the color list. Click on the dropdown arrow to see the entire list of Harmony Rules based on the selected base color.

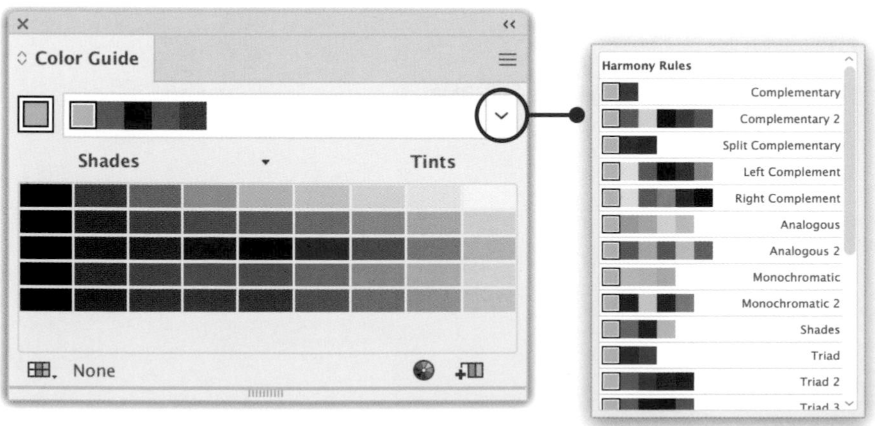

Color Guide Options

Show Color Families

Clicking on the dropdown menu gives you options for the Color Guide panel so you can choose how the color harmonies will family together. Choose from showing Shades/Tints of the colors, Warm/Cool colors, or Muted/Vivid renditions of the colors. The base color harmonies remain the same down the vertical center of the Color Guide and the family variations are shown on each side.

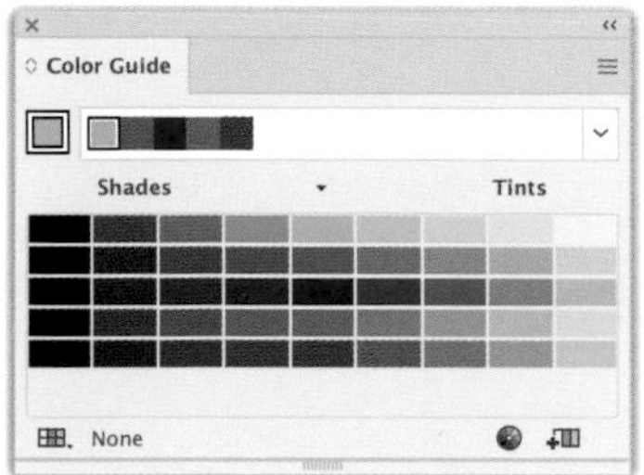

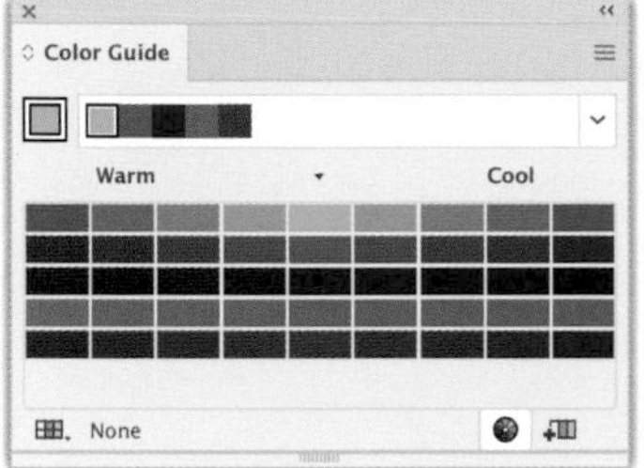

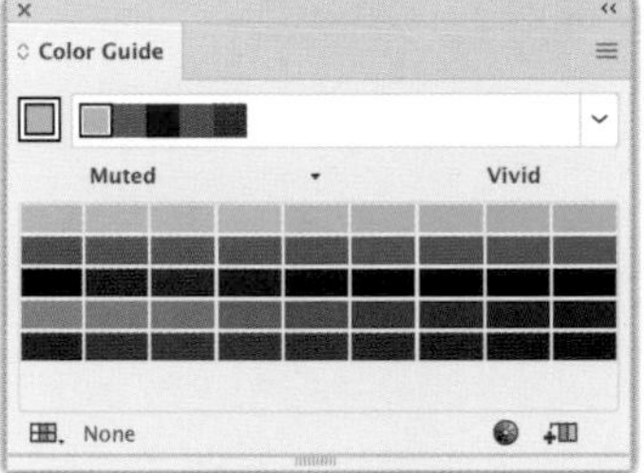

Color Guide Options

Clicking on the dropdown menu gives you options for the Color Guide panel. You can set the number of steps (color variations) from the original color set. Steps are the number of color options on each side of the selected set, from 3 to 20 Steps. Variation determines how much different these color will be from the original set of colors.

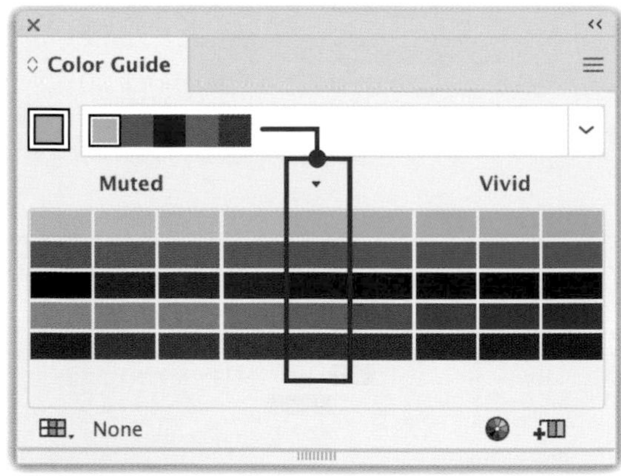

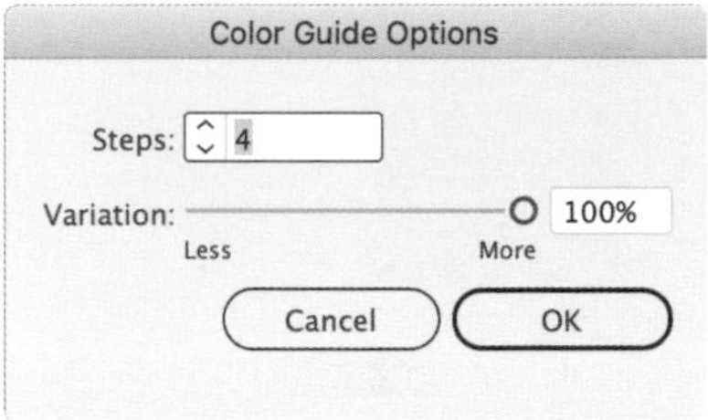

Add Colors to Swatches Panel

To add colors to the Swatches panel, click the add swatches icon in the lower right of the Color Guide panel. This will add the colors displayed in the color harmony bar at the top of the panel. If you want to add more colors based on you choice of Shades/Tints, Warm/Cool, or Muted/Vivid colors, select individual colors from the variations. Click on any color variation and hold Command (Mac) or Ctrl (PC) to select nonconsecutive colors. Or, click on a color, hold Shift and click on another color to select all the colors in between. Click on the Add to Swatches icon to add the selected colors to the Swatches panel.

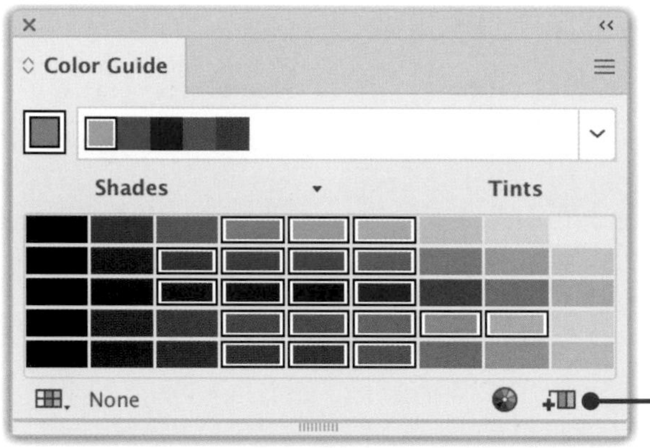

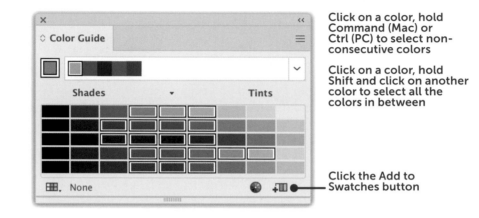

Click on a color, hold Command (Mac) or Ctrl (PC) to select non-consecutive colors

Click on a color, hold Shift and click on another color to select all the colors in between

Click the Add to Swatches button

Gradient Panel

A gradient is a graduated blend of two or more colors or tints of the same color. You can use gradients to create color blends, add volume to vector objects, and add a light and shadow effects to your artwork. In Illustrator, you can create, apply, and modify a gradient using the Gradient panel, the Gradient tool, or the Control panel.

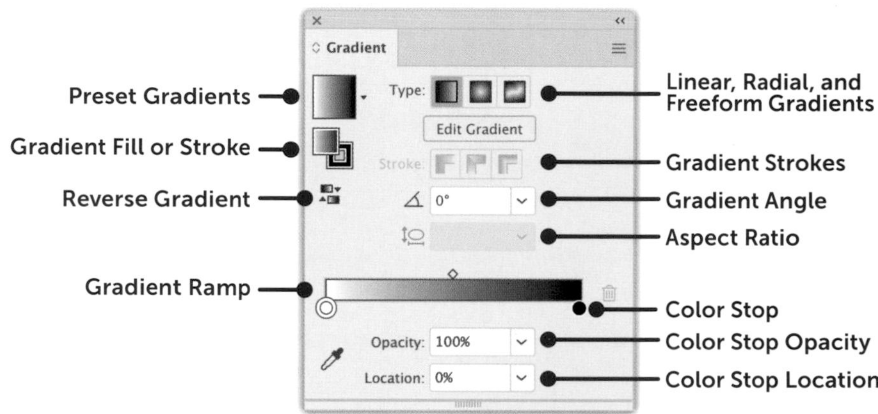

You can create or modify a gradient using the Gradient tool or the Gradient panel. Use the Gradient tool when you want to create or modify gradients directly in the artwork and view the modifications in real time.

Apply a Predefined Gradient

When you open the Gradient panel to apply a gradient to a fill or stroke of an object, the black and white gradient is the default. Select the object you want to add a gradient to, select the fill or stroke icon in the Gradient panel, and click on the white and black gradient icon to set the gradient. Once the gradient has been applied, you can edit it.

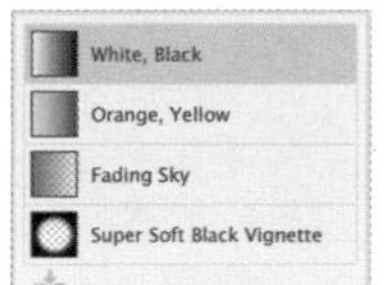

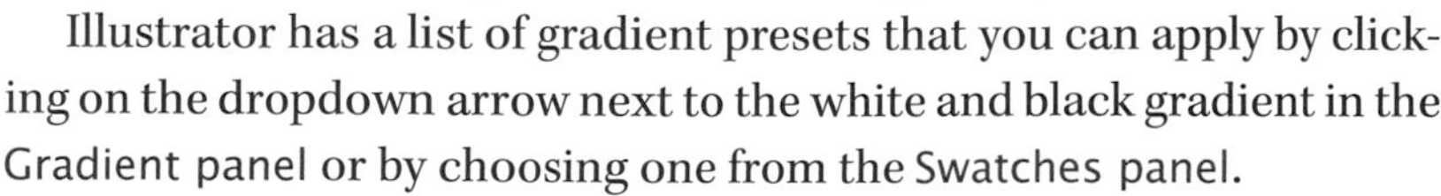

Illustrator has a list of gradient presets that you can apply by clicking on the dropdown arrow next to the white and black gradient in the Gradient panel or by choosing one from the Swatches panel.

Editing and Creating Gradients

To create a new gradient, you must edit an existing gradient and save it as a new gradient once the colors and attributes have been changed. There is no create new gradient command like there is to create a new color swatch.

Edit Colors

To edit the colors in the gradient, double-click on the color stop under the gradient ramp. This will bring up three options for choosing or editing color: create a color from the sliders, choose a color from the swatches, or choose a color by using the Eyedropper tool and selecting an area of color.

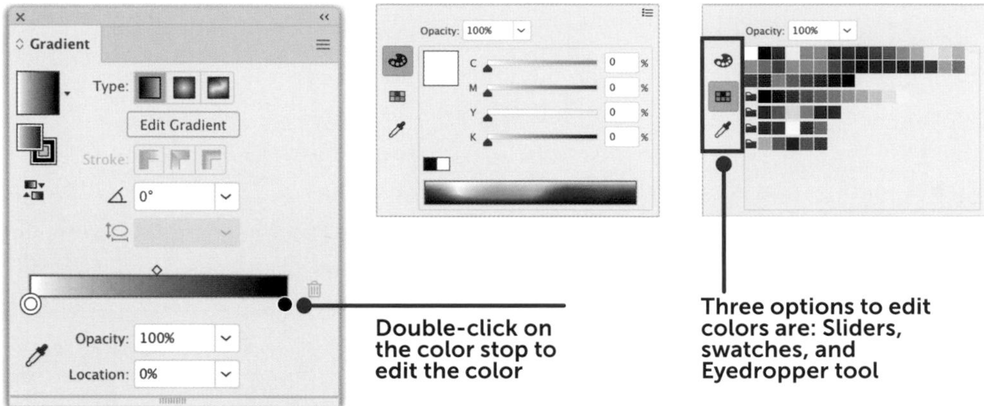

Add Colors

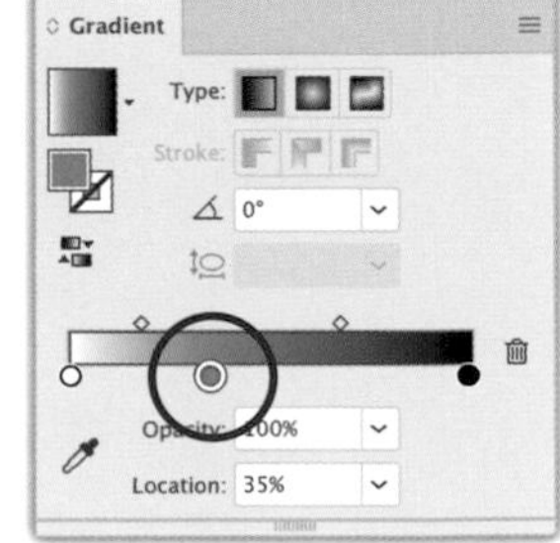

To add a color to the gradient ramp, click under the ramp to add a color stop. You can add as many color stops as will fit in the ramp. Slide the colors on the ramp to adjust the color blends. You can also reposition the colors in a different order on the ramp. To duplicate colors already on the ramp, hold Option (Mac) or Alt (PC) and drag the color stop to a new location to create a copy of the same color.

To control the precise location of the color stop on the ramp, click on the color stop and enter in the value in the Location field at the bottom of the panel.

Opacity

Changing the opacity on a gradient color will make the color more transparent, allowing any other content to show through when a gradient is placed over another object. To change the opacity of a color on the gradient, select the color stop and set an opacity from the Opacity dropdown menu or enter in a value and press return.

Location

The location controls two attributes on the gradient ramp: the location of the color stops and the blend rate between two color stops (indicated by the diamond above the color ramp). The left color stop location is set at 0% and the right color stop is set at 100% by default. To add a color stop in the middle of the ramp, click under the ramp, then set the location to be 50% .

To set the blend rate between colors, click on the diamond above the ramp. The default blend is set at 50% between two color stops to provide a smooth transition between them. To edit the blend rate, slide the diamond on the gradient ramp toward one color stop. You can also choose a value from the drop-down menu or enter in a value and press return. A location diamond will appear between color stops, so the more colors that are added to the gradient ramp, the more location diamonds will appear to control the blend rate between each color stops.

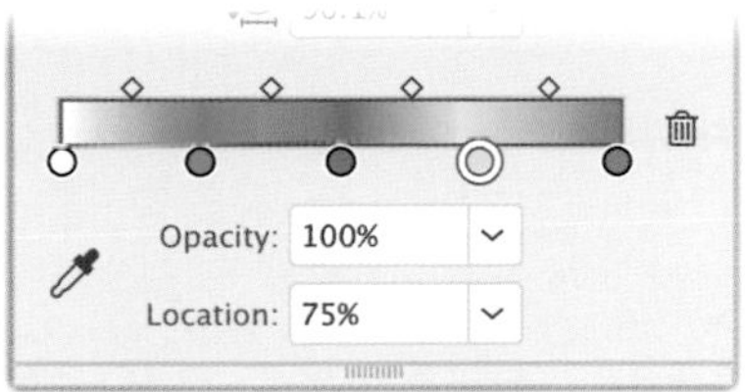

Delete Colors

To delete a color from the gradient ramp, select the color stop and drag it off the ramp. The color stop will be removed and the remaining colors will blend together.

Gradient Types

You can create three types of gradients in Illustrator:

Linear: This blends colors from one point to another in a straight line.
Radial: This blends colors from one point to another point in a circular pattern.
Freeform: This creates a smooth blend of color based on stops within a shape. The Freeform gradient is created by adding color stops and blending to shade the area around the line.

Editing Linear / Radial Gradient Attributes

Gradients can be modified to control the start and stop color locations, the angle of a linear gradient, and the aspect ratio of a radial gradient.

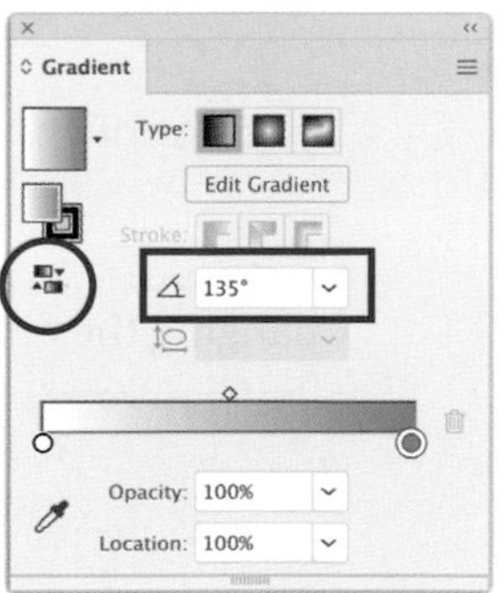

Gradient Angle

To modify the angle of a linear gradient from the standard 0-degree angle, click on the angle dropdown menu for preset angles or enter a value in the field and press return.

Gradient Direction

To change the direction of a linear or radial gradient click on the reverse gradient icon. The reverse gradient switches the start and stop colors' positions.

Aspect Ratio

When drawing a shape, the aspect ratio is set to 100%, which means if a circle or square is drawn, the radial gradient will be the same width and height. When a shape is not the same width and height, the radial fill will not match the ratio of the shape. To change the aspect of a radial gradient, click on the aspect ratio dropdown menu or enter a value and press return.

Circle with a radial gradient

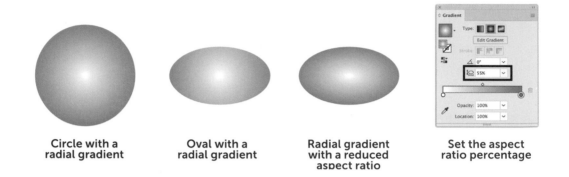

Oval with a radial gradient

Radial gradient with a reduced aspect ratio

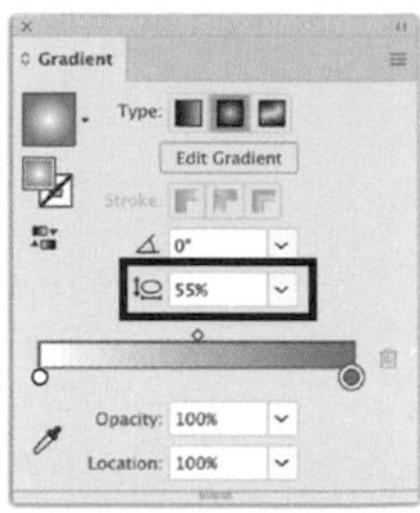

Set the aspect ratio percentage

Gradient Tool

The Gradient tool (shortcut is G) can be used to edit linear or radial gradients directly on an object. Select the object you want to edit and choose the Gradient tool. The Gradient Annotator will appear on the object, which will allow you to edit the gradient attributes directly on the shape. The Gradient Annotator can be turned on and off in the View menu by choosing Hide/Show Gradient Annotator.

To edit the colors of the gradient, the same gradient ramp as in the Gradient panel appears on the shape. Double-click on the color stops to edit the colors, click under the ramp to add colors, or drag a color stop off the ramp to delete the color. Use the location diamonds to control the blend rate of the color stops.

To change the start and stop locations of the gradient, select the black circle (the start end). Select the black square to control the stop point of the gradient.

To change the angle of the gradient, hover just off the black square (stop point) and the rotate icon will appear to allow rotation of the gradient ramp.

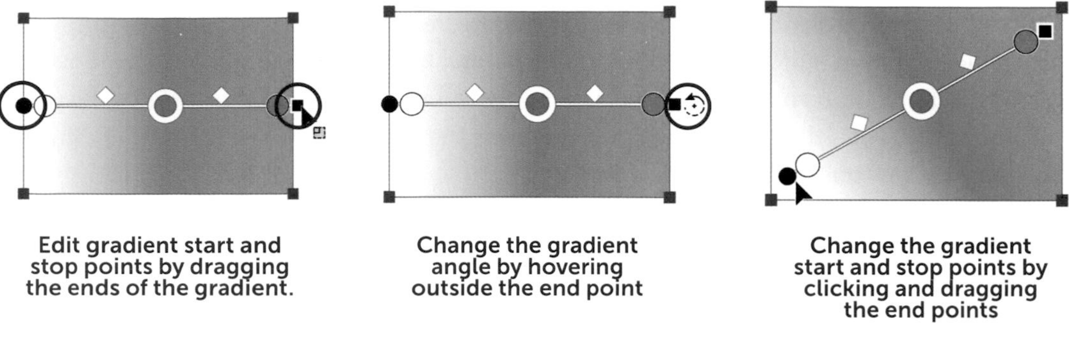

Edit gradient start and stop points by dragging the ends of the gradient.

Change the gradient angle by hovering outside the end point

Change the gradient start and stop points by clicking and dragging the end points

On a radial gradient, the Gradient tool can be used to edit the start and stop points and angles, as well as the focal point of a sphere.

Pull on the gradient start and stop points on the left double circle to edit the gradient start and stop. A dotted line shows the overall gradient area for adjusting.

To change the aspect ratio, click on the black circle on top of the circle and pull in or out from the center of the shape.

Double-click on the shape to change the focal point of the gradient to create a sphere. Double-click in the center of the shape to set the focal point back to its default (center) position.

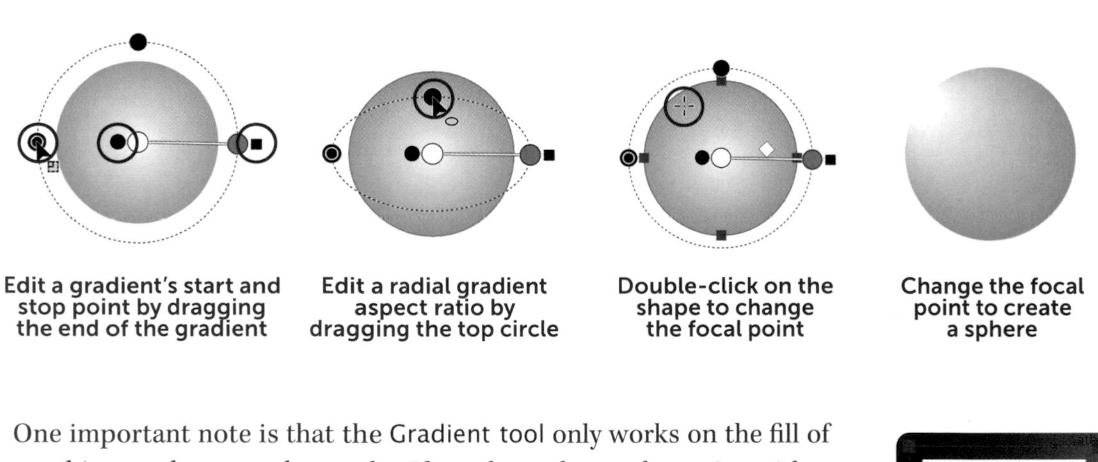

Edit a gradient's start and stop point by dragging the end of the gradient

Edit a radial gradient aspect ratio by dragging the top circle

Double-click on the shape to change the focal point

Change the focal point to create a sphere

One important note is that the Gradient tool only works on the fill of an object and not on the stroke. If you have the stroke active with a gradient applied, the Gradient tool will show that the operation is not available.

Apply a Gradient to a Stroke

When applying a stroke to a path, there are three choices. A gradient applied within the path, a gradient applied along the path, and a gradient applied across the path.

- **Within**: A gradient **within** the path has the gradient begin and end with the ends of the open path. When the gradient is applied with a closed shape, the edges of the shape are

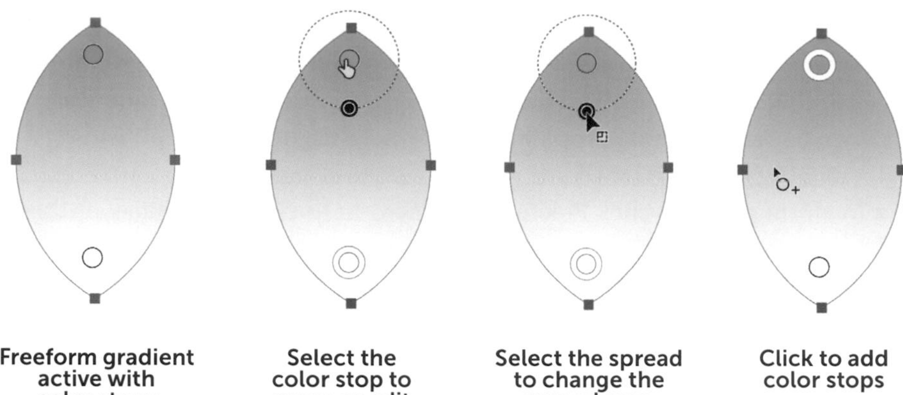

Freeform gradient active with color stops
Select the color stop to move or edit
Select the spread to change the spread area
Click to add color stops

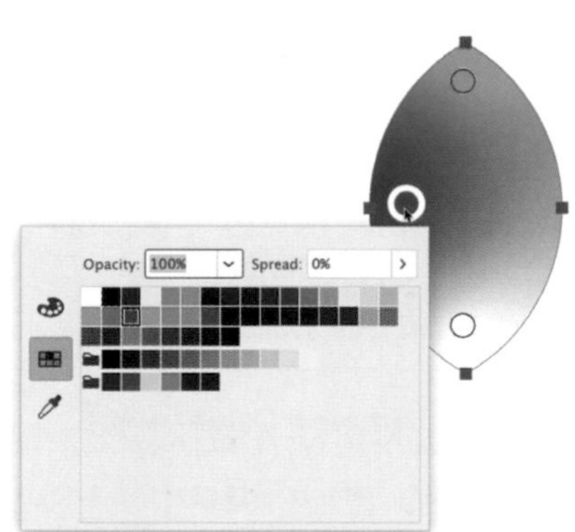

Edit stop colors: To edit a color stop, double-click on it to call up the color options. A panel will open and you can choose from the Color, Swatches, or Color Picker. Opacity of the color can also be set here or in the Gradient panel.

Reposition stop colors: Color stops can be rearranged by dragging them around to different locations in the shape. Color stops can also be dragged to the edge of the shape as long as the color stop is touching the shape. If the color stop is dragged off the shape it will be deleted.

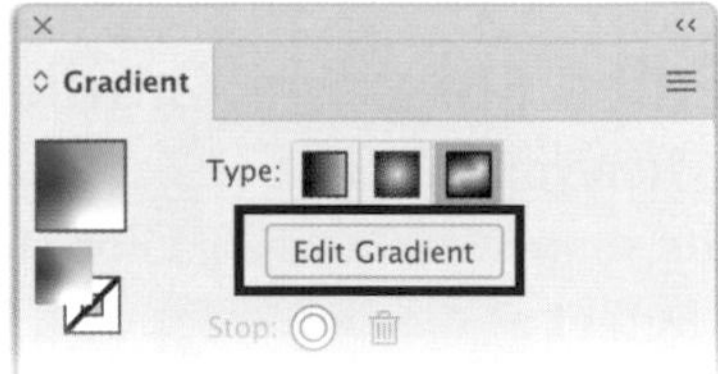

Edit an existing Freeform gradient: Once you click off a shape that has a Freeform gradient applied to it, you have to select the shape and click on the Edit Gradient button in the Gradient panel or Properties panel to get back into editing the gradient for the shape.

Edit a Freeform Gradient Using Lines

When you select an object and click on the Freeform gradient type, choose Lines for the gradient type. A Freeform gradient using lines flows along the lines and not around the points. This can give a smoother transition between color stops. Using the Lines method of Freeform gradients will not allow the Spread feature to be used like it could be around the points.

Click anywhere in the object to create the first color stop, or click on an existing color stop. Click to create the next color stop or connect it to an existing color stop. A straight line connects the first and the second color stops. When adding more color stops or clicking on additional color stops, the straight line changes into a curved line to accommodate the shape. Add color stops to the end of the path or along the path.

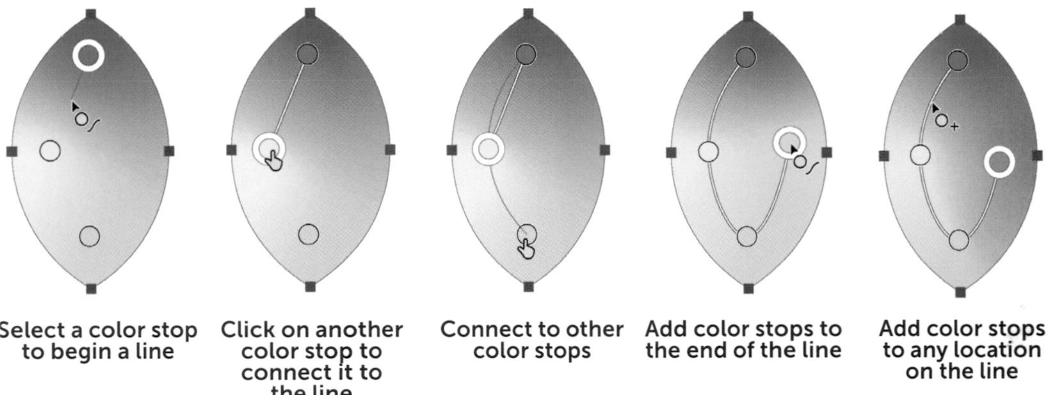

Select a color stop to begin a line | Click on another color stop to connect it to the line | Connect to other color stops | Add color stops to the end of the line | Add color stops to any location on the line

You can create multiple individual line segments in an object. To create a new set of connected color stops independent of the other color stops, drag the cursor outside the object and bring it back to the object and then click anywhere to create the first color stop for the new line. To connect two color stops that are at the end of a line, drag the line segments and join them together.

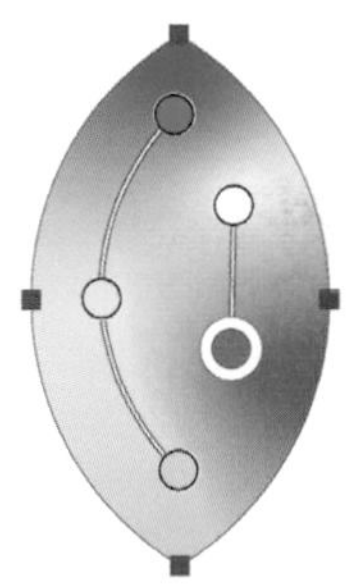

Create a new gradient line by dragging the cursor out the shape, then back in and click

Color Libraries in AI

Color libraries in Illustrator have preset colors, including ink libraries such as the Pantone matching systems. These libraries are accessed via the Swatches panel by clicking on the Library icon at the lower right or selecting it from the dropdown menu under Open Swatch Library.

When a swatch library is opened, it appears in a new floating panel. You can drag swatches from a swatch library into the Swatches panel for use in your document. You cannot add swatches to delete swatches from, or edit the swatches in any of the libraries panels.

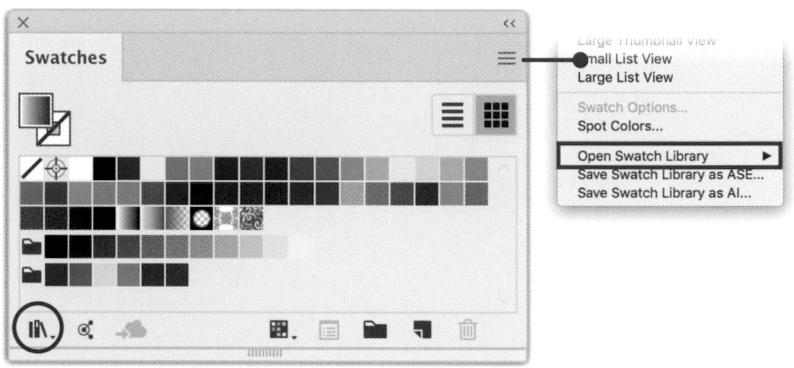

Spot Colors

Access spot colors from the Pantone Matching system, Toyo, DIC, Focoltone, HKS, and Tru-Match in the Color Books section of the Swatches Library.

Saving Swatches

To save swatches to be used in other Illustrator files or in other Adobe Applications, you can create and save custom swatches from the Swatches panel.

AI or ASE files

There are two different ways to save swatches. From the dropdown menu of the Swatches panel you can choose from Save Swatches Library as ASE and Save Swatches Library as AI.

ASE stands for Adobe Swatch Exchange and is used to save color from Illustrator to be used in other Adobe applications. Some limitations apply to saving swatches as an .ase file. Any tints of a color, patterns, or gradients will not be saved in an ASE color library.

AI Swatch Library is used to find colors from a previous Illustrator file. AI Swatch Libraries can contain tints of a color, patterns, and gradients, but they can only be used in Illustrator documents.

All the swatches in the Swatches panel will be saved in the Library if you choose Save as AI. Any tints of a color, patterns, or gradients will not be saved if ASE is selected. Name the Swatches Library file in the save window, and save the Library in a location that is accessible. The default location is to save the library is in the same location as all the other Illustrator libraries.

Importing Swatches

To bring swatches into Illustrator, access the color library menu from the Swatches panel. Choose Open Swatch Library > Other Library... at the bottom of the list of Swatch Libraries. Locate the Swatch Library and click Open. The swatches will open in a new panel independent of the Swatches panel. The colors can be applied from this window or be dragged into the Swatches panel if they need to be edited.

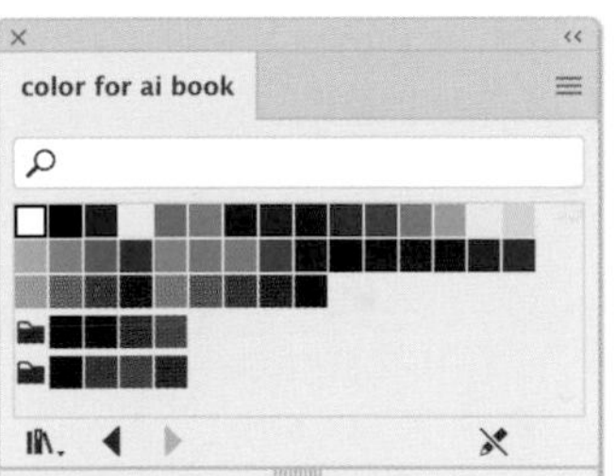

Color in CC Libraries

A Creative Cloud Library is a collection of design assets. Several types of design assets can be added to a Creative Cloud Library. With Adobe Illustrator, assets can be Colors, Color Themes, Brushes, Character Styles, Graphics, and Text.

Creative Cloud Libraries help you organize, browse, and access creative assets. For example, you may want to create a Creative Cloud Library that contains all the components most frequently used in a specific project.

The Libraries panel is under the Window menu. To have access to the content of the Libraries outside of Illustrator, choose View on Website from the Libraries panel dropdown menu. You can go to creativecloud.adobe.com, log in using your Adobe ID and password, click on Your Work at the top left, then choose from your Libraries on the left side.

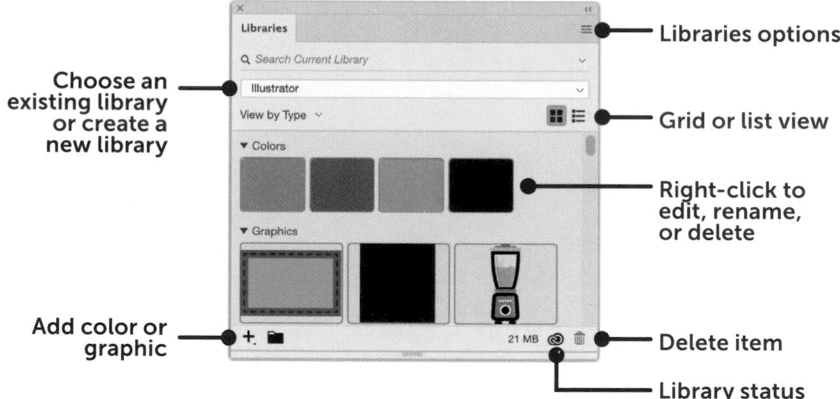

Creative Cloud Color Storage and Usage

Add Colors

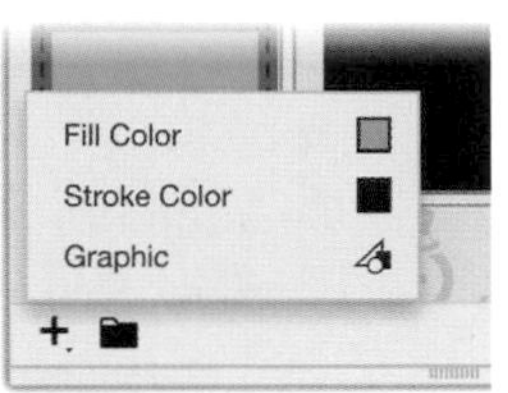

To add colors to the Libraries panel, create an object and apply a stroke and/or fill. Select the object and click on the + at the lower right of the Libraries panel. A window will allow you to add the fill color, stroke color, or the object to the Libraries panel.

Once the color has been added, hover the mouse pointer on an available color asset to view the name and hexadecimal code of the color. To change the name of the swatch, hover over the lower portion of the swatch and double-click on the HEX color to edit the name. To change or edit the color in the Libraries panel, double-click on the color swatch. This opens the Color Picker. Change the swatch color and click OK. You can also right-click on the color and choose Edit as well as other options.

To add colors to the Libraries panel through the Swatches panel, click on the dropdown menu on the Swatches panel and choose New Color Swatch. Create the color you want, check the box at the lower left (Add to my Library), choose the library you want to add it to from the dropdown menu, and click OK.

Apply Colors

To use colors that are in the Libraries panel, create an object and target the fill or stroke in the toolbar and select the color from the Libraries panel. Colors in the Libraries panel are non-global colors. If you edit a color in the Libraries, the color in the object will not reflect that color edit.

Colors in the Libraries panel can be added to the Swatches panel by selecting the color swatch, right-click, and choosing Add Color to Swatches.

Delete Colors

To delete a color from the Libraries panel, select the color and click the trash can icon in the lower right, or right-click on the color and choose Delete.

Share, Collaborate, Import, and Export Libraries

To share colors in your library with other people or team members, click on the dropdown menu from the Libraries panel. Here you can select from a list of options.

To export a current library, select it and choose Export Name Of Library. This will save the library to be imported into a library for another person or onto another computer.

To import a library, click on the folder icon next to Select Library and click Import. Navigate to a saved library, select it, and click Open.

When you choose Collaborate, a web browser will open to your Adobe Creative Cloud account and a dialog box will appear. Add in the email address for the person you want to collaborate with, and choose from options that allow them to View or Edit the Library. Then click Invite.

To share a library, choose Share Link, and a web browser will open to your Adobe Creative Cloud account. A dialog box will appear. A link can be copied and shared with other people. Link options allows others to Follow your library and to Save to your Creative Cloud account.

CC Libraries in Other Adobe Applications

The Libraries panel in other Abode applications may be listed under the Window menu in CC Libraries. To open the Libraries panel in other Adobe Applications, choose the library from the dropdown list. From there you can access the library you saved your colors to and now have access to the colors that were created in Illustrator—without having to re-create them in your current application.

Recolor Artwork

In Illustrator, your vector creations may get very complicated. Selecting the shapes to change or edit the colors could be a tedious task. Recolor Artwork makes editing existing colors easy without the need to select and isolate the fills or strokes that the colors are applied to. You can also recolor your vector artwork by applying color harmonies to the artwork, changing the number of colors used in the creation, and adjusting the saturation of the colors.

Recolor Artwork Panel

Select your artwork and open the Recolor Artwork panel. Choose Edit > Edit Colors > Recolor Artwork or click the Recolor button at the bottom of the Properties panel.

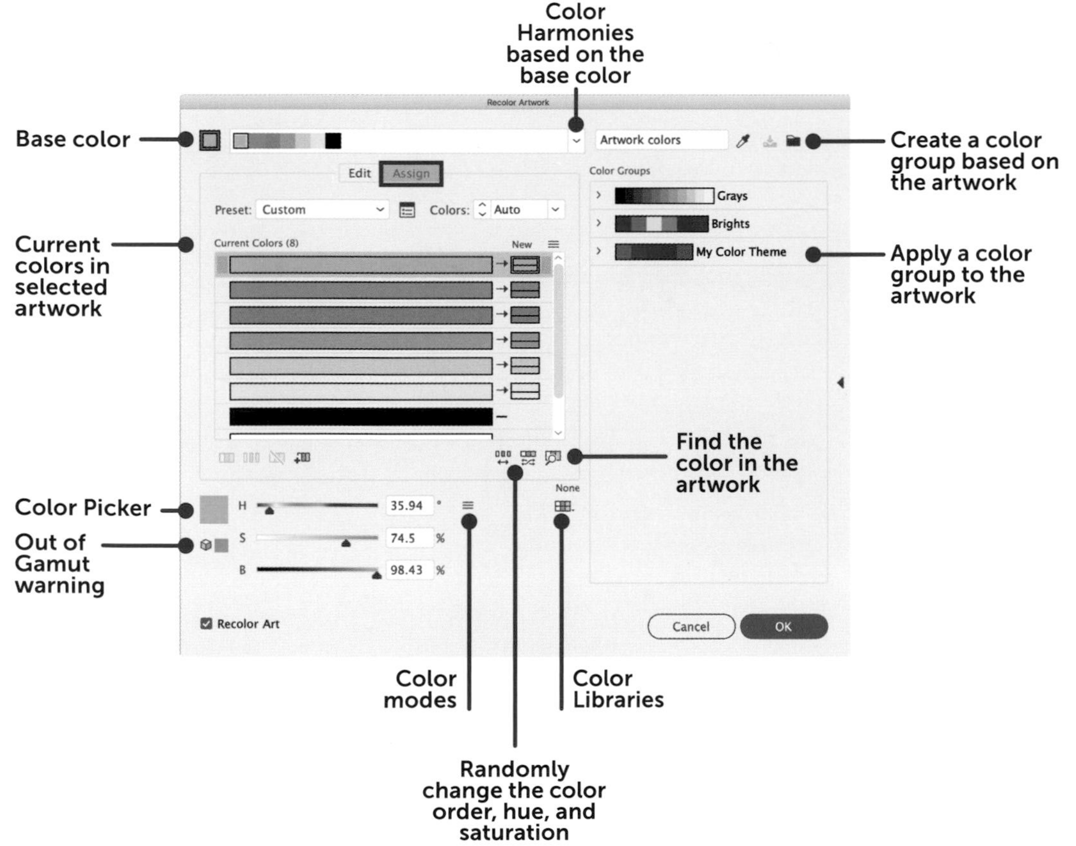

Assign Colors

When the Recolor Artwork is open, the mode is in Assign. This mode allows you to see the number of colors used in the artwork, listed under Current Colors. You can save new color groups to your Swatches panel by selecting the artwork colors and making a Color Group. You can add or reduce the number of colors used, and change colors in the creation. The hue, saturation, and brightness sliders allow you to change the adjustments on the colors in the artwork and generate color harmonies based on your current base color choice.

Color Harmonies

At the top of the panel are the squares of colors starting with the base color, which is what the color harmonies are based on. The base color is the one that appears on the top of the list in the Current Colors section. Click on the drop-down menu at the right of the color harmonies bar to access all the color harmonies for your base color.

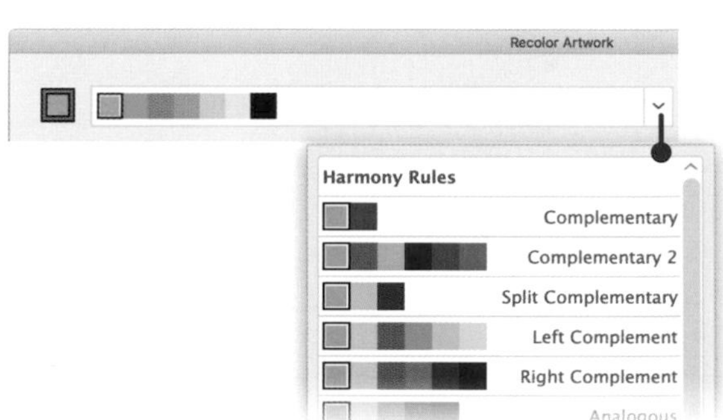

Once a new color harmony is chosen, the artwork will replace the current colors with the selected colors from the harmonies.

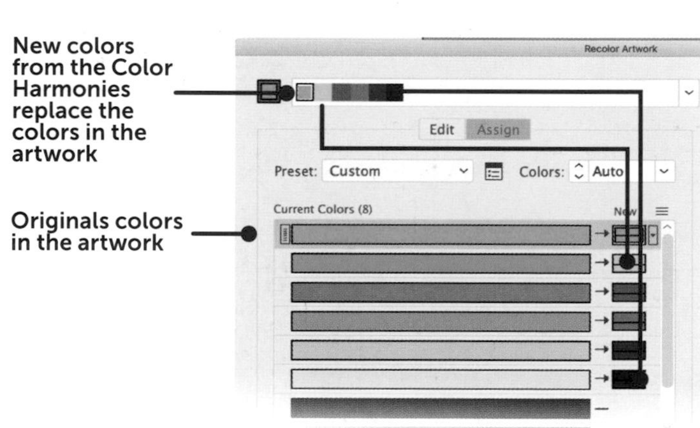

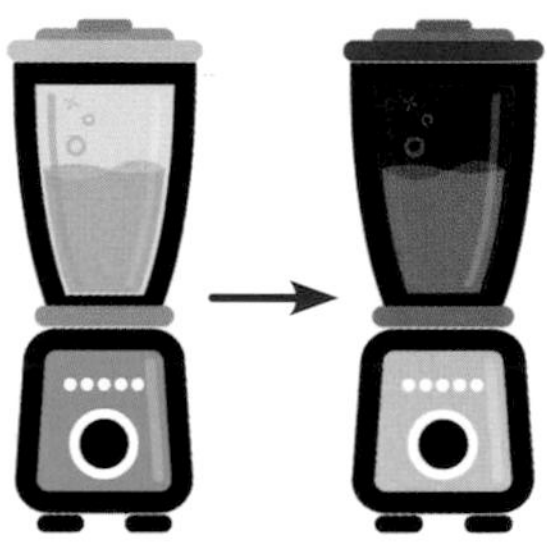

Originals colors in the artwork now mapped to a new set of colors from the color harmonies

Color Groups

Color groups are created by choosing color harmonies, then clicking on the Color Group folder to add it to the list. The list of color groups can be selected to remap the artwork to the new Color Group colors.

Hue, Saturation, and Brightness

The hue, saturation, and brightness (HSB) sliders can be used to select a color from the Current Colors list to edit the hue, saturation or brightness of that color. Each color from the list can be selected and edited individually with the HSB sliders.

Editing Existing Colors

The Edit option in Recolor artwork allow you to use the color wheel to edit your color choices. Use the Link Color Harmony option to have colors change together.

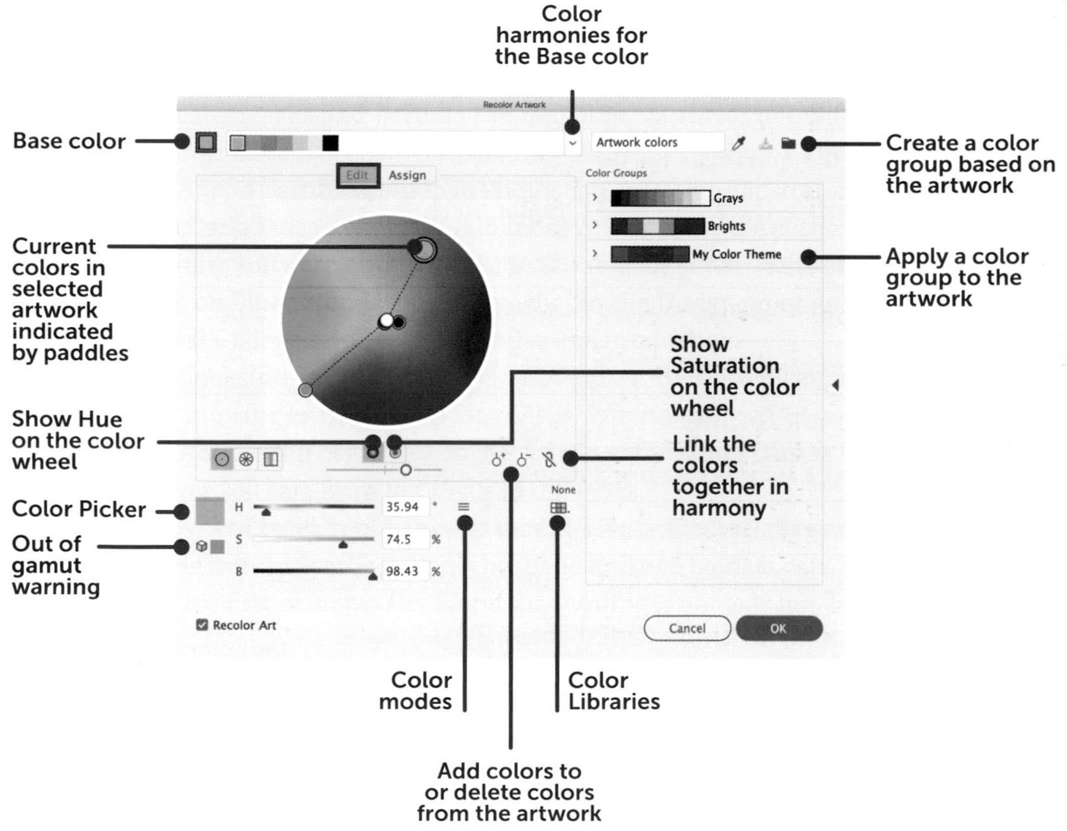

Color Wheel

The color wheel allows dynamic changes to an applied color and allows you to keep the colors linked together to adjust the hue or saturation overall. You can choose new colors while keeping all the other colors in harmony.

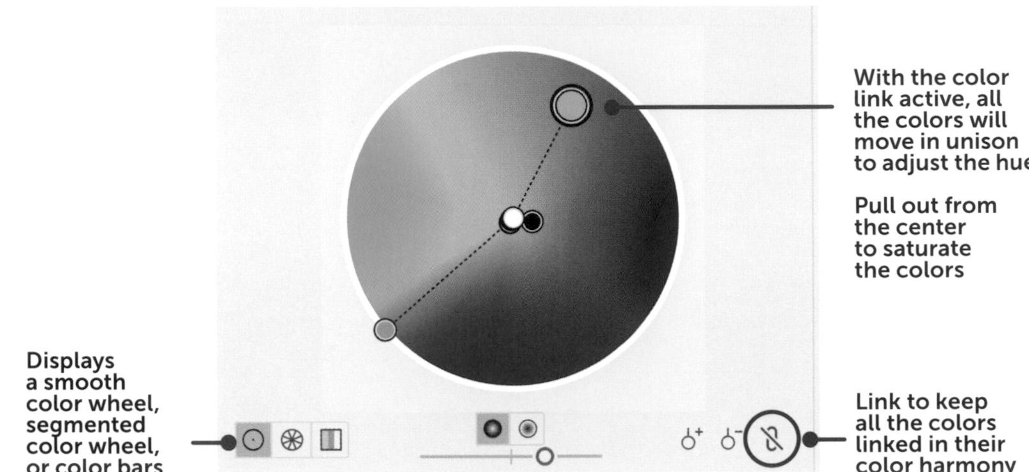
With the color link active, all the colors will move in unison to adjust the hue
Pull out from the center to saturate the colors
Displays a smooth color wheel, segmented color wheel, or color bars
Link to keep all the colors linked in their color harmony

By linking the colors and rotating the paddles around the wheel, you can very quickly and easily recolor your artwork while keeping the original color harmonies.

8 Output

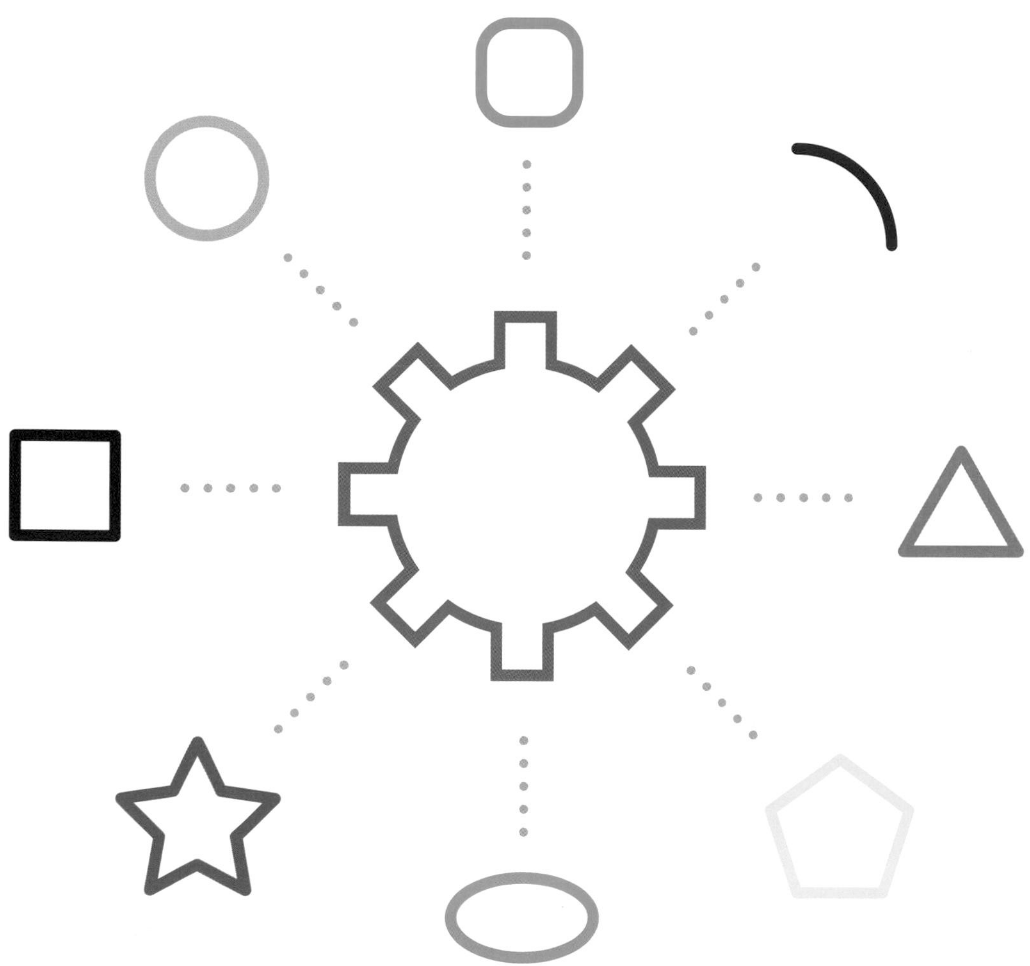

PDF

The Portable Document Format (PDF) is the most common way to share your finished creations. Adobe Acrobat Reader is the free application that can open a PDF. Those who want to edit PDFs or add extra functionality to them can use Adobe Acrobat Pro, the powerful sibling to Reader. If you don't know if you have Acrobat Reader, you can open a PDF in a web browser and use the built in PDF reader.

The advantage of sharing a document as an Acrobat PDF rather than an Illustrator document is that PDFs retains the layout and text formatting of the application without the recipient needing to have the author's applications or fonts.

Presets

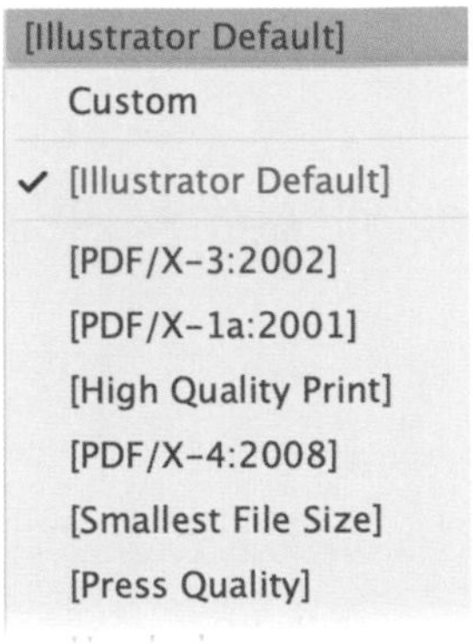

Illustrator can save directly to a PDF format using File > Save > Adobe PDF. You'll be prompted for a name and a location where you can save the PDF. The next dialog box has several choices to make regarding the quality and color rendering options. If you cancel at either stage, no PDF is made.

[Illustrator Default]
Custom
✓ [Illustrator Default]
[PDF/X-3:2002]
[PDF/X-1a:2001]
[High Quality Print]
[PDF/X-4:2008]
[Smallest File Size]
[Press Quality]

The Presets for saving a file are in a dropdown menu of the Save As PDF dialog box labeled Adobe PDF Presets. The Presets in brackets ("[" and "]") are built-in and cannot be changed. You can customize the settings by choosing Custom from the list of Presets to use later. And there are many things that you may want to change! When working with a printing company, some will often suggest which preset to choose or supply a custom preset they have made to suit their workflow.

One note on choosing Presets based on what "sounds" right. If you have any doubt about the Presets, saving your file with the Illustrator Default will be the best overall option. My recommendation is to never used Smallest File Size, even if a small file size is your main goal. Smallest File Size removes most of the information from a file, including the fonts you have used, leaving the end recipient with a file that does not look like the one originally created. High Quality Print is a good choice if you are printing a file on a home printer and want good quality output but are not looking for print-shop quality. Press Quality is a good choice if you are preparing a file to be printed by a print shop that higher-end equipment and great paper quality.

PDF/X is the industry standard for printing graphics. This preset renders the file, images, fonts, and colors so they produce professional quality results. These files also contain all the information in the file needed to edit the file if necessary. The different numbered version PDF/X (1a, 3, and 4) of the PDF/X standards are newer standards with revisions to the color rendering.

General Options

When you save a file as an Illustrator PDF, the first set of options are the General options.

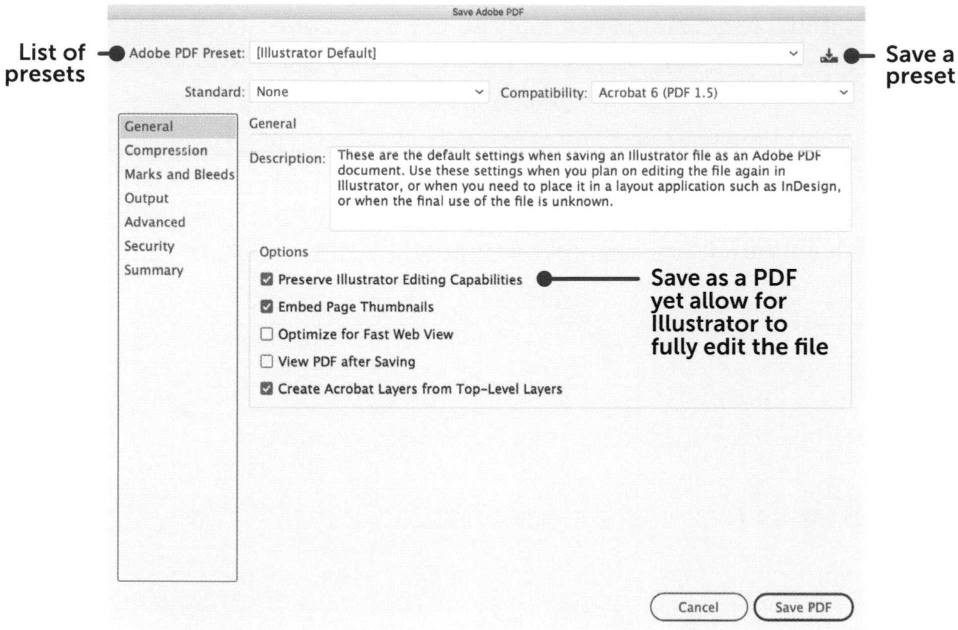

Here you can choose the presets if the Illustrator default is not what you need. The checkboxes allow you to Preserve Illustrator Editing Capabilities, which keeps the entire file editable in Illustrator but it appears as a PDF for other users who may not have Illustrator or have no need to edit the file. This is a good option to have checked if any editing it going to be done in the file.

Embed Page Thumbnails creates an image of each artboard in the file and displays it as a page thumbnail when the PDF is opened in Adobe Acrobat or Adobe Reader.

Optimize for Fast Web View speeds up the preview and reading process if the PDF is to be displayed on the web.

Create Acrobat Layers takes the top-level layers of the Illustrator file and retains the layers in the PDF as they were built.

Compression

Compression mainly pertains to images embedded or linked in the Illustrator file. .The Smallest File Size preset greatly reduces the image quality down to produce an image that is only suitable for on-screen display and not for print. More print-friendly settings for images would be ZIP compression rather than JPEG, since ZIP is lossless compression, unlike JPEG which literally discards data and creates compression lines in the images. The Downsampling option will take any image with a resolution higher than what is specified and reduce the image resolution to a lower specified resolution. This may reduce the quality, but it also will reduce the file size. A good rule of thumb for image resolution is if the images are going to be printed professionally, 300 ppi (pixels per inch) is needed. 150 ppi is acceptable for home printing, 72 ppi is low resolution and only suitable for display on the web or a screen.

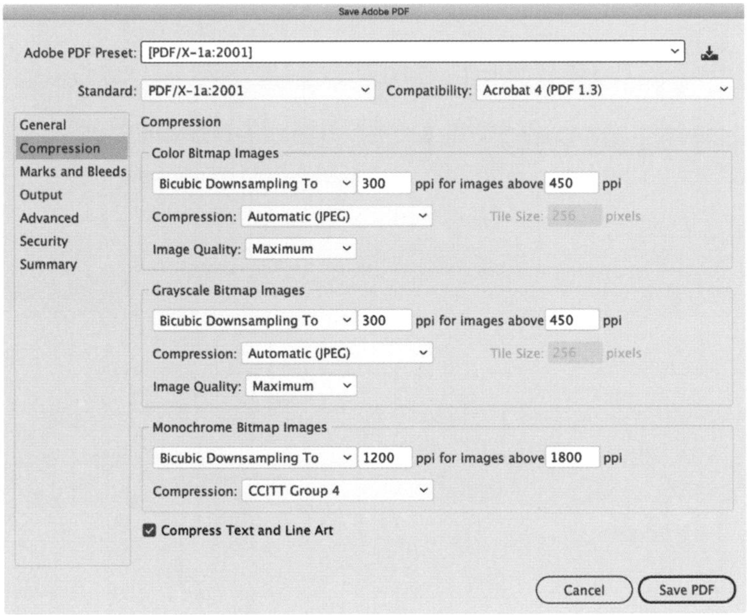

Marks and Bleeds

Beyond the edge of the artboard is the bleed and the Crop Marks that determine where the paper will be cut after the final printing. Some printers may also want more information like filename or date (Page Information). Color bars can be applied to show the density of the ink when the file goes on press.

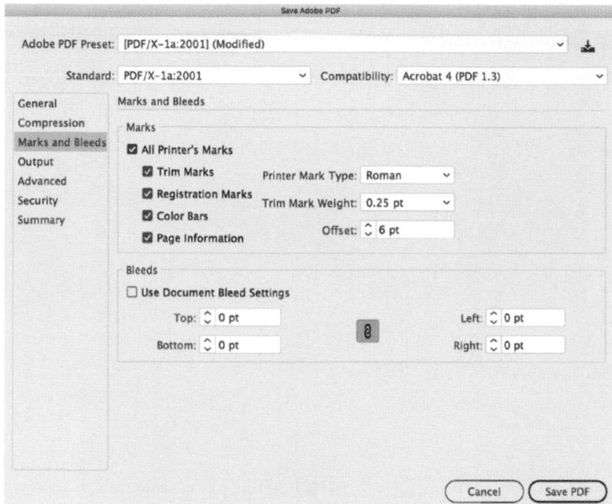

Use only the settings that your print shop asks for! If you've set your document's bleed to their specs, you can check Use Document Bleed Settings in the Document Setup.

Output

Color output is controlled in the Output options. The key choice here is the Destination. If the PDF were to be viewed on screen only, sRGB is the profile that represents standard displays. For print, choose the profile indicated by your print shop—they may even supply one to install—and choose it here. With a Destination profile chosen, specify the Color Conversion: Convert to Destination (Preserve Numbers).

A CMYK Destination, ensures that all RGB content is converted to the printer's desired profile while maintaining any CMYK choices you may have made in the Illustrator file.

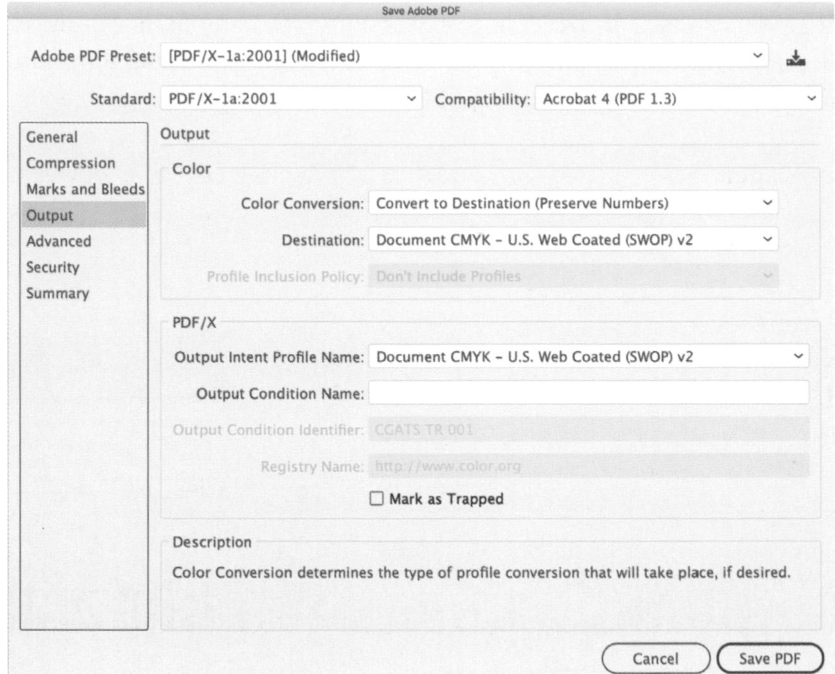

It is possible that you will be asked to do no Color Conversion so that the printer may do it themselves with the most up-to-date profile. Profile Inclusion Policy is also a decision for the print shop, because some of their software may not like to have unnecessary data embedded in the PDF.

Advanced

Subset Fonts When Percent of Characters Used Is Less Than (X%) sets the threshold for embedding complete fonts based on how many of the font's characters are used in the document. If the percentage of characters used in the document for any given font is exceeded, then that specific font is completely embedded. Embedding complete fonts increases file size, but if you want to make sure you completely embed all fonts, enter 0%.

Security

With Document Open Password, you can encrypt the PDF so it cannot be opened without a password that you choose. These settings are set in Adobe Acrobat, not in Illustrator. If you open a PDF in Illustrator, there may be an Open Password or Editing Password needed to open or edit the file.

The Permissions portion is less secure, but it attempts to prevent certain actions (printing or extracting content) unless a password is provided. However, there is software that circumvents this.

Summary

The Summary shows the setting options of the PDF. Each option in the Summary can be expanded to see the details of the chosen settings.

The Warnings section shows possible issues with the compatibility of the settings with the chosen preset and options.

Save Summary... allows you to save a digital version of the summary for reference.

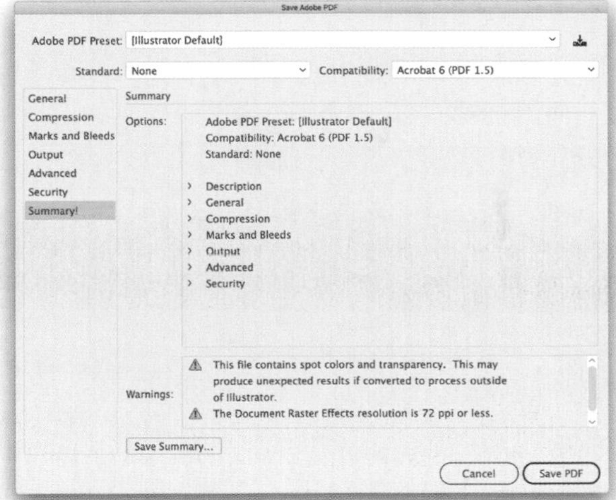

Asset Export

The Export for Screens panel is a more efficient way to generate assets of different sizes and file formats in one easy-to-use panel. Through the Asset Export panel you can quickly generate and export your creations for web and mobile workflows in several different file formats; PNG-8 and PNG-24, JPG, SVG, and PDF. The size of the files can be customized to create different versions of the artwork in multiple sizes for web use or for iOS and Android devices.

Adding Assets

Adding any of your creations to the Asset Export panel (Window Asset Export) is as simple as selecting your artwork and dragging it into the Asset Export window. Note: either group the artwork into a single asset before dragging it into the Asset Export window, or hold down Option (Mac) or Alt (PC) and drag it in. If the artwork is not grouped, all the individual shapes will appear as separate assets, which is probably not what you want.

Add as many assets to the export window from the current document as you need. The Asset Export panel does not act like a library, so if you switch over to another open document, the panel will not show assets from outside the current document. If you have several assets to export, it may be easier to have them all in one document.

Export Settings

To control the setting of your export, click on the Asset Export panel dropdown menu to dial in the specific setting for each format. You can edit the export settings for PNG-8, PNG-24, JPEG quality, SVG, and PDF.

File Formats

PNG, JPG, SVG, and PDF are the supported file types to export from the Asset Export panel.

Portable Network Graphic (PNG) comes in two flavors, PNG-8 (256 colors) and PNG-24 (about 16.7 million color), sometimes just labeled at PNG. PNG files render as pixels even though the original artwork is a vector. They are favored for many graphics, specifically vector artwork, type, logos, and illustrations since they render without any compression and little loss of quality. Another big feature they offer is a transparent background, which makes the content of the file easy to use on any layout for web or mobile devices. A PNG-8 produces a relatively small file since the maximum number of colors is 256. PNG-24s are larger files because the number of colors contains 16.7 million possibilities. PNG files are not ideal for photographic images, since a PNG-8 can only render limited colors.

PNG-8 files will only render transparency when there is no color

PNG-24 files will render translucency creating a true transparent effect

One more thing to consider when using PNG-8 and PNG-24 formats is how to render drop shadows. All PNGs support background transparency, and there is one case where this may become an issue: that is if your graphic has a drop shadow that fades into that transparency. A PNG-8 will render the drop shadow with a halo around it since the transparency only exists when there is no color in the file. A PNG-24 will render the drop shadows as translucent.

Joint Photographic Expert Group (JPEG, shortened to JPG) files are used for exporting images. They are not ideal for graphics, logos, or type. JPGs support full color range, which is why they are ideal for photographs. But they do not support transparency. JPGs are also good for producing small file sizes due to their ability to compress a file, however, the more a JPG is compressed, the more the file degrades in quality. Some people refer to this as pixelization since you notice the pixels with more compression. JPG files are referred to as having "lossy" compression do to this attribute.

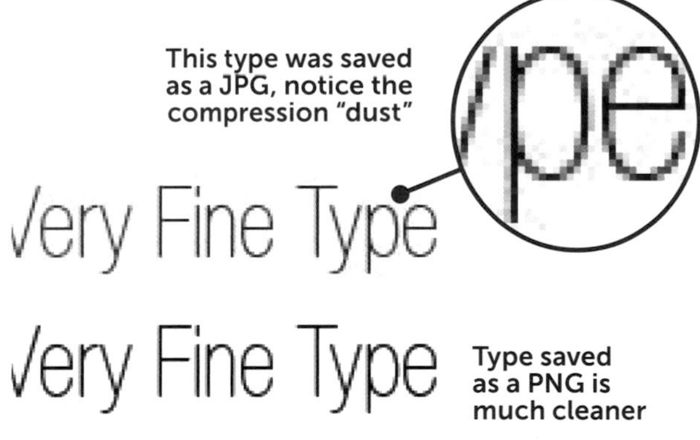

JPG files that are highly compressed will get JPG compression lines as the file format tries to group and convert similar colors into a single color to save file size. When a file is highly compressed, these lines can become visible and the degradation in quality is quite noticeable. Typically, we don't use a JPG export from Illustrator since most, if not all of the artwork is vector-based rather than a photograph.

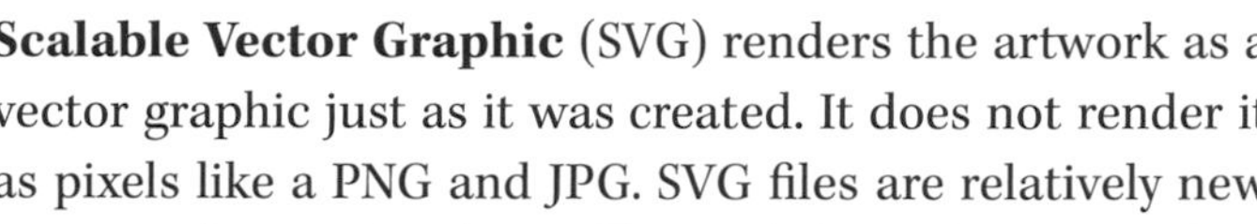

Scalable Vector Graphic (SVG) renders the artwork as a vector graphic just as it was created. It does not render it as pixels like a PNG and JPG. SVG files are relatively new compared to PNG and JPG files, and they offer some advantages over them as well.

Scalability is one of the major advantages of SVG files because they are resolution independent. This means they retain the same quality no matter what resolution or size they are being displayed at. On a display where a JPG might appear blurry, an SVG will still look high-quality.

Editing capability is another unique feature. You can create and edit an SVG in Illustrator, whereas a PNG or JPG has to be editing in a pixel-based editing software even though the original file may have started in Illustrator.

Performance of an SVG file is much faster on loading since there is no file that needs to be download. No file? An SVG is all numbers that spell out the curves and lines of the vector artwork. This makes your website or mobile app work faster for visitors, improving the user experience.

Style control is another benefit of using an SVG file. You can control properties such as fill color, stroke color, sizing, and more through CSS without having to create and export multiple files.

Portable Document Format (PDF) is a file type that maintains editing capabilities of the Illustrator file while providing the recipient with a printable and viewable file that has all the quality and functionality of the original file.

Export Assets

Now it's time to export your assets in the formats and sizes you need for your specific end user or device. Select the asset you want to export and choose the settings, file format(s), and sizes for the files. When iOS formatting is chosen, the files sizes display as .5x, 1x, 2x, etc. of the size of the original artwork. When Android is selected, that format sizing is labeled as ldpi, mdpi, hdpi, xldpi, etc.

Add scale and formats to the list by clicking on the + Add Scale section. Click the export button, and choose the location of the files, and a folder will be created with internal folders reflecting the different scales.

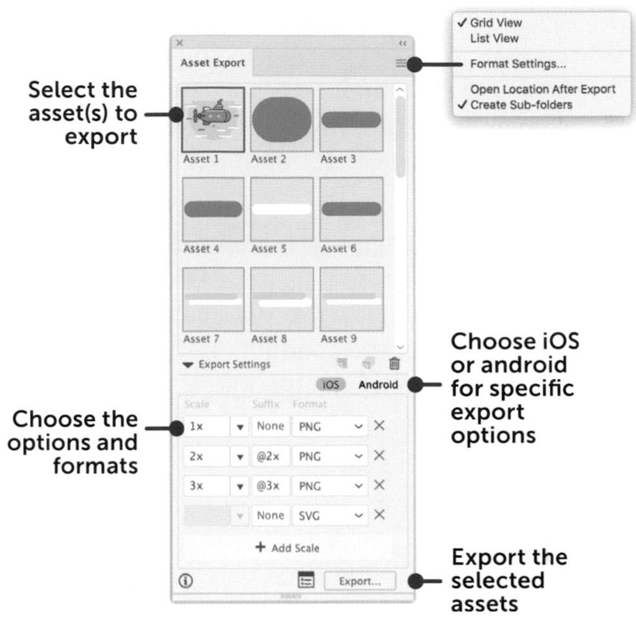

Updating Assets

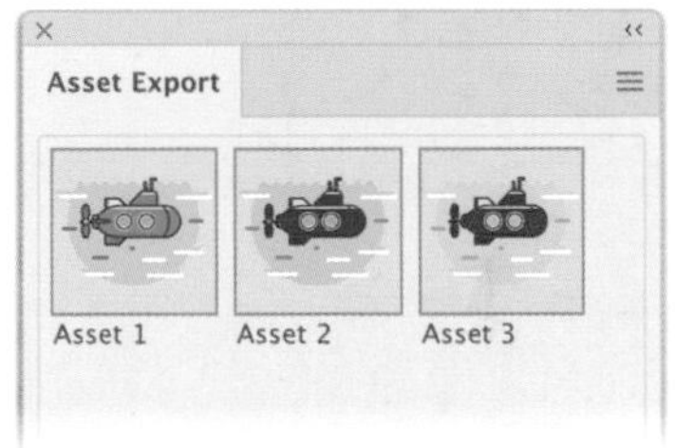

Updating the assets in Illustrator does not require you to reload them into the Asset Export panel. Illustrator is smarter than that. Any updates that are applied to the artwork will immediately be reflected in the Asset Export panel. This pertains to all grouped or ungrouped artwork so if you update artwork with multiple objects and each object is in the Asset Export panel, all the pieces will update regardless of whether or not the final artwork is grouped. If you want to create a version of the same artwork, duplicate the artwork on the artboard, make the edits, and drag that file into the Asset Export panel. This will give you a unique version of that artwork.

When artwork is deleted from the file, the version of the Asset will also be deleted from the Asset Export panel. Assets only exist in the panel if they are in the file.

Print

Now that all the hard work is done, let's print out the file to see what it looks like. Choose File > Print to open the printer dialog box. The printer information at the top of the dialog box will reflect the printer that your device is connected to. If there are multiple printers, choose the correct printer from the printer dropdown menu.

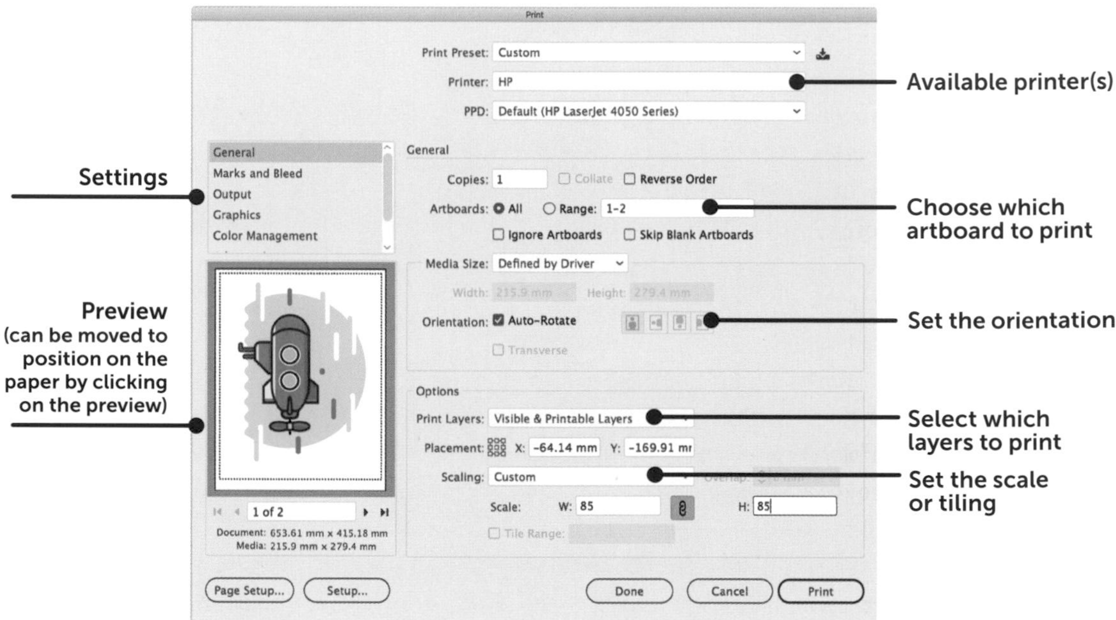

Choose the number of copies and which artboards to print. If there are multiple artboards and you want to a print a specific one, you can use ranges such as 1-4 or 2, 3, 5 to select the artboards you want to print. Set the rotation of the paper in the Orientation field if the printer presets allow. Options let you print all layers or only layers that are visible, which is quite handy if you have instructions on a layer that do not print. Set the scale of the artwork and see the size in the Preview window. The dashed line in the Preview shows the printable area and the artwork position on the paper. Click on the Preview to move the position of the artwork on the paper where you'd like. Preview will also offer the ability to select how other artboards and artwork is positioned on the paper.

Package

When you are done with your Illustrator creation, you may want to send it off to a client or other recipient. If the end user does not have Illustrator, a PDF will work just fine. But for our colleagues, you may need to supply your "native" files with a copy of the fonts as well as any linked graphics just in case edits need to be made to the file and all its assets. That is what Package is for.

A Copy of Everything

When an Illustrator project is complete and saved, we may wish or need to share it and its assets with others. The assets used in the file may not be in one central location; they could be on several hard drives and in many folders. Luckily, if we don't have any missing links or fonts, we can make a package that leaves our assets in place and makes copies of them all consolidated in a folder.

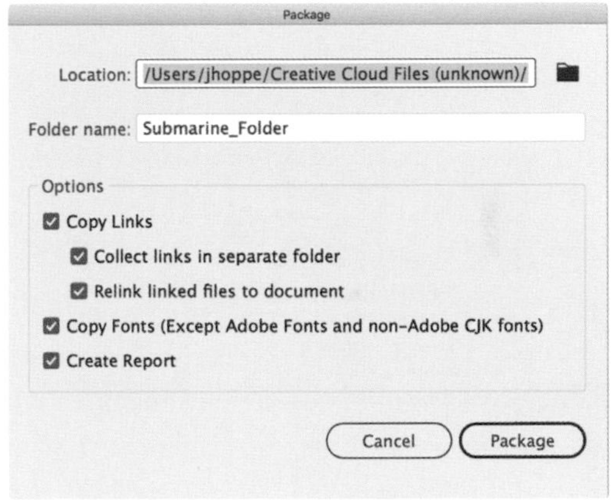

To create the package folder, choose File > Package.... A simple dialog box will appear with a few options. Check all the options to collect and package all the available content in the file.

The dialog that appears either after or instead of the printing instructions is the most significant. In it, you name the package, choose where to save it, and designate its contents. I almost always check the first five checkboxes. I want the package to have all the assets necessary to successfully open it. All the fonts, except Asian language (CJK—Chinese, Japanese, Korean) fonts and those from the Typekit service, will be copied to a subfolder called "Document fonts."

Copy Links creates a folder called "Links" in which all placed images and graphics are copied. Relink linked files to document ensures that those graphics know they're linked to the copy of your Illustrator document.

Copy Fonts copies the fonts that are available. CJK characters are excluded. Any fonts that are activated through Adobe Type are not copied since they can be activated upon opening of the Illustrator file.

Finally, Create Report will include a text file regrading the file.

When you hit the Package button, you get to read the last dialog box: a message from Adobe's legal department warning that one cannot share fonts with those who do not have a license to use them. That dialog has an important checkbox labeled Don't Show Again. A happy dialog box will show at the end if you want to see all your hard work in a nice neat folder.

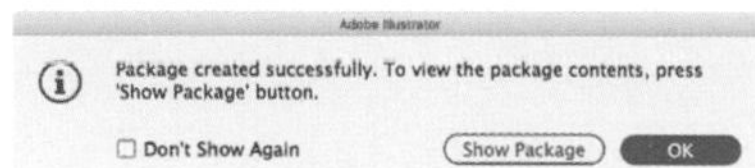

Appendix

The Adobe applications allow us to customize keyboard shortcuts. Use Edit > Keyboard Shortcuts..., create a new Set based on the defaults, and you can tweak or invent shortcuts for the commands you use most. What follows are many of Illustrator's default shortcuts. I'm using this convention, as I have throughout the book: Mac version/ PC (Windows) version.

Illustrator Keyboard Shortcuts

Tools

Selection	**V**
Direct Selection	**A**
Magic Wand	**Y**
Lasso	**Q**
Artboard	**Shift+O**
Pen	**P**
Add Anchor Point	=
Delete Anchor Point	-
Anchor Point	**Shift+C**
Curvature Tool	**Shift+`**
Line Segment	\
Rectangle	**M**
Ellipse	**L**
Paintbrush	**B**
Blob Brush	**Shift+B**
Pencil	**N**
Shaper Tool	**Shift+N**
Symbol Sprayer	**Shift+S**
Column Graph	**J**
Slice	**Shift+K**
Perspective Grid	**Shift+P**
Perspective Selection	**Shift+V**
Type	**T**
Touch Type	**Shift+T**
Gradient	**G**
Mesh	**U**

Shape Builder	**Shift+M**
Live Paint Bucket	**K**
Live Paint Selection	**Shift+L**
Rotate	**R**
Reflect	**O**
Scale	**S**
Width	**Shift+W**
Warp	**Shift+R**
Free Transform	**E**
Eyedropper	**I**
Blend	**W**
Eraser	**Shift+E**
Scissors	**C**
Hand	**H**
Zoom	**Z**
Toggle Fill/Stroke	**X**
Default	**D**
Swap Fill/Stroke	**Shift+X**
Color	**,**
Gradient	**.**
None	**/**
Toggle Screen Mode	**F**
Show/Hide All Palettes	**Tab**
Show/Hide All But Toolbox	**Shift+Tab**
Increase Diameter	**]**
Decrease Diameter	**[**
Symbolism Tools—Increase Intensity	**Shift+}**
Symbolism Tools—Decrease Intensity	**Shift+{**
Toggle Drawing Mode	**Shift+D**
Presentation Mode	**Shift+F**

Menu Commands

Preferences	
General	**Cmd+K (Mac) / Ctrl+K (PC)**
Units	**Cmd+, (Mac) / Ctrl+, (PC)**
Hide Others	**Opt+Cmd+H (Mac) / Alt+Ctrl+H (PC)**
Quit Illustrator	**Cmd+Q**
File	
New	**Cmd+N (Mac) / Ctrl+N (PC)**
New from Template	**Shift+Cmd+N (Mac) / Shift+Ctrl+N (PC)**
Open	**Cmd+O (Mac) / Ctrl+O (PC)**
Browse in Bridge	**Opt+Cmd+O (Mac) / Alt+Ctrl+O (PC)**
Close	**Cmd+W (Mac) / Ctrl+W (PC)**
Save	**Cmd+S (Mac) / Ctrl+s (PC)**
Save As	**Shift+Cmd+S (Mac) / Shift+Ctrl+S (PC)**
Save a Copy	**Opt+Cmd+S (Mac) / Alt+Ctrl+S (PC)**
Revert	**Opt+Cmd+Z (Mac) / Alt+Ctrl+Z (PC)**
Place	**Shift+Cmd+P (Mac) / Shift+Ctrl+P (PC)**
Export for Screens	**Opt+Cmd+E (Mac) / Alt+Ctrl+E (PC)**
Save for Web (Legacy)	**Opt+Shift+Cmd+S (Mac) / Alt+Shift+Ctrl+S (PC)**
Package	**Opt+Shift+Cmd+P (Mac) / Alt+Shift+Ctrl+P (PC)**
Other Script	**Cmd+F12 (Mac) / Ctrl+F12 (PC)**
Document Setup	**Opt+Cmd+P (Mac) / Alt+Ctrl+P (PC)**
File Info	**Opt+Shift+Cmd+I (Mac) / Alt+Shift+Ctrl+I (PC)**
Print	**Cmd+P (Mac) / Ctrl+P (PC)**

Edit	
Undo	**Cmd+Z (Mac) / Ctrl+Z (PC)**
Redo	**Shift+Cmd+Z (Mac) / Shift+Ctrl+Z (PC)**
Cut	**Cmd+X (Mac) / Ctrl+X (PC)**
Copy	**Cmd+C (Mac) / Ctrl+C (PC)**
Paste	**Cmd+V (Mac) / Ctrl+V (PC)**
Paste in Front	**Cmd+F (Mac) / Ctrl+F (PC)**
Paste in Back	**Cmd+B (Mac) / Ctrl+B (PC)**
Paste in Place	**Shift+Cmd+V (Mac) / Shift+Ctrl+V (PC)**
Paste on All Artboards	**Opt+Shift+Cmd+V (Mac) / Alt+Shift+Ctrl+K (PC)**
Check Spelling	**Cmd+I (Mac) / Ctrl+I (PC)**
Color Settings	**Shift+Cmd+K (Mac) / Shift+Ctrl+K (PC)**
Keyboard Shortcuts	**Opt+Shift+Cmd+K (Mac) / Alt+Shift+Ctrl+K (PC)**

Object	
Transform Again	**Cmd+D (Mac) / Ctrl+D (PC)**
Move	**Shift+Cmd+M (Mac) / Shift+Ctrl+M (PC)**
Bring to Front	**Shift+Cmd+] (Mac) / Shift+Ctrl+] (PC)**
Bring Forward	**Cmd+] (Mac) / Ctrl+] (PC)**
Send Backward	**Cmd+[(Mac) / Ctrl+[(PC)**
Send to Back	**Shift+Cmd+[(Mac) / Shift+Ctrl+K (PC)**
Group	**Cmd+G (Mac) / Ctrl+G (PC)**
Ungroup	**Shift+Cmd+G (Mac) / Shift+Ctrl+G (PC)**
Selection	**Cmd+2 (Mac) / Ctrl+2 (PC)**
Unlock All	**Opt+Cmd+2 (Mac) / Alt+Ctrl+2 (PC)**
Selection	**Cmd+3 (Mac) / Ctrl+3 (PC)**
Show All	**Opt+Cmd+3 (Mac) / Alt+Ctrl+3 (PC)**
Join	**Cmd+J (Mac) / Ctrl+J (PC)**

Average	Opt+Cmd+J (Mac) / Alt+Ctrl+J (PC)
Edit Pattern	Shift+Cmd+F8 (Mac) / Shift+Ctrl+F8 (PC)
Make	Opt+Cmd+B (Mac) / Alt+Ctrl+B (PC)
Release	Opt+Shift+Cmd+B (Mac) /Alt+Ctrl+B (PC)
Make with Warp	Opt+Shift+Cmd+W (Mac) / Alt+Shift +Ctrl+W (PC)
Make with Mesh	Opt+Cmd+M (Mac) / Alt+Ctrl+M (PC)
Make with Top Object	Opt+Cmd+C (Mac) / Alt+Ctrl+C (PC)
Live Paint Make	Opt+Cmd+X (Mac) / Alt+Ctrl+X (PC)
Clipping Mask Make	Cmd+7 (Mac) / Ctrl+7 (PC)
Clipping Mask Release	Opt+Cmd+7 (Mac) / Alt+Ctrl+7 (PC)
Compound Path Make	Cmd+8 (Mac) / Ctrl+8 (PC)
Compound Path Release	Opt+Shift+Cmd+8 (Mac) / Alt+Shift+Ctrl+K (PC)
Type	
Create Outlines	Shift+Cmd+O (Mac) / Shift+Ctrl+O (PC)
Discretionary Hyphen	Shift+Cmd+- (Mac) / Shift+Ctrl+- (PC)
Em Space	Shift+Cmd+M (Mac) / Shift+Ctrl+M (PC)
En Space	Shift+Cmd+N (Mac) / Shift+Ctrl+N (PC)
Thin Space	Opt+Shift+Cmd+M (Mac) / Alt+Shift Ctrl+M (PC)
Show Hidden Characters	Opt+Cmd+I (Mac) / Alt+Ctrl+I (PC)
Select	
All	Cmd+A (Mac) / Ctrl+A (PC)
All on Active Artboard	Opt+Cmd+A (Mac) / Alt+Ctrl+A (PC)
Deselect	Shift+Cmd+A (Mac) / Shift+Ctrl+A (PC)
Reselect	Cmd+6 (Mac) / Ctrl+6 (PC)
Next Object Above	Opt+Cmd+] (Mac) / Alt+Ctrl+] (PC)
Next Object Below	Opt+Cmd+[(Mac) / Alt+Ctrl+[(PC)

Effect	
Apply Last Effect	**Shift+Cmd+E (Mac) / Shift+Ctrl+E (PC)**
Last Effect	**Opt+Shift+Cmd+E (Mac) / Alt+Shift+Ctrl+E (PC)**

View	
Preview	**Cmd+Y (Mac) / Ctrl+Y (PC)**
GPU Preview	**Cmd+E (Mac) / Ctrl+E (PC)**
Overprint Preview	**Opt+Shift+Cmd+Y (Mac) / Alt+Shift+Ctrl+Y (PC)**
Pixel Preview	**Opt+Cmd+Y (Mac) / Alt+Ctrl+Y (PC)**
Zoom In	**Cmd+= (Mac) / Ctrl+= (PC)**
Zoom Out	**Cmd+- (Mac) / Ctrl+- (PC)**
Fit Artboard in Window	**Cmd+0 (Mac) / Ctrl+) (PC)**
Fit All in Window	**Opt+Cmd+0 (Mac) / Alt+Ctrl+0 (PC)**
Actual Size	**Cmd+1 (Mac) / Ctrl+1 (PC)**
Hide Edges	**Cmd+H (Mac) / Ctrl+H (PC)**
Hide Artboards	**Shift+Cmd+H (Mac) / Shift+Ctrl+H (PC)**
Show Template	**Shift+Cmd+W (Mac) / Shift+Ctrl+W (PC)**
Hide Bounding Box	**Shift+Cmd+B (Mac) / Shift+Ctrl+B (PC)**
Show Transparency Grid	**Shift+Cmd+D (Mac) / Shift+Ctrl+D (PC)**
Show Rulers	**Cmd+R (Mac) / Ctrl+R (PC)**
Change to Artboard Rulers	**Opt+Cmd+R (Mac) / Alt+Ctrl+R (PC)**
Hide Gradient Annotator	**Opt+Cmd+G (Mac) / Alt+Ctrl+G (PC)**
Show Text Threads	**Shift+Cmd+Y (Mac) / Shift+Ctrl+Y (PC)**
Smart Guides	**Cmd+U (Mac) / Ctrl+U (PC)**
Show Perspective Grid	**Shift+Cmd+I (Mac) / Shift+Ctrl+I (PC)**
Hide Guides	**Cmd+; (Mac) / Ctrl+; (PC)**
Lock Guides	**Opt+Cmd+; (Mac) / Alt+Ctrl+; (PC)**
Make Guides	**Cmd+5 (Mac) / Ctrl+5 (PC)**

Release Guides	**Opt+Cmd+5 (Mac) / Alt+Ctrl+5 (PC)**
Show Grid	**Cmd+' (Mac) / Ctrl+' (PC)**
Snap to Grid	**Shift+Cmd+' (Mac) / Shift+Ctrl+' (PC)**
Snap to Point	**Opt+Cmd+' (Mac) / Alt+Ctrl+' (PC)**
Window	
Minimize Window	**Cmd+M (Mac) / Ctrl+M (PC)**
Align	**Shift+F7 (Mac) / Ctrl+F7 (PC)**
Appearance	**Shift+F6 (Mac) / Ctrl+F6 (PC)**
Attributes	**Cmd+F11 (Mac) / Ctrl+F11 (PC)**
Brushes	**F5**
Color	**F6**
Color Guide	**Shift+F3**
Gradient	**Cmd+F9 (Mac) / Ctrl+F9 (PC)**
Graphic Styles	**Shift+F5**
Info	**Cmd+F8 (Mac) / Ctrl+F8 (PC)**
Layers	**F7**
Pathfinder	**Shift+Cmd+F9 (Mac) / Shift+Ctrl+F9 (PC)**
Stroke	**Cmd+F10 (Mac) / Ctrl+F10 (PC)**
Symbols	**Shift+Cmd+F11 (Mac) / Shift+Ctrl+F11 (PC)**
Transform	**Shift+F8**
Transparency	**Shift+Cmd+F10 (Mac) / Shift+Ctrl+F10 (PC)**
Character	**Cmd+T (Mac) / Ctrl+T (PC)**
OpenType	**Opt+Shift+Cmd+T (Mac) / Alt+Shift+Ctrl+T (PC)**
Paragraph	**Opt+Cmd+T (Mac) / Alt+Ctrl+T (PC)**
Tabs	**Shift+Cmd+T (Mac) / Shift+Ctrl+T (PC)**
Illustrator Help	**F1**

Miscellaneous Shortcuts

Switch Selection Tools	**Opt+Cmd+Tab (Mac) / Alt+Ctrl+Tab (PC)**
Point Size Up	**Shift+Cmd+. (Mac) / Shift+Ctrl+. (PC)**
Point Size Down	**Shift+Cmd+, (Mac) / Shift+Ctrl+, (PC)**
Font Size Step Up	**Opt+Shift+Cmd+. (Mac) / Alt+Shift+Ctrl+. (PC)**
Font Size Step Down	**Opt+Shift+Cmd+, (Mac) / Alt+Shift+Ctrl+, (PC)**
Kern Looser	**Shift+Cmd+] (Mac) / Shift+Ctrl+] (PC)**
Kern Tighter	**Shift+Cmd+[(Mac) / Shift+Ctrl+[(PC)**
Tracking	**Opt+Cmd+K (Mac) / Alt+Ctrl+K (PC)**
Clear Tracking	**Opt+Cmd+Q (Mac) / Alt+Ctrl+Q (PC)**
Spacing	**Opt+Shift+Cmd+O (Mac) / Alt+Shift+Ctrl+O (PC)**
Uniform Type	**Shift+Cmd+X (Mac) / Shift+Ctrl+X (PC)**
Highlight Font	**Opt+Shift+Cmd+F (Mac) / Alt+Shift+Ctrl+F (PC)**
Highlight Font (Secondary)	**Opt+Shift+Cmd+M (Mac) / Alt+Shift+Ctrl+M (PC)**
Left Align Text	**Shift+Cmd+L (Mac) / Alt+Shift+Ctrl+L (PC)**
Center Text	**Shift+Cmd+C (Mac) / Alt+Shift+Ctrl+C (PC)**
Right Align Text	**Shift+Cmd+R (Mac) / Alt+Shift+Ctrl+R (PC)**
Justify Text Left	**Shift+Cmd+J (Mac) / Alt+Shift+Ctrl+J (PC)**
Justify All Lines	**Shift+Cmd+F (Mac) / Alt+Shift+Ctrl+F (PC)**
Toggle Auto Hyphenation	**Opt+Shift+Cmd+H (Mac) / Alt+Shift+Ctrl+H (PC)**
Toggle Line Composer	**Opt+Shift+Cmd+C (Mac) / Alt+Shift+Ctrl+C (PC)**

Subscript	**Opt+Shift+Cmd+= (Mac) / Alt+Shift+Ctrl++ (PC)**
Superscript	**Shift+Cmd+= (Mac) / Shift+Ctrl++ (PC)**
Lock Others	**Opt+Shift+Cmd+2 (Mac) / Alt+Shift+Ctrl+2 (PC)**
Hide Others	**Opt+Shift+Cmd+3 (Mac) / Alt+Shift+Ctrl+3 (PC)**
Repeat Pathfinder	**Cmd+4 (Mac) / Ctrl+4 (PC)**
Average & Join	**Opt+Shift+Cmd+J (Mac) / Alt+Shift+Ctrl+J (PC)**
New Symbol	**F8**
Add New Fill	**Cmd+/ (Mac) / Ctrl+/ (PC)**
Add New Stroke	**Opt+Cmd+/ (Mac) / Alt+Ctrl+/ (PC)**
New Layer	**Cmd+L (Mac) / Ctrl+L (PC)**
New Layer with Dialog	**Opt+Cmd+L (Mac) / Alt+Ctrl+L (PC)**
Switch Units	**Opt+Shift+Cmd+U (Mac) / Alt+Shift+Ctrl+U (PC)**
New File (No Dialog)	**Opt+Cmd+N (Mac) / Alt+Ctrl+N (PC)**
Close All	**Opt+Cmd+W (Mac) / Alt+Ctrl+W (PC)**
Cut (Secondary)	**F2**
Copy (Secondary)	**F3**
Paste (Secondary)	**F4**
Zoom In (Secondary)	**Cmd++ (Mac) / Ctrl++ (PC)**
Navigate to Next Document	**Cmd+` (Mac) / Ctrl+` (PC)**
Navigate to Previous Document	**Shift+Cmd+` (Mac) / Shift+Ctrl+` (PC)**
Navigate to Next Document Group	**Opt+Cmd+` (Mac) / Alt+Ctrl+` (PC)**
Navigate to Previous Document Group	**Opt+Shift+Cmd+` (Mac) / Alt+Shift+Ctrl+` (PC)**

Index

A

Add Anchor Point tool, 196
Add shape mode
  Pathfinder panel, 159
  Shape Builder tool, 164
Adobe Acrobat Reader, 338
Adobe Color Themes panel, 27, 312–315
Adobe fonts, 278–280
Adobe Swatch Exchange (ASE), 328
Align panel, 231–233
  Distribute Objects section, 232
  Distribute Spacing section, 233
alignment
  blend, 217
  object, 231, 233
  stroke, 168
alternate characters, 256
anchor points
  adding, 196
  converting, 31, 188, 196
  deleting, 44, 196
  handles created from, 196
  moving, 187–188, 195, 196
  preferences for, 7, 106
  selected vs. non-selected, 187
Animated Zoom, 15, 117
anti-aliasing, 236
Appearance of Black preferences, 120
Appearance panel, 166–171
  Blend modes, 171
  effects, 243, 244, 250
  opacity settings, 170, 239
  sewn patch project and, 69–73
  shape attributes in, 286
  Stroke panel, 166–170
  Swatch panel, 171
apple drawing project, 16–18
Arc tool, 18, 32, 56, 144, 190
arcs, drawing, 18, 32, 56, 144
area type
  converting to point type, 271
  options for working with, 272
  resizing/reshaping, 269–270
Area Type tool, 262
Artboard panel, 14, 126
Artboard tool, 14, 127, 128
artboards, 126–128
  aligning to, 233
  creating new, 14
  duplicating, 128
  layouts for, 128
  options panel for, 127
  printing, 98, 128, 348
  rectangles converted to, 128
artwork
  pixel-perfect, 235–236
  recoloring, 333–336
ascenders, 260
aspect ratio, 322
Asset Export panel, 96, 344–347
assets
  adding, 344
  exporting, 347
  updating, 347
attributes
  character, 263–266, 281
  paragraph, 267, 281
  shape, 286

B

backpack project, 91–93
baseline for type, 258
baseline shift, 8, 107, 265
black, appearance of, 120
Black and White mode, 176–177
bleed, document, 125
Blend Options dialog box, 79, 89, 215–216
Blend tool, 215, 218, 219
  blended objects, 215–220
  editing the spine of, 216–218
  expanding, 219–220
  multiple objects as, 218–219
  options for creating, 215–216
  releasing, 219
blending modes, 93, 171, 241–242
blog of author, 151, 158
boiling water icon, 57–58
Break Link to Symbol option, 222
breaking links
  to graphic styles, 249, 290
  to symbols, 222
Brightness setting, 9
Bring to Front option, 82

C

campfire project, 86–88
camping gear project
  backpack creation, 91–93

- tent creation, 83–85
- *See also* outdoor scene project

Chamfer corners, 182, 184
Character panel, 263
character styles, 281, 284–285
- creating, applying, and editing, 285
- setting options for, 284

Character Styles panel, 284
characters, 256
- adjusting individual, 276
- formatting styles for, 265
- setting attributes of, 263–266

circles
- drawing, 31, 35, 39, 140–141
- transforming, 151

Clear Overrides option, 283
Clipboard preferences, 118–119
clipping masks, 252–254
- creating, 252–253
- editing, 137, 253–254
- opacity masks vs., 240
- releasing, 254

closed paths, 208
closed shapes, 190, 191
cloud icon project, 35–36
CMYK color mode, 292, 299, 305
coffee pot icon project, 53
Collaborate option, 332
color
- applying, 21, 25, 177, 331
- blend and base, 241
- CMYK, 292, 299, 305
- device limitations, 295–296
- editing, 307, 335–336
- explanation of, 294–300
- fill and stroke, 171
- global vs. local, 305
- gradient, 320–321
- opacity of, 308, 320
- recolor options, 333–336
- RGB, 292, 298–299, 305
- sampling, 309–311
- selecting, 17, 25–27
- spot, 292–293, 304, 306, 328
- themes based on, 312–315
- tinting, 308

color.adobe.com website, 315
color families, 317
color groups, 302, 335
Color Guide panel, 316–318
color harmonies, 316, 317, 334
color libraries, 27, 328–329, 331
color modes, 125, 305
color palette, 79
Color panel, 25–26, 303, 307
Color Picker, 25, 303, 307
color profiles, 297
color space, 294–295
color themes, 312–315
- creating, 312
- exploring, 313
- online options, 315
- saving, 314

color variations, 317
color wheel, 336
compound paths, 49, 162, 211–214
- creating, 211–212
- editing, 212
- releasing, 212
- type as, 213–214

compound shapes, 162
compression settings, 340
Constrain transform option, 189, 205
Convert Anchor Point tool, 196, 197, 198
converting
- anchor points, 31, 188, 193, 196
- area and point type, 271
- lines to shapes, 49
- rectangles to artboards, 128

Copy command, 40, 42–43
corner points, 31, 188, 196
corner widgets, 36, 142–143, 180, 181–184
corners
- editing, 142–143, 181–184
- rounded, 53, 83, 89, 142, 181, 184
- scaling, 205–206
- stroke styles for, 167–168

Creative Cloud Libraries, 330–332
crop marks, 341
Crop shape mode, 161
Curvature tool, 180, 192–194
- blend spine edited with, 216
- converting points on paths with, 193
- curved lines drawn with, 193–194
- straight lines drawn with, 194

curved lines
- Curvature tool for drawing, 193–194
- Pen tool for drawing, 195, 197
- straight lines with, 197–198

D

dashed lines, 169
data recovery preferences, 118
Delete Anchor Point tool, 196
deleting
- anchor points, 44, 193, 196
- colors, 331

effects, 244
gradient colors, 321
graphic styles, 290
guides, 228
lines, 165
shapes, 33, 164
styles, 283

descenders, 260

Direct Selection tool
editing corners with, 20, 36, 183–184
finer editing with, 204–205
joining lines with, 192
moving points on paths with, 195
reshaping objects with, 180, 187–188
Selection tool vs., 184, 187

Display dropdown menu, 266

Distribute Objects settings, 232

Distribute Spacing settings, 233

Divide shape mode, 160

dividing paths, 210–211

documents
creating new, 124–125
preferences specific to, 102
saving in Illustrator, 97

dotted lines, 169

Draw Inside mode
clipping mask creation, 252, 253
outdoor scene project, 87, 90, 91
plaid fabric project, 66
sewn patch project, 74
texture application, 48–49

Draw Normal mode, 49

drawing
arcs, 18, 32, 56, 144
circles, 31, 35, 39, 140–141
curves, 193–194
grids, 145–147
lines, 19, 37, 39, 41, 143
ovals, 16
polygons, 43
rectangles, 19, 140
shapes, 19, 140–142, 144–147
spirals, 144–145
squares, 41

duplicating
artboards, 128
lines, 37, 40
objects, 234–235
shapes, 33, 35, 46, 80, 148

dynamic symbols, 223

E

editing, 180
blend spines, 216–218
clipping masks, 137, 253–254
colors, 307, 335–336
compound paths, 212
corners, 142–143, 181–184
effects, 244
gradients, 28, 319, 321–322
groups, 136
Live Shapes, 185–186
opacity masks, 240
paragraph styles, 283
patterns, 226
precision, 152
shapes, 148–152, 187–200
symbols, 221–222

effects, 243–245
applying, 243, 244
expanding, 250–251
modifying or deleting, 244
Outer Glow, 71
scaling, 207, 244–245
Transform, 64–66, 71
turning off, 244
types of, 243
Zig Zag, 57, 59, 69, 70, 75

Effects menu, 243

egg and avocado icons, 61

Ellipse tool, 16, 31, 33, 39, 140–141

ellipses
drawing, 140–141
Live Shape Properties, 185
transforming, 151

end caps for strokes, 167

Essentials Classic Workspace, 11

Even-Odd rule, 211

Exclude shape mode, 160

Expand Appearance function, 246

Expand function, 246

Expand shapes button, 162

expanding
blended objects, 219–220
compound shapes, 162
effects, 250–251
strokes, 208–209

Export for Screens panel, 344

exporting
assets, 96, 344–347
libraries, 331
text, 277

Eyedropper tool, 309, 310, 311

F

file formats, 344–346
JPG, 345–346
PDF, 346

- PNG, 344–345
- SVG, 346

File Handling preferences, 118–119
fill color, 171, 177
Fill Line checkbox, 143
finding missing fonts, 279
flattening layers, 137
flipping shapes, 148, 203
fonts, 256
- activating, 279
- Adobe, 278–280
- missing, 279
- outline, 277
- packaging, 280
- setting, 263
- *See also* type

formatting, character, 265
Free Distort option, 189, 205
Free Transform tool, 180, 189, 205
freeform gradients, 321, 325–327
- applying, 325
- editing, 325–327

freeform lines, 199

G

gamut, color, 296
General preferences, 6, 102–104
global colors, 26, 305
global preferences, 102
global rulers, 227
Glyphs panel, 256
GPU Performance preferences, 10, 117
Gradient Annotator, 322
Gradient panel, 28–29, 319–327
Gradient tool, 322–323
gradients, 28–29, 319–327
- applying to strokes, 323–324
- color adjustments to, 320–321
- editing, 28, 319, 321–323
- freeform, 325–327
- predefined, 319
- types of, 321

graphic styles, 248–249, 286–290
- applying, 248, 287
- breaking the link to, 249, 290
- creating and saving, 248, 286
- deleting, 290
- groups with, 287
- libraries of, 289
- merging multiple, 287
- removing, 248
- renaming, 290
- replacing, 249, 290
- scaling, 288–289
- type with, 288

Graphic Styles panel, 248, 286, 287, 289
Grayscale mode, 178
grids, 230
- hiding, 52, 230
- pixel, 236
- polar, 146–147
- preferences for, 110–111
- rectangular, 145–146
- setting up, 51, 63
- snapping to, 51, 63, 230
- transparency, 238
- *See also* guides

Group Selection tool, 205, 212
groups, 153–154
- color, 302
- graphic style, 287
- Knockout, 239–240
- layer, 136
- shape, 42
- symbols vs., 155

guides, 227–230
- creating, 228
- options for, 228
- preferences for, 110, 228
- rulers and, 227
- showing/hiding, 228
- Snap to Point function, 228
- *See also* grids; Smart Guides

Guides & Grid preferences, 110–111

H

Hand tool, 15
handles
- creating from points, 196
- moving or changing direction of, 197
- preferences for, 7, 106

HEX colors, 307
hiding
- grids, 52, 230
- guides, 228
- rulers, 227

Home screen display, 6
hot air balloon project, 78–81
- Blend Options, 79
- drawing creation, 78
- Live Paint mode, 79–80
- Pathfinder Unite mode, 80–82

hyphenation
- preferences for, 113
- setting controls for, 267

I

Illustrator
- keyboard shortcuts, 351–360
- saving documents in, 97

Illustrator Effects, 243
Image Trace panel, 47–48, 175–178
- Black and White mode, 176–177
- Color mode, 178
- Grayscale mode, 178

importing
- libraries, 331
- swatches, 329
- text, 268

instances, symbol, 155
Intersect shape mode, 17, 159
Inverted Round corners, 182, 184
Isolation mode, 154, 253

J

jasonhoppe.com blog, 151, 158
joining
- lines, 191–192
- paths, 199, 210
- shapes, 82, 83, 191

JPG file format, 345–346

K

kerning type, 258–259, 264
Keyboard Increment value, 6, 103
keyboard shortcuts, 351–360
- customizing, 122–123, 351
- for menu commands, 354–358
- for tools, 352–353

kitchen icons project, 50–61
- boiling water, 57–58
- coffee pot, 63
- egg and avocado, 61
- grid setup, 51–52
- spatula and spoon, 54–56
- toaster with toast, 59–60

Knockout Group, 239–240

L

Lab color space, 294–295
Language options, 108, 266
Laskevitch, Steve, 294
layers, 130–138
- benefits of using, 133
- creating new, 133, 135
- explained, 130
- managing, 131
- merging and flattening, 137
- moving objects to, 135
- naming, 132
- options for, 134
- releasing items to, 138
- reordering, 135

Layers panel
- clipping masks in, 137, 253, 254
- editing groups in, 136
- expanded effects in, 250, 251
- Isolation mode in, 154
- labeled illustration of, 129
- locating objects in, 136
- locking/unlocking objects in, 154
- merging items in, 137
- moving objects in, 135
- options for working in, 134
- overview of, 130–131
- reordering layers in, 135
- selecting objects in, 136, 154

layouts, artboard, 128
leading, type, 259, 264
libraries
- color, 27, 328–329, 331
- Creative Cloud, 330–332
- exporting/importing, 331
- graphic style, 289
- sharing, 332
- swatch, 328, 329

Libraries panel, 330, 331, 332
Line Segment tool, 19, 143, 190
line segments
- Line Segment tool for drawing, 19, 143
- Pencil tool for drawing, 200

line spacing, 259
Line tool, 37, 39, 41, 54, 143
linear gradients, 321, 324
lines
- curved, 193–194, 195
- dashed, 169
- deleting, 165
- dotted, 169
- drawing, 19, 37, 39, 41, 143
- duplicating, 37, 40
- freeform, 199
- grouping, 38
- joining, 191–192
- moving points on, 196
- round capping, 21, 32, 37, 39, 41
- shapes created from, 49
- straight, 194, 195
- transforming, 152, 201–207

linking text containers, 274
links
- to graphic styles, 249, 290
- to symbols, 222

Live Paint Bucket tool, 79, 172
Live Paint Objects, 172–174
- creating and filling, 172
- expanding shapes as, 174
- Gap options for, 173
- hot air balloon project, 79–80

Live Paint Selection tool, 172, 173
Live Paintbrush tool, 172
Live Shapes, 185–186
local colors, 305
Locate Object command, 136
locking/unlocking
- guides, 228
- objects, 154

M

masks
- clipping, 137, 240, 252–254
- opacity, 240, 242

math calculations, 234
measurement units
- document sizes and, 125
- preferences for, 109
- rulers and, 227
- type and, 258

menu command shortcuts, 354-358
Merge Selected option, 137
Merge shape mode, 161
merging
- graphic styles, 287
- layers, 137

Minus Back shape mode, 161
Minus Front shape mode, 20, 22, 33, 36, 84, 86, 159
missing fonts, 279
miter limit, 210
monitor color limitations, 295, 296
moon icon project, 33
Move dialog, 201
moving
- anchor points, 187–188, 195, 196
- guides, 228
- objects, 235

Multiply Blend mode, 93, 171, 241

N

naming/renaming
- graphic styles, 290
- layers, 132

New Character Style dialog box, 284
New Paragraph Style dialog box, 281–282
non-global colors, 305
Normal Blend mode, 171

O

offset paths, 29, 44, 210
opacity
- changing, 239
- color-related, 308, 320
- selecting objects by, 242
- setting, 93, 170, 239, 308

opacity masks, 240, 242
open paths, 208
open shapes, 190, 191
outdoor scene project, 77–95
- assembly process, 94–95
- backpack creation, 91–93
- campfire creation, 86–88
- camping tent creation, 83–85
- pine trees creation, 89–90

Outer Glow effect, 71
Outline Path option, 247
Outline shape mode, 161, 191
Outline Stroke option, 208–209
outline type, 277, 288
output options, 96–98, 338–350
- asset export, 96, 344–347
- file formats and, 344–346
- packaging projects, 349–350
- PDF file settings, 338–343
- printing files, 98, 348
- saving files, 97, 338–343

ovals, drawing, 16, 140–141
overrides, style, 283

P

packaging
- fonts, 280
- projects, 349–350

panels, expanding/collapsing, 12
paragraph attributes, 267
Paragraph panel, 267, 281
paragraph styles, 281–283
- applying, 282–283
- attributes for, 281
- creating new, 281–282
- editing and deleting, 283
- loading, 283

Paragraph Styles panel, 281, 283
paragraph type, 261, 269–270
Pathfinder panel, 158–163, 180
- Add or Unite mode, 16, 21, 35, 81, 159
- Expand button, 162
- Intersect mode, 17, 159
- Minus Front mode, 20, 22, 33, 36, 84, 86, 159
- options available in, 163
- Pathfinders Section, 160–161

Shape Modes Section, 158–160

paths, 208–214
- compound, 49, 162, 211–214
- direction of, 197, 198
- dividing, 210–211
- joining, 199, 210
- moving, 187–188
- offset, 29, 44, 210
- open vs. closed, 208
- selecting, 195
- strokes as, 208
- type on, 262, 270, 273

Pattern Options panel, 224, 225, 226

patterns, 224–226
- creating, 62, 224–225
- editing and scaling, 226
- saving, 225

PDF files, 97, 338–343, 346
- Advanced options, 343
- Compression options, 340
- General options, 339
- Marks and Bleeds options, 341
- Output options, 342
- presets for saving, 338
- Security options, 343

Pen tool, 180, 194–199
- adding and deleting points with, 196
- connecting open paths with, 199
- curved lines drawn with, 195, 197
- handles created and moved with, 196–197
- moving points on lines with, 196
- path direction changes with, 197, 198
- project examples using, 45, 51
- selecting between tools in, 194
- straight lines drawn with, 195, 197–198

Pencil tool, 180, 199–200
- freeform lines drawn with, 199
- line segments drawn with, 200
- overview of options in, 199

Performance preferences, 10, 117

Perspective Distort option, 189, 205

Photoshop Effects, 243

picas and points, 258

pie charts, 151

pine trees project, 89–90

pixel grid, 236

pixel-perfect artwork, 235–236

plaid fabric project, 62–67
- Draw Inside mode, 66
- grid setup, 63
- Recolor Artwork panel, 67
- Transform effect, 64–66

plug-ins, 114

PNG file format, 344–345

point type, 261
- converting to area type, 271
- resizing, 269

points and picas, 258

Polar Grid tool, 146

Polygon tool, 83, 89, 141

polygons
- drawing, 43, 141
- Live Shape Properties, 186
- transforming, 151

positioning objects, 233

preferences, 6–10, 102–120
- Appearance of Black, 120
- File Handling & Clipboard, 118–119
- General, 6, 102–104
- GPU Performance, 10, 117
- Guides & Grid, 110–111
- Hyphenation, 113
- Plug-ins & Scratch Disks, 114
- Selection & Anchor Display, 7, 105–106
- Slices, 113
- Smart Guides, 111–112
- Type, 8, 107–108
- Units, 8, 109
- User Interface, 9, 115–116

presets
- document, 124
- gradient, 319
- PDF file, 338
- tracing, 175

Preview Mode setting, 125

Preview panel, 98

printer limitations, 295–296

printing artboards, 98, 128, 348

process colors, 292

Properties panel, 149
- Appearance options, 25
- Character section, 263
- Edit Clipping Path icon, 253
- Flip Vertical button, 46
- Quick Actions section, 47–48

R

radial gradients, 321, 322, 323

rain icon project, 37–38

Raster Effects setting, 35

Recolor Artwork panel, 333–336
- assigning colors in, 334–335
- backpack creation project, 93
- editing colors in, 335–336
- labeled illustration of, 333
- plaid fabric project, 67
- sewn patch project, 75–76

Rectangle Grid tool, 145

Rectangle tool, 19, 41, 66, 140
rectangles
- converting to artboards, 128
- drawing, 19, 66, 140
- filling with color, 73
- grid creation, 145–146
- Live Shape Properties, 185
- transforming, 150

Redefine Style option, 283
Reflect tool, 180, 203
Release To Layers command, 138
releasing
- blended objects, 219
- clipping masks, 254
- compound paths, 212
- guides, 228
- items to layers, 138
- text threads, 274

removing
- graphic styles, 248
- text threading, 274
- *See also* deleting

renaming. *See* naming/renaming
reordering
- layers, 135
- swatches, 302

Replace Spine option, 217
replacing
- blend spines, 217
- graphic styles, 249, 290
- missing fonts, 279
- symbols, 222

resizing. *See* sizing/resizing
resolution settings, 119, 124, 125, 340
Reverse Gradient button, 29
Reverse Spine option, 218
RGB color mode, 292, 298–299, 305
Rotate tool, 22, 23, 39–40, 42, 180, 201
rotation, 201–202
- of grouped lines, 23
- of shapes, 42, 148

rounded corners, 53, 83, 89, 142, 181, 184
Rubber Band preview, 194
rulers, 227

S

sampling color, 309–311
sans serif type, 257
saving
- color themes, 314
- graphic styles, 248, 286
- Illustrator files, 97
- library files, 289
- patterns, 225
- PDF files, 97, 338–343
- swatches, 328, 329

Scale Corners option, 181, 205, 206
Scale Stroke & Effects checkbox, 18, 46, 89, 94, 207, 244, 288
Scale tool, 180, 202–203, 288
scaling, 202–203
- corners, 205–206
- effects, 207, 244–245
- graphic styles, 288–289
- objects, 235
- patterns, 226
- strokes, 94, 207
- type, 265
- *See also* sizing/resizing

Scissor tool, 210
scratch disks, 114
Screen Blend mode, 171, 241
security options, 343
selecting
- colors, 17, 25–27
- objects, 136, 153–154
- paths, 195
- text, 270

Selection & Anchor Display preferences, 7, 105–106
Selection tool
- Direct Selection tool vs., 184, 187
- duplicating objects with, 22, 35
- selecting objects with, 135, 153

Send to Back option, 60, 82, 84, 92
serif type, 257
sewn patch project, 68–76
- Appearance panel options, 69–70
- Draw Inside mode, 74
- effects applied to, 71
- Recolor Artwork panel, 75–76
- stroke building, 71–73

Shape Builder tool, 33, 164–165, 180
shapes
- blending, 89
- compound, 162
- converting lines to, 49
- deleting, 33, 164
- dividing, 211
- drawing, 19, 140–142, 144–147
- duplicating, 33, 35, 46, 81, 148
- editing, 148–152, 187–200
- expanding, 162, 174
- filling with color, 73, 177
- flipping, 148
- grouping, 42
- joining, 82, 83, 191
- Live Shapes, 185–186

open vs. closed, 190, 191
resizing, 148
rotating, 42, 148
transforming, 150–152, 201–207
vector, 140

sharing
libraries, 332
PDF files, 338

Shear tool, 38, 204
shearing objects, 204
shortcuts. *See* keyboard shortcuts
Size/Leading control, 8, 107
sizing/resizing
shapes, 148
text areas, 269–270
type, 263–264
See also scaling

Skew tool, 86
slices, 113
Smart Guides, 229–230
labels displayed with, 229
preferences for, 111–112
turning on, 187

Smooth Color blend, 215, 219
smooth points, 31, 58, 188, 196
Snap to Grid option, 51, 63, 230
Snap to Point function, 228
snowflake icon project, 41–43
spatula and spoon icons, 54–56
Specified Distance blend, 216
Specified Steps blend, 215
Spiral tool, 144
spirals, drawing, 144–145
spot colors, 292–293, 304, 306, 328
spray can project, 19–23
squares, drawing, 41
Star tool, 34, 75, 141–142
stars
drawing, 34, 75, 141–142
transforming, 152

stitch effect, 70
straight lines
Curvature tool for drawing, 194
curved lines with, 197–198
Pen tool for drawing, 195, 197

Stroke panel, 18, 19, 32, 53, 166–170
strokes
alignment of, 168
arrowheads and tails for, 170
color swatch for, 171
corner style options, 167–168
dashed and dotted line, 169
end cap options, 167
expanded, 208–209
gradients applied to, 323–324
outline, 208–209
paths as, 208
rounded corner, 53, 83, 89
scaling, 94, 207
sewn patch project, 69–73

styles
character, 281, 284–285
deleting, 283
graphic, 248–249, 286–290
loading, 283
overrides for, 283
paragraph, 281–283
stroke corner, 167–168

sun icon project, 39–40
SVG file format, 346
swatches
creating, 303–304
importing, 329
reordering, 302
saving, 328, 329

Swatches panel, 301–306
adding colors to, 307, 311, 318
color groups used in, 302
creating swatches for, 303–304
editing colors in, 307
labeled illustration of, 301
libraries accessed via, 328
options available in, 302
patterns added to, 224, 225
reordering swatches in, 302
saving swatches from, 328, 329
selecting colors in, 17, 25, 26–27
tints added to, 308

Symbol Sprayer tool, 222, 223
symbols, 221–223
breaking links to, 222
creating and using, 221
dynamic, 223
editing, 221–222
groups vs., 155
instances of, 155
replacing, 222
tools for, 222–223

Symbols panel, 155, 221–222

T

tent creation project, 83–85
text
character attributes, 263–266
character styles, 281
containers for, 262
creating, 261–262

exporting, 277
graphic styles on, 288
importing, 268
linking containers of, 274
options for working with, 271–275
outline fonts for, 277
paragraph attributes, 267
paragraph styles, 281–283
preferences for, 107–108
resizing areas of, 269–270
selecting, 270
wrapping, 275
See also type

text box, 262
texture, adding, 47–49
themes, color, 312–315
threaded text, 274
thumbnails
layer, 134
swatch, 302
tinting colors, 308
toaster icon project, 59–60
tools
displaying hidden, 11
keyboard shortcuts for, 352–353
Touch Type tool, 276
tracing presets, 175
Tracking control, 8, 107
tracking type, 259, 265
Transform effect, 64–66, 71
Transform panel
Live Shapes edited in, 185–186
math calculations performed in, 234
precision editing in, 152, 233, 234
Rectangle Properties section, 182
Scale Stroke & Effects checkbox, 18, 46, 89, 94
scaling objects in, 94, 235
transforming shapes in, 150–152
transformations, 180
corner scaling, 205–206
movement, 201
reflection/flip, 203
rotation, 201–202
scaling, 202–203
shape and line, 150–152, 201–207
shearing, 204
stroke & effect scaling, 207
tools for, 204–205
transparency, 238–239
Knockout Group, 239–240
showing grid for, 238
Transparency panel, 238
Trim shape mode, 160
type
area, 269–270, 271, 272
as compound path, 213–214
converting, 271
graphic styles on, 288
outlined, 277, 288
on a path, 262, 270, 273
point and paragraph, 261, 269–270
setting preferences for, 8, 107–108
terminology for working with, 256–260
See also text
Type tool, 57, 261
selecting type with, 270
text creation with, 261–262
typefaces. *See* fonts

U

UI Scaling setting, 9, 116
Ungroup command, 153
Unite shape mode, 16, 21, 35, 81
Units preference settings, 8, 109
updating assets, 347
User Interface preferences, 9, 115–116

V

vector shapes, 140

W

water droplet project, 31–32
weather icons project, 30–49
cloud, 35–36
moon and stars, 33–34
rain, 37–38
snowflake, 41–43
sun, 39–40
texture, 47–49
water droplet, 31–32
wind, 44–46
Web safe colors, 307
widgets, corner, 36, 142–143, 180, 181–184
wind icon project, 44–46
workspaces
choosing, 11–12, 121
creating new, 12, 121

X

x-height of letters, 260

Z

Zig Zag effect, 57, 59, 69, 70, 75
Zoom tool, 15